The Author

Research staff at CERN from 1969 to retirement in 2006. Member and, from 1992, leader of the Gas Detectors Development group initiated by Georges Charpak (1992 Nobel Laureate in Physics). From 2006 to 2018, associated with the TERA foundation (Fondazione per Adroterapia Oncologica).
Presently, Honorary Member of CERN.

aker and regular attendee of major conferences
Sauli has been Adjunct Professor at Northeast-
n) and Bicocca University (Milano). In 2005,
Doctor Honoris Causa at Université de Haute
was appointed Editor of Nuclear Instruments
ics Research. He is the author of several books
cles devoted to radiation detectors. Amongst
ents are the Multi-step Chamber and the Gas
described in this book.

Organizer, invited spe
and workshops, Fabio
ern University (Bosto
he was awarded the
Alsace. In 2000, he
and Methods in Phys
and hundreds of arti
his original developm
Electron Multiplier,

Contents

Preface

In his previous book, *Gaseous Radiation Detectors* (Cambridge University Press, 2014), the author describes the basic physics of collection and multiplication of charges released by ionizing radiation in a gas, as well as the design, performance and applications of current gaseous detectors. The present work covers in more detail the development of advanced devices, collectively named "Micro-Pattern Gaseous Detectors", aimed at the improvement of localization accuracy and rate capability, principally in view of applications in particle physics experimentation. The content is distilled from information provided in a very large number of publications, conference proceedings and dedicated workshops.

The author aimed to outline, in the different chapters, the most basic observations, measurements and applications of the various families of detectors. In light of the rapid evolution of the technologies, the reader is encouraged to follow the more recent works of the various research groups, with the help of the references provided. An abridged list of relevant literature on gaseous phenomena and detectors is provided in the "Further Readings" section at the end of the volume.

Promising new approaches to the development of better performing detectors are discussed in the last chapter.

Fabio Sauli was research staff at CERN from 1969 to his retirement in 2006, a member and then leader of the Gas Detectors

Development group initiated by Georges Charpak and 1992 Nobel Laureate in Physics. He collaborated with the Italian TERA foundation (Fondazione per Adroterapia Oncologica) from 2006 to 2018 and is presently Honorary Member of CERN. He is the author of several books and hundreds of articles devoted to radiation detectors.

Acronyms

ALICE	A Large Ion Collider Experiment at CERN
ASIC	Application-Specific Integrated Circuit
BAND	Boron Array Neutron Detector
CAT	Compteur A Trous
CBM	Compressed Baryonic Matter experiment
CCD	Charge Coupled Device camera
CENBG	Centre d'Etudes Nuclaires de Bordeaux-Gradignan (France)
CERN	European Organization for Nuclear Research, Geneva (Switzerland)
COCA-COLA	Coated Cathode Conductive Layer detector
CMS	Compact Muon Solenoid
CNAO	Centro Nazionale di Adroterapia Oncologica, Pavia (Italy)
CNESM	Close-contact Neutron Emission Surface Mapping
COG	Centre-of-gravity
COMPASS	COmmon Muon Proton Apparatus for Structure and Spectroscopy
CRID	Cherenkov Ring Imaging Detector
CVD	Chemical Vapour Deposition
DESY	Deutsches Elektronen-Synchrotron, Hamburg (Germany)
DGEM	Double-GEM
DLC	Diamond-like Carbon

DME	Dimethyl ether, CH_3OCH_3
EDXRF	Energy Dispersive X-ray Fluorescence imaging
ENEA	Agenzia nazionale per le nuove tecnologie, Frascati (Italy)
EPID	Electronic Portal Imaging Device
EF	Ethyl ferrocene $Fe(C_5H_4)_2C_2H_5$
FTM	Fast Timing Micropattern Gas Detector
FWHM or fwhm	Full Width at Half Maximum
GEM	Gas Electron Multiplier
GEMGrid	GEM Grid hybrid device
G-GEM	Glass GEM
GIF	Gamma Irradiation Facility at CERN
GPD	Gas Pixel detector
GSI	GSI Helmholtzzentrum fur Schwerionenforschung, Darmstadt (Germany)
HBD	Hadron Blind Detector
IBF	Ions Back-flow
IHEP	Institute of High Energy Physics, Beijing (China)
ILC	International Linear Collider
ILL	Institut Laue-Langevin, Grenoble (France)
IMRT	Intensity Modulated Radiation Therapy
INAF	Istituto Nazionale di Astrofisica, Roma (Italy)
INTEGRAL	International Gamma Ray Astrophysics Laboratory
ITO	Indium Tin Oxide
JET	Joint European Torus laboratory, Culham, Oxfordshire (UK)
JINR	Joint Institute for Nuclear Research, Dubna (Russia)
JLAB	Thomas Jefferson National Laboratory, Newport News VA (USA)

KAIST	Korea Institute of Science and Technology, Seoul (RK)
LA	Liquid Argon
LEM	Large Electron Multiplier
LHC	Large Hadron Collider at CERN
LHM	Liquid Hole Multiplier
MICRODOT	Micro-dot Detector
MICROWELL	Micro-well Detector
MGD	Micro-groove Detector
MHSP	Micro-hole and Strip Plate
MICROMEGAS	Micro-mesh Gaseous Structure
MIPA	Micro-pin Array
MPGD	Micro-Pattern Gas Detector
MSAC	Multi-Step Avalanche Chamber
μ-PIC	Micro-pixel detector
μ-RWELL	Micro-resistive Well Detector
M^3-PIC	Micro-mesh Micro-pixel Detector
MSGC	Microstrip Gas Counter
MWPC	Multiwire Proportional Chamber
NEXT	Neutrino Experiment with a Xenon Time Projection Chamber
NITPC	Negative Ions TPC
PACEM	Photon-Assisted Cascaded Electron Multiplier
PRAXyS	Polarimeter for Relativistic Astrophysical Sources
PRF	Pad Response Function
PPC	Parallel Plate Chamber
PPAC	Parallel Plate Avalanche Counter
PRR	Proton Range Radiography
PSI	Paul Sherrer Institute, Villigen (Switzerland)
QE	Quantum Efficiency
RHIC	Relativistic Heavy Ions Collider, Brookhaven, Upton NY (USA)

RICH	Ring Imaging Cherenkov Counter
RMS or rms	Root Mean Square
RPWELL	Resistive Plate WELL
RWELL	Resistive WELL
SGC	Small Gap Chamber
SMEX	Small Explorer
SLAC	Stanford Linear Accelerator Center, Stanford CA (USA)
SRPWELL	Segmented Resistive Plate WELL
TDR	Technical Design Report
TEA	Triethyl amine, $(C_3H_5)_3N$
TGD	Thick Grove Detector
TGEM	Triple-GEM
TH-GEM	Thick GEM
TMAE	Tetrakis dimethyl amino ethylene, $C_2[(CH_3)_2N]_4$
TPC	Time Projection Chamber
TRD	Transition Radiation Detector
T2K	Tokay to Kamioka Experiment
UV	Ultra-Violet
WELL	Well detector

Chapter 1

One Century of Gaseous Detectors

1.1 Early Gaseous Counters

The discovery that gaseous media are ionized by radiation can be
tracked back to the late years of the nineteen century, in the course of
the early studies on natural radioactivity. Extensively used by Pierre
and Marie Curie, the ionization (or ion) chamber consists of a gas-
filled vessel with charge-collecting electrodes, Figure 1.1. Electrons
and ions created in the gas by the interaction of radiation with the
medium migrate under the effect of an applied electric field and are
collected at the electrodes; the measured current depends on the
nature of the radiation and is proportional to its intensity.

With a scheme avoiding edge distortions, and in the absence of
charge losses due to recombination or electron capture, the device
has a linear response to the radiation flux over many orders of
magnitude. With a variety of design and fillings (including liquids),
ion chambers have been widely used, and are still the tool of choice
for intense beam calibration and monitoring purposes (Sugaya et al.,
1996; Giordanengo et al., 2013).

The ionization chamber has unitary gain, and can therefore
detect only high radiation fluxes. With a similar design, the Parallel
Plate Avalanche Counter (PPAC) permits to increase the amount
of detectable charge, applying between the electrodes a potential
difference high enough to induce ionizing collisions between primary
electrons and gas molecules; starting with the primary electrons, the
avalanche process continues through the gap until the charges, electron and ions, reach the electrodes, anode and cathode, respectively.

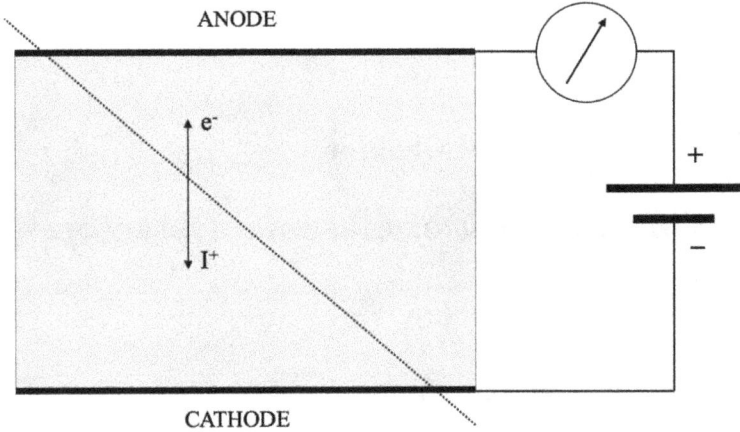

Figure 1.1:   Schematics of the ionization chamber.

While permitting to detect small amounts of charge, the PPAC does not have a response proportional to the primary ionization, since the total charge created depends on the distance of the multiplying electrons from the anode.

Rather prone to discharges, standard PPACs have been over the years of limited use. The observation, however, that a very narrow gap permits to reach high and stable gains is the basis for the design of the MICROMEGAS devices, to be described in Chapter 4.

In their early works at the beginning of previous century, Ernest Rutherford, Hans Geiger and collaborators found that the use as the anode of a thin wire, coaxial with a cylindrical cathode, permits to amplify the primary charge, while preserving a proportional response in most of the sensitive volume. Figure 1.2 shows schematically the original setup, with the reaction vessel (on the left) used to introduce, with the gas flow, the radioactive molecules into the counter (on the right) (Rutherford and Geiger, 1908).

Owing to the radial structure of the electric field in the cylinder, electrons smoothly drift through the sensitive volume towards the anode; only on approaching the wire do they start experiencing ionizing collisions with the gas molecules in the increasing field. This results in an avalanche-like charge multiplication process, permitting to amplify the primary ionization and allowing the detection of low

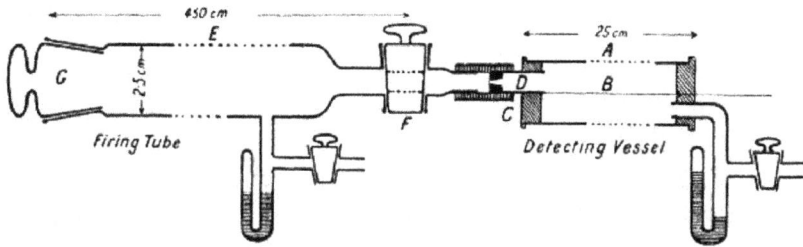

Figure 1.2:   The wire proportional counter.
Source: Rutherford and Geiger (1908).

ionization events with the simple electrical sensors available at the time. Operated at a wide range of pressures and gases, the counter provides a signal proportional to the primary ionization charge; the ratio between detected and primary charges defines its gain.

With a low-pressure gas filling, the counters permit to reach a very high gain mode of operation, the so-called Geiger–Müller regime. Through a process of photon production and reabsorption, the primary avalanche spreads along the anode wire, resulting in a large increase in the total charge, until the process is damped by external resistive circuitry.

Wire proportional counters and Geiger counters were, and still are, largely used for radiation detection and monitoring. Proportional counters, which are very successful in terms of efficiency and energy resolution, provide, however, positional information only corresponding to their size; arrays of counters permit the detection over large areas.

Developed in the late fifties of the previous century, pulsed spark chambers provided a visual image of the particles' trajectories, and were used for decades to study cosmic rays and interaction yields in particle physics experiments. Initially implemented with photographic recording systems, spark chambers evolved into digital devices, with the coordinates of the sparks recorded and analysed electronically. Large-volume pulsed devices, such as the streamer chambers, consented to acquire detailed images of rather complex events.

For a more detailed description of these early systems see for example (Rice-Evans, 1974).

1.2   Multiwire Proportional Chambers

Despite their considerable impact in particle physics experimentation, spark chambers suffer from an intrinsic operating rate limitation, a few hertz at best, seriously hindering the realization of systems capable of withstanding the ever-increasing particle fluxes provided by the new generations of accelerators. A more powerful device was introduced at the end of the sixties with the invention of the multiwire proportional chamber (MWPC) (Charpak *et al.*, 1968).

The MWPC consists of a set of parallel, regularly spaced thin anode wires between two cathode planes. On application of a symmetric difference of potential between anodes and cathodes, electric field lines and equipotentials develop as shown in Figure 1.3. Charges released by ionization in the gas volume drift towards the electrodes, electrons to the anodes and ions to the cathodes. On approaching the thin wires, the electrons, accelerated by the increasing field, undergo inelastic collisions with the gas molecules, starting an avalanche; the process continues until the amplified electron charge reaches the

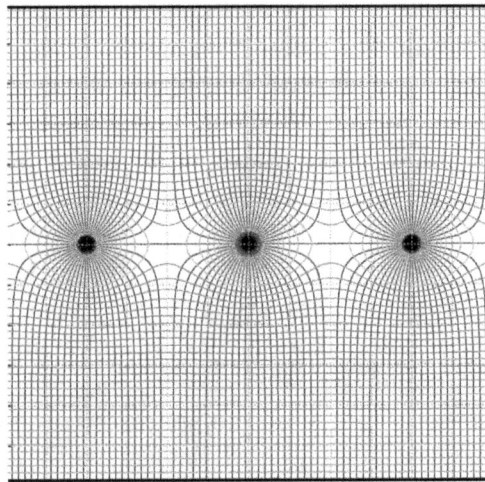

Figure 1.3:   Field lines and equipotentials in the MWPC.

anodes. The multiplication process initiates typically at a distance of few wire radii, ensuring a proportional response to the ionization released in most of the gas volume. With a proper choice of geometry and gases, proportional gains up and above 10^6 can be achieved, permitting the detection of small amounts of primary charges.

Efficiently operating at radiation fluxes up to 10^4 mm^{-2}s^{-1}, MWPCs were quickly adopted to instrument the new generations of detectors in particle physics experiments. Limited originally in localization properties by the wire spacing, 1 or 2 mm, the development of the induced cathode signals readout method permitted to reach position accuracies well below a millimetre (Charpak *et al.*, 1970).

The subsequent exploitation of the drift time measurement improved the localization even further, and permitted to realize large-volume, high-rate detectors at moderate costs (Walenta *et al.*, 1971). Noticeable examples of these advanced devices are the cylindrical drift chambers operated around colliding beam accelerators and the large-volume Time Projection Chambers (TPCs) (Nygren and Marx, 1978). Large-area, UV-photon-sensitive wire chambers were used as detectors in particle identification systems based on the Cherenkov Ring Imaging (RICH) technology (Ypsilantis, 1981). MWPCs of various designs have also been successfully used in other applied fields, medicine and biology.

The significant contribution of the new technology to the field of particle physics was acknowledged with the award to Georges Charpak of the 1992 Nobel Prize for physics.

Despite the performance improvements offered by wire systems, the progress with particle physics accelerators and the new quest for rare events, requiring operation of the detectors at very high particle flux, eventually exceeded the MWPC rate capability, limited by the long collection time of positive ions created in the avalanche process. A new family of innovative devices, collectively named micro-pattern gaseous detectors (MPGDs), introduced in the late nineties, permitted to overcome some of the weaknesses of wire systems, in particular their rate limitations. A large CERN-based international research project on the new devices, RD51, initiated in 2009, with at the time of writing close to ninety participating institutions and around 600 members, contributes to the research, and collects and distributes the information with the help of regular meetings and workshops (Titov and Ropelewski, 2013). The search is still motivated by the increasing requirements of particle physics experiments, and in particular by the luminosity upgrade of CERN's Large Hadron Collider to be completed in 2021. An even cursory keyword search on the field easily provides several thousand related publications, making a comprehensive coverage of the subject a rather demanding issue. The author apologizes in advance for improper quotes and involuntary omissions.

Chapter 2

Micro-Strip Gas Counters

2.1 Introduction

Despite their good performances and large impact on particle physics experimentation, the gaseous detectors making use of thin anode wires as amplifying elements outlined in the previous chapter suffer from a number of limitations. The displacements and instabilities due to electrostatic forces set a limit to the maximum length of unsupported wires, particularly for small wire spacings; in typical operating conditions, the maximum stable length of the anodes is around 90 cm for 2 mm wire spacing and ~10 cm for 1 mm, unless the wire grids are stabilized by unwieldy and inefficiency-causing internal supports (Majewski and Sauli, 1975). For large chamber sizes, the gravitational sagging of the electrodes adds to the stability problem and requires the use of heavy support frames (Majumdar and Mukhopadhyay, 2007).

A more fundamental limitation of wire structures results from the slow retrograde motion of positive ions generated at the anodes by the avalanching process, and the long time required to neutralize them at the cathodes. At high radiation fluxes, this accumulation in the gap of positive space charge induces a decrease of the electric field around the wires, with a consequent reduction of gain. For common wire chamber geometry, the gain drop begins at a particle flux around $10^4 \, \mathrm{mm}^{-2} \, \mathrm{s}^{-1}$, easily exceeded in experiments (Breskin *et al.*, 1975). Last but not least, detectors built with thousands of thin wires are mechanically rather fragile and prone to failures in case of accidental discharges.

7

An innovative detector, developed in 1988 by Anton Oed at the Institute Laue Langevin (ILL) in Grenoble, permits to overcome some of the abovementioned limitations. Appropriately named micro-strip gas counter (MSGC), the device consists of a set of thin parallel metallic strips engraved on an insulating substrate and alternatively connected as thin anodes and wider cathodes, Figure 2.1 (Oed, 1988). On application of a difference of potentials between adjacent strips, and with an overlaying gas volume delimited by a drift electrode, the electric field lines converge to the thin anodes creating a field structure evocative of that of MWPCs: electrons released by ionization in the gas volume drift towards the anode and multiply in an avalanche process, Figure 2.2. Using high-resolution, industrially available photolithographic technologies, the distance

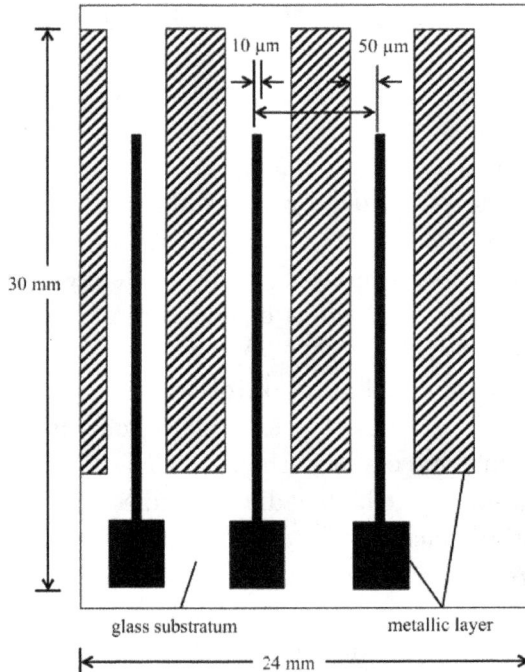

Figure 2.1:   MSGC schematics, with thin anodic strips alternating with wider cathodes.

Source: Oed (1988).

Figure 2.2: Electric field lines and equipotentials in the MSGC.

between strips, or pitch, can be reduced to a fraction of a millimetre, 100 or 200 μm being a common choice (Oed *et al.*, 1991, 1989). Figure 2.3 shows Anton Oed wearing a T-shirt well illustrating his invention.

Owing to the short distance between strips, the majority of the positive ions released in the avalanches are quickly collected and neutralized at the cathodes, minimizing the space-charge-induced field distortions encountered in wire detectors at high radiation fluxes. Adding to the narrow pitch, the fast clearance of the charge results in an increase by almost two orders of magnitude in the rate capability as compared to wire chambers.

Originally aimed at instrumenting a neutron spectrometer at the ILL, owing to their promising performances in terms of rate capability and resolution, MSGCs were soon planned for use as tracking detectors in large experiments and for other applications, see Section 2.8. A problem appeared, however, with the long-term reliability of the devices, as the fragile electrodes could be easily damaged by accidental discharges induced by manufacturing defects

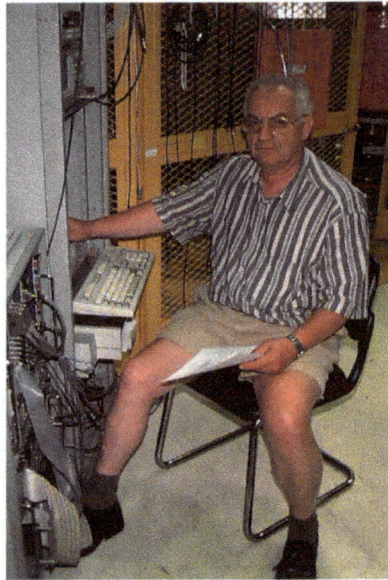

Figure 2.3:   Anton Oed (ca. 1989, courtesy himself).

or large energy releases due to heavily ionizing radiation in harsh operating conditions, Section 2.7. While the low-amplification factors required for detection of neutrons permitted to reliably operate the ILL spectrometer for many years, despite improvements due to the contribution of many research groups, the technology was essentially abandoned for detection of the smaller charges released by fast particles requiring higher gains. The efforts helped, however, to sprout the development of many alternative and innovative devices, collectively named micro-pattern gas detectors (MPGDs), described in the following chapters.

2.2   MSGC Manufacturing

The micro-strip fabrication process is a high-resolution version of those used for manufacturing printed circuits, also exploiting photolithographic technologies developed by the semiconductor and precision optical industry, providing sub-micron accuracies for the width and position of the strips. In their early and most widespread implementation, MSGCs were manufactured on borosilicate or soda-lime glass plates coated with a thin (few μm) chromium layer. A variety of other supports and metals have been, however, used in an effort to improve the performances of the detector, and particularly their rate capability and long-term operating lifetime under high radiation fluxes. Thin foil polymer substrates permitted also to realize larger sensitive areas at low cost, as well as non-planar devices. For a summary of manufacturing technologies, see for example (Sauli, 1999).

Two major techniques are used for metal coating of insulating substrates: vacuum depositions and sputtering. In the first method, the metal is heated up to evaporation with resistances, high-frequency induction coils, electrons or laser beams, and condensates on the substrate. Metal sputtering is instead produced by bombardment of the target in a cavity by plasma or by argon ions; in magnetron sputtering, a magnetic field is added to increase the plasma ionization and accelerate the process. The adherence to the support of a sputtered layer is generally better due to the higher energy of the metallic particles; in vacuum deposition, the same result can be obtained with the initial deposition of a thin adhesion underlayer of another metal, such as titanium or chromium.

To ensure adherence and good manufacturing characteristics, supports must have optical surface quality and planarity; the requirements are met by several commercial glasses, available at low cost and large sizes from $30\,\mu$m thickness and above, such as the borosilicate DESAG D-263 and the alkali-free AF-45.[1] Other rigid substrates have been used: quartz, silicon, ceramics and sapphire.

Most glass and ceramic plates available with sufficiently good surface quality have a very high bulk resistivity, of the order of $10^{16}\,\Omega \cdot$ cm, corresponding to a surface resistivity of about 10^{18} Ω/square for a 100-μm-thick plate. Such high values may cause gain instabilities, due to surface polarization and the charging up of the insulator between strips under irradiation. Substrates with bulk resistivity in the range between 10^{10} and 10^{12} $\Omega \cdot$ cm (surface

---

[1]Schott Advanced Optics, Grünenplan/Eschershausen, Germany.

resistivity between 10^{12} and $10^{14}\,\Omega$/square) have been found to be adequate to reduce of the gain shifts observed for insulating supports at high radiation fluxes; this is achieved either using special semi-conducting (or electronic) glasses, or by appropriate surface treatments of various insulators, to be discussed later.

A variety of thin polymer foil supports have also been used, offering several advantages over glass: flexible and permitting to realize non-planar detectors, they have low density and atomic number, reducing multiple scattering, photon and neutron conversion backgrounds in the detector. The availability of polymer materials in a wide range of bulk resistivities also allows reducing charging-up problems (see Section 2.4).

The basic photolithographic etching method used to manufacture MSGCs is schematically illustrated in Figure 2.4. The metal-coated

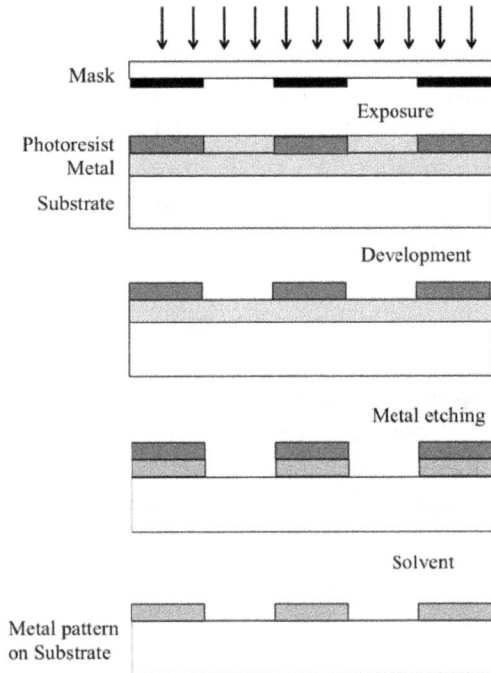

Figure 2.4:   Wet etching photolithographic processing used to manufacture an MSGC.

substrate is laminated with a photosensitive resin foil and exposed to an UV light source through a mask duplicating the desired strip pattern. The exposed areas of the photoresist are then stripped in an alkaline solution, allowing the removal of the excess metal by immersion in acid baths (wet etching) or by exposure to chemically active ions (plasma etching). Finally, a solvent removes the residual non-exposed resin.

The acids used for wet etching tend to be rather aggressive for the support, and might modify its electrical properties, particularly in the case of the thin resistive coatings described in the following sections. In plasma etching, instead, the non-protected metal is removed either by sputtering with an argon ion plasma or with a reactive chlorine plasma; both conditionings are less aggressive for the substrate.

A variant of the previous technique has been used to manufacture MSGC with gold strips, Figure 2.5. A thin (100 Å) underlayer of

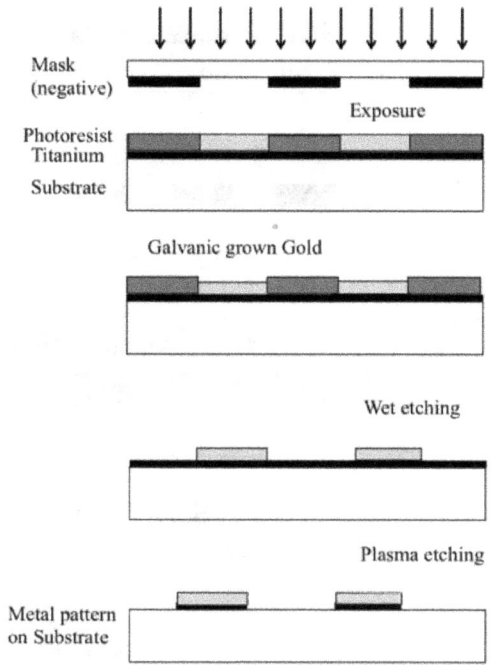

Figure 2.5:   Processing to manufacture MSGCs with gold strips.

titanium is vacuum evaporated on the substrate, and after layering the photoresist the plate is exposed to UV light through a negative mask. Channels are open in the locations of the strips, uncovering the titanium layer; a thick gold layer is then galvanically grown on top of the underlayer. The non-exposed photoresist is then dissolved, and the unwanted titanium layer removed by plasma etching. The process appears not to modify the surface resistivity of the glass, and moreover results in slightly rounded edges of the strips, beneficial for safer chamber operation due to smaller fields at edges.

In an alternative manufacturing method named lift-off, Figure 2.6, a negative photoresist film is first deposited on the

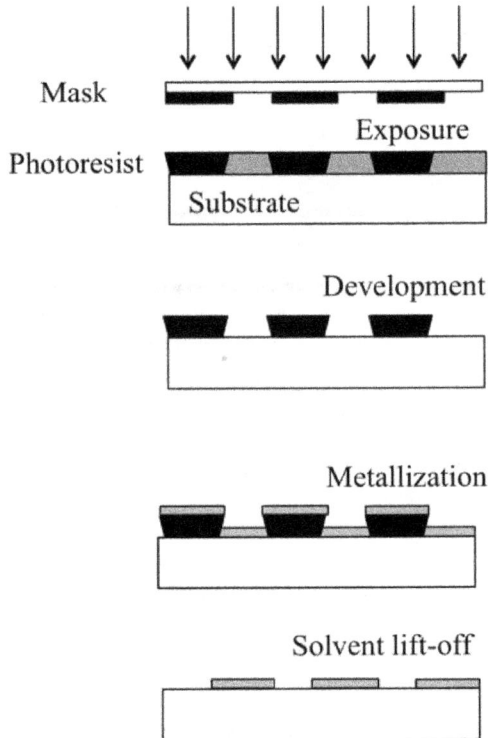

Figure 2.6:   Lift-off processing.

substrate, exposed to the UV light source through a mask and developed, thus presenting open channels with the desired strips. A metal layer is then deposited over the whole surface with one of the described methods, and the resin removed with a solvent carrying away the overlaying metal; this leaves the metal strips in the previously exposed regions. There is no hard etching of the metal layer, and the process is therefore less aggressive for the substrate.

While chromium has been used for most of the early MSGCs, to limit signal losses due to its high resistance, gold, copper or aluminium are better choices for long MSGC strips because of their higher conductivity. To avoid electro-migration or diffusion in the support, aluminium is sometimes loaded with a few percent of other materials such as copper, silicon or titanium.

Several firms produce, using direct laser writing, the masks of the desired size and precision; they are manufactured on soda-lime glass with thin chromium strips using one of the procedures described above. The photoresist is exposed to a high-resolution electron or laser beam directly driven by a computer reproducing the desired strip pattern.

The quality of the manufactured plates affects the performance of the detectors. The uniformity of gain depends on the precision in the width and thickness of the strips; the maximum achievable gain is defined by the voltage that can be safely applied to the strips before discharges occur, and this is closely related to the shape and smoothness of the edges of the strips and to the insulation properties of the substrate between anodes and cathodes. Interruptions in the continuity of strips can be found manually with a resistance meter; for large productions, the test can be automatized with a computer-controlled multi-head probe.

Figure 2.7 is an example of an MSGC plate manufactured on thin borosilicate glass, with alternating 100-μm-wide cathode and 10-μm anode strips, at a 200-μm pitch. The close-up of Figure 2.8 illustrates the excellent definition of the strips' edges, essential for the correct performance of the detector.

Figure 2.7:   View of a typical MSGC plate near the anode strips' end.
Source: Picture CERN.

Figure 2.8:   Close-up view of the strips, at 200-μm pitch, showing the very good quality of the artwork.
Source: Picture CERN.

2.3 Basic MSGC Operation

The micro-pattern devices described above rely on the charge amplification in the high electric field around the thin anode strips. Figure 2.9 shows the computed electric field structure for a typical MSGC geometry on application of a difference of potential between

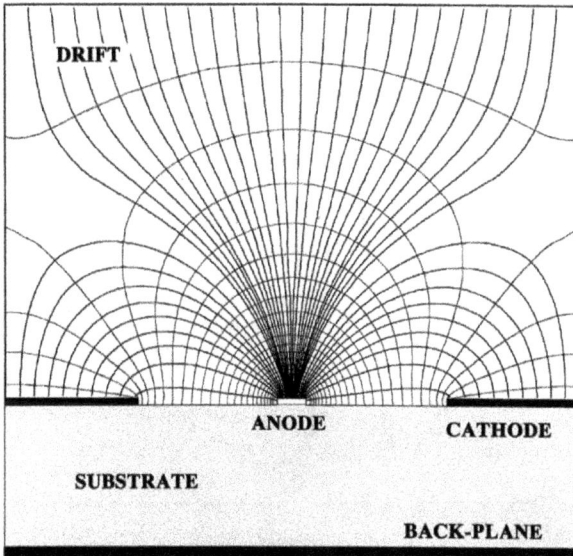

Figure 2.9: Electric field lines and equipotential close to the substrate.

alternating strips, with field lines converging from the negative cathode strips and the upper drift electrode on the thin positive anode strips; to permit an easier connection to the readout electronics, anodes are usually kept at ground potential. The backplane, not always present in MSGC structures, can be grounded to ensure screening from external noise. With a suitable choice of voltages and filling gas, electrons released in the drift gap by ionizing encounters move towards the anodes, where they multiply in the high electric field. The neighbouring cathodes swiftly collect the majority of positive ions created in the avalanche process, thus reducing the problems due to accumulation of space charge in the gap.

MSGCs have been built in a wide range of geometries and tested under various conditions. For laboratory studies, MSGC plates are usually assembled in a gas-tight vessel, with a gap-defining drift electrode, power supply inputs and signal output connections. For reasons to be discussed in Section 2.4, basic measurements are realized at low source rates and after a settling time that can be of several minutes.

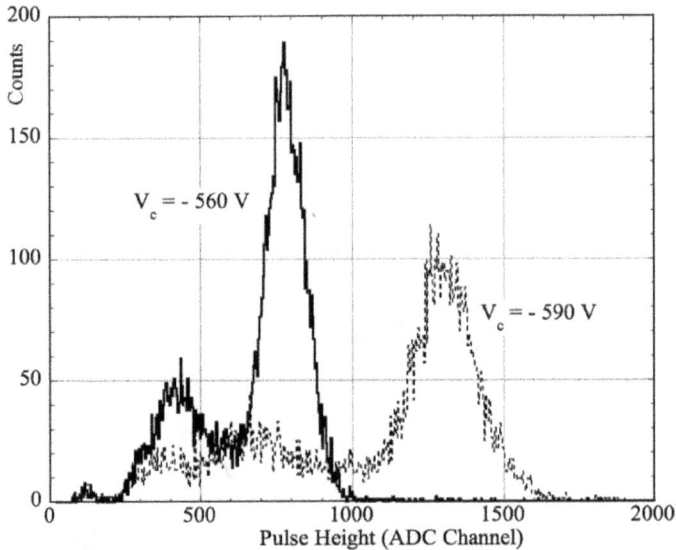

Figure 2.10:   Pulse height spectra recorded for a ^{55}Fe X-ray source at two values of the cathode strips' voltage.

Source: Barr *et al.* (1998).

Figure 2.10 is a typical pulse height spectrum recorded with a 200-μm pitch MSGC at two values of gains exposed to a ^{55}Fe X-ray source, showing the characteristic shape of the 5.9-keV main peak and the $\sim$ 3-keV escape (Barr *et al.*, 1998). Figure 2.11 is an example of pulse height spectrum measured with the same device, with a 3-mm drift gap, for a minimum ionizing particles' beam (Bachmann *et al.*, 1998).

Using as sensors charge amplifiers of calibrated sensitivity, and taking into account the total primary ionization induced by the radiation, one can estimate from the recorded pulse height spectra the proportional multiplication factor, or gain, of the detector.

A more accurate method for estimating the gain consists of the simultaneous measurement of the total detector current and of the X-ray conversions rate, see, for example, Sauli (2014, Chapter 7.2); this technique provides a result that is less affected by the value of the (often unknown) detector capacitance and various inter-strip signal couplings. An example of gain measured on the same MSGC plate with different gas fillings is given in Figure 2.12 (Beckers *et al.*, 1994);

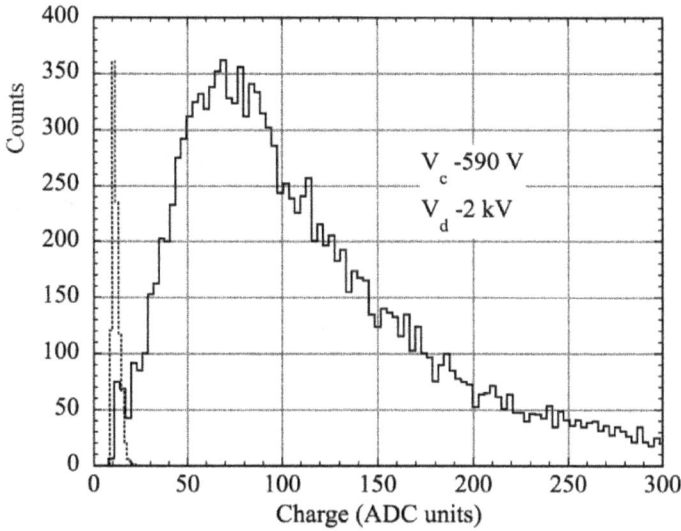

Figure 2.11: Pulse height recorded for minimum ionizing particles. The small peak is due to noise.

Source: Bachmann *et al.* (1998).

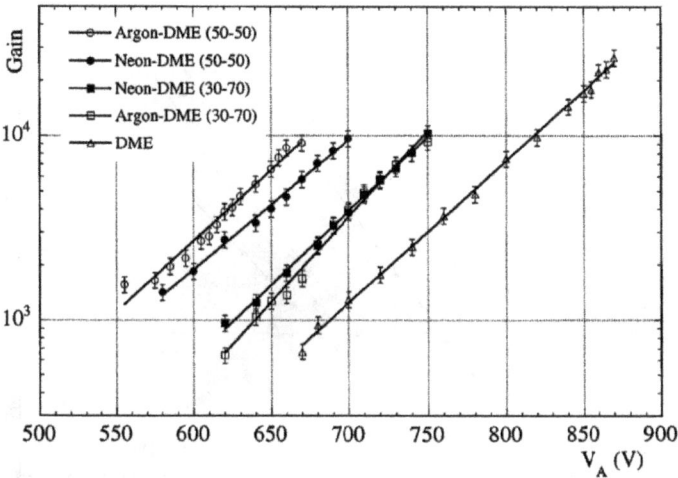

Figure 2.12: Examples of MSGC gains as a function of voltage in several gas mixtures.

Source: Beckers *et al.* (1994).

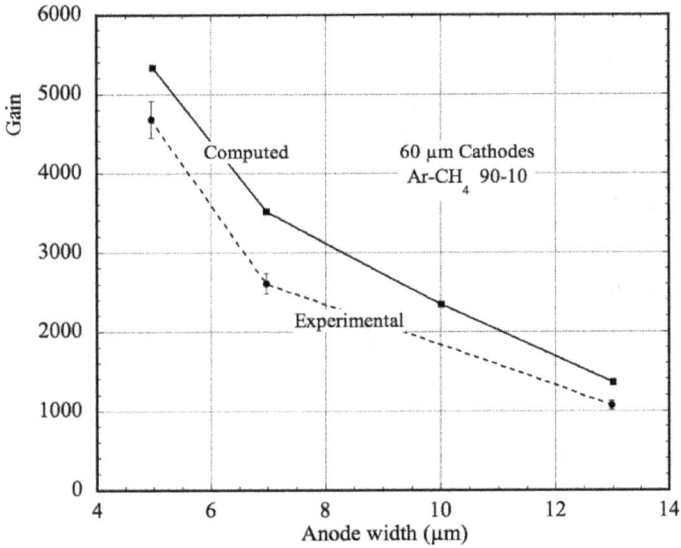

Figure 2.13:   Computed and measured dependence of gain from the anode strips' width.

Source: Beckers *et al.* (1994).

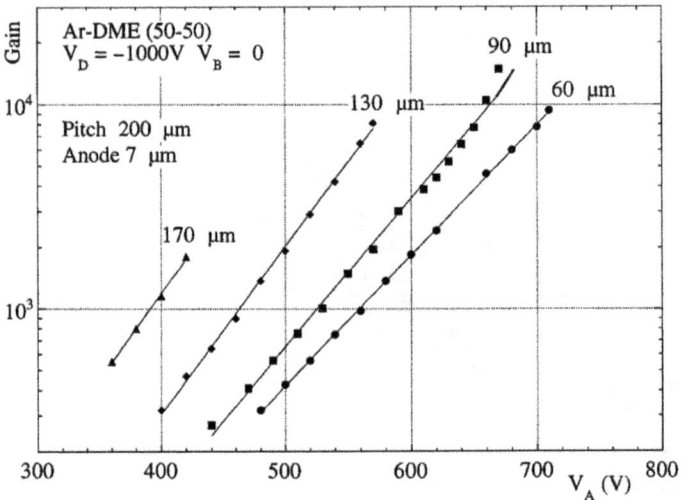

Figure 2.14:   Dependence of gain from voltage in a range of cathode strips' width for 7-μm-wide anodes.

Source: Bouclier *et al.* (1995).

the almost perfect exponential dependence from voltage confirms the good proportional response of the counter.

For a given pitch, the observed gain depends on the anode width, Figure 2.13; the cathode width also affects the gain performance, with a tendency to permit higher gains before discharge at an optimum value, around $90 \, \mu$m, as seen in Figure 2.14 (Bouclier *et al.*, 1995).

2.4 Choice of the Substrate

As mentioned in the introduction, a gain shift after application of voltage (usually a decrease) is observed on most micro-strip structures manufactured on insulating supports, with a time constant that depends on irradiation rates. Figure 2.15 (Bouclier *et al.*, 1992) is an example of this behaviour, measured at low-irradiation rates on an MSGC manufactured on borosilicate glass. The gain decrease, with a time constant of hours, is correlated to an increase of the plate resistivity, that can be deduced from the current recorded as a function of time in the absence of radiation; it can be attributed

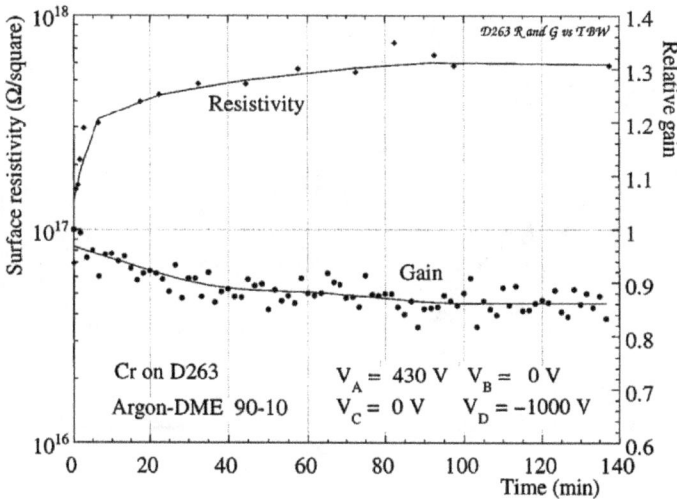

Figure 2.15:   Time dependence of gain and surface resistivity after power on for a glass substrate.

Source: Bouclier *et al.* (1992).

to the migration in the glass of ions, possibly of sodium, under the effect of the electric field, before reaching equilibrium. The process is commonly named polarization.

A different process is observed irradiating a detector after the initial setting time at increasing source rates: the gain decreases within minutes, reaching a rate-dependent plateau, Figure 2.16 (Bouclier *et al.*, 1994). The gain drop is observed to occur locally in the irradiated spot, while the rest of the chamber remains unaffected, pointing at a process of surface charging up as responsible; this makes the use of the device in non-uniform radiation fields rather problematic.

The described surface charging-up processes and ensuing field modifications have been described both mathematically and experimentally (Fang *et al.*, 1995).

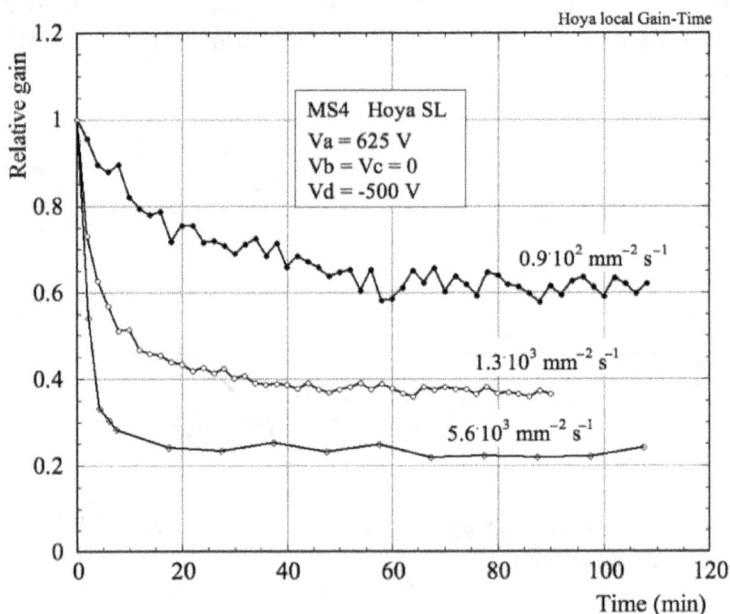

Figure 2.16:   Substrate charging up: time dependence of gain after application of voltage, at increasing radiation fluxes.

Source: Bouclier *et al.* (1993).

Use of a custom-made electron-conducting glass[2] with a resistivity in the range 109–1012 Ω · cm, originally developed for use with spark counters and named Pestov glass from its designer, moderates or eliminates the observed gain shifts up to high radiation fluxes (Minakov *et al.*, 1993). The time evolution of gain for micro-strip detectors exposed to a high radiation flux, manufactured on electronic conducting glass with bulk resistivity between 10^{10} and 10^{12} Ω · cm is compared in Figure 2.17 (Pestov and Shekhtman, 1994). The absence of gain-modifying processes up to very high radiation fluxes for electronic glasses is also shown in Figure 2.18 (Bouclier *et al.*, 1992).

While promising, bulk-conducting glass is expensive and fragile to handle, particularly in thin layers. Similar performances can be obtained reducing only the surface resistivity to between 10^{14} and 10^{15} Ω/square. Several methods of conditioning insulating supports have been explored to obtain values in this range. Early tests with

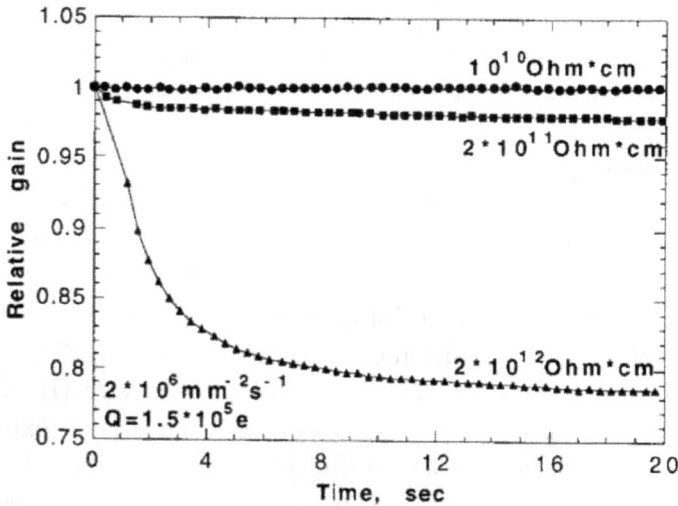

Figure 2.17: Time dependence of gain measured for MSGCs manufactured on substrates having different resistivity.

Source: Pestov and Shekhtman (1994).

---

[2]A material is said to have electronic or ohmic conductivity when the current is roughly proportional to the applied voltage.

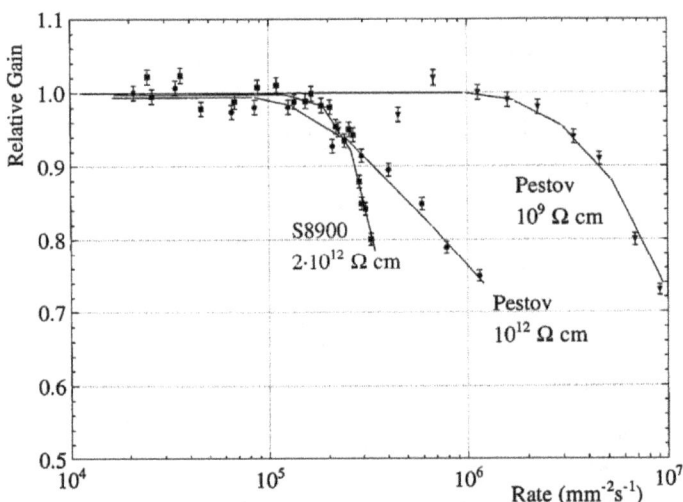

Figure 2.18:   Rate dependence of gain for substrates of different resistivity. *Source*: Bouclier *et al.* (1992).

phosphor and boron implantation in quartz and silicon oxide were reported (Budtz-Jørgensen, 1992); doubts exist, however, on the long-term stability of the implants, most ions being rather mobile in amorphous glass. Deposition by sputtering over the insulating support of an electron-conducting glass layer is an intrinsically simpler and more reliable technique to control surface resistivity (Savard *et al.*, 1993; Bishai *et al.*, 1997; Cho *et al.*, 1997).

Better uniformity and stability over large areas can be obtained by chemical vapour deposition of diamond-like carbon (DLC) layers, chemically treated to provide the required resistivity (Hollenstein, 1994). Detectors made with this technology have been extensively tested both in the laboratory and in particle beams (Bouclier *et al.*, 1996; Cicognani *et al.*, 1997; Barr *et al.*, 1998), and constituted the baseline for the realization of the large detector arrays foreseen for use in particle physics experiments (Bellazzini *et al.*, 2001). The controlled resistivity layer, with typically 10^{15} Ω/square, can be deposited on the plate before or after photolithographic processing with similar results.

Figure 2.19 shows the measured dependence of gain from the soft X-ray flux of a micro-strip plate on D-263 glass, bare or coated with a

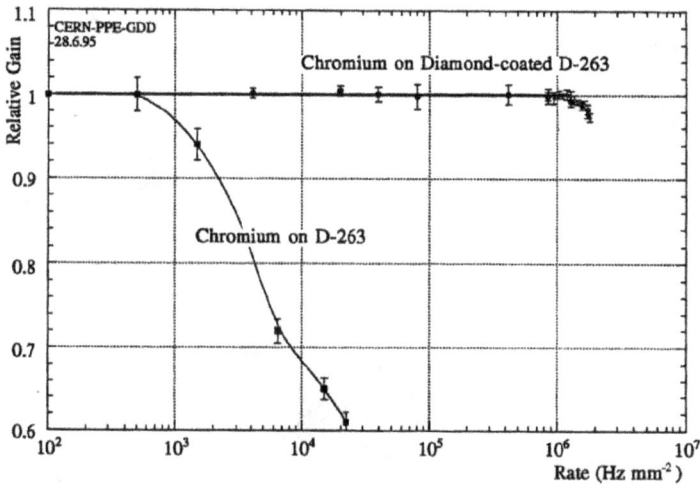

Figure 2.19: Rate dependence of gain for MSGCs manufactured on bare and diamond-coated glass.

Source: Bouclier *et al.* (1996).

CVD diamond-like layer before manufacturing (Bouclier *et al.*, 1996). Substantially similar results have been obtained by overcoating the MSGC plates after manufacture, an intrinsically safer procedure since it avoids problems due to non-uniformity or inclusions in the DCL coating; it was found, however, that the thin layer could be damaged by local discharges (Boimska *et al.*, 1998).

The rigidity of the substrate permits the construction of very light, self-supporting detectors; a medium-sized MSGC design, extensively tested in beam conditions, is shown schematically in Figure 2.20 (Bohm *et al.*, 1995). Thin frames with holes for the gas inlet-outlet are directly glued on the glass plate; the top drift electrode, also manufactured on thin metal-coated glass, completes the structure and ensures gas tightness. To prevent discharges induced by the high field at the ends of the anode strips, they may be coated with a thin insulating or passivation layer (Angelini *et al.*, 1996). Large-value resistors connecting individual or grouped cathode strips to the high-voltage supply provide an additional protection.

Attempts have been made to use thin polymer foils as supports (Stahl *et al.*, 1990, 1994; Bouclier *et al.*, 1992; Dixit *et al.*, 1994).

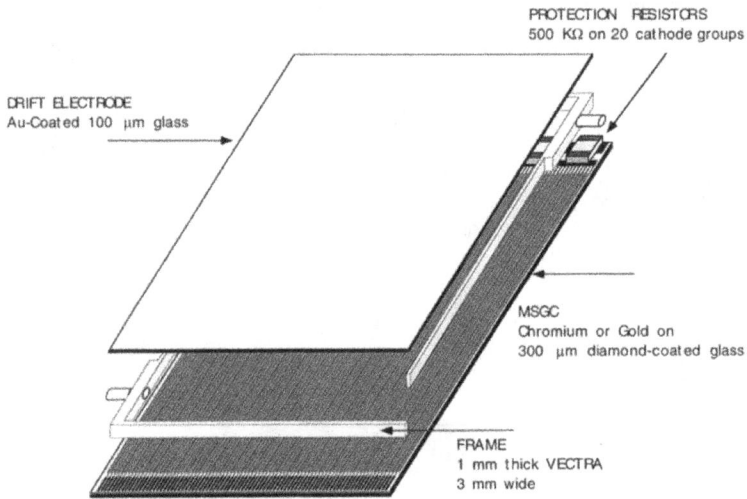

PROTECTION RESISTORS
500 KΩ on 20 cathode groups

DRIFT ELECTRODE
Au-Coated 100 μm glass

MSGC
Chromium or Gold on
300 μm diamond-coated glass

FRAME
1 mm thick VECTRA
3 mm wide

Figure 2.20:   Schematics of the light MSGC manufacturing assembly.
Source: Bohm *et al.* (1995).

Polarization and charging up can be avoided by choosing a material with moderate bulk resistivity, or having the surface conductivity increased by ion implantation. In general, however, because of modest surface quality and poor metal adherence, the results were only moderately successful. The technology for manufacturing metal patterns on polymer supports has been largely improved in subsequent years owing to the development of new families of micro-pattern detectors described in the following chapters.

The use of plastic substrates permits to realize non-planar detector geometry, suitable for colliding beam experiments. Good results have been obtained with tedlar, a polyvinyl fluoride foil,[3] owing to its moderate resistivity, $\sim 10^{12} \, \Omega \cdot cm$, and adequate metal adherence properties. The time dependence of gain, a function of rate, measured on a small prototype, is given in Figure 2.21 (Bouclier *et al.*, 1992); a semi-cylindrical MSGC prototype built on the plastic support is shown in Figure 2.22.

---

[3]Du Pont de Nemours Inc, USA.

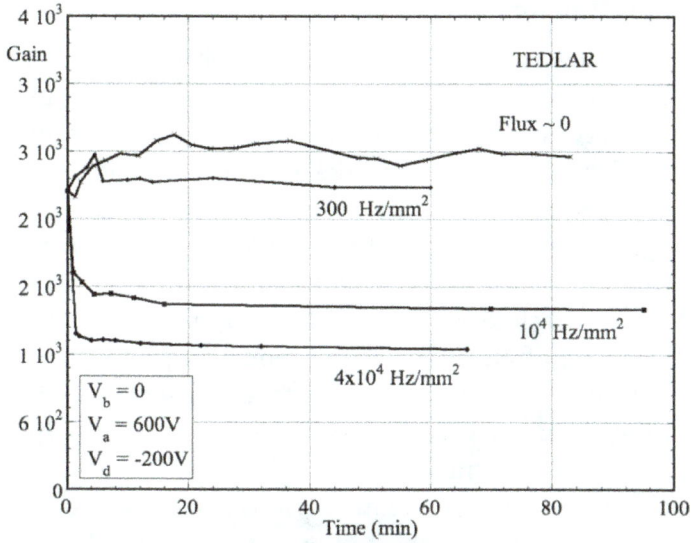

Figure 2.21: Time dependence of gain under different iradiation rates for a plastic substrate.

Source: Bouclier *et al.* (1992).

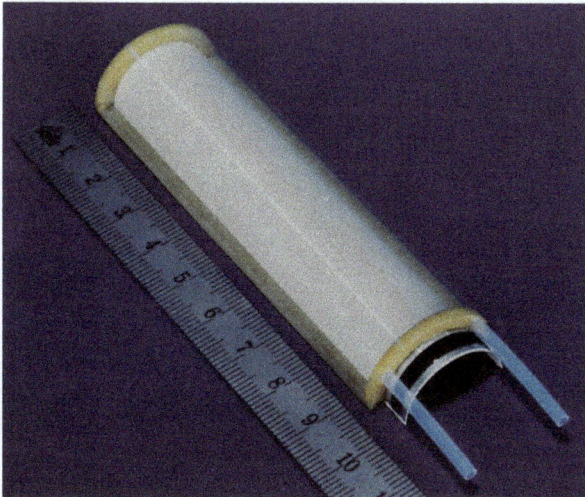

Figure 2.22: A small semi-cylindrical MSGC prototype realized on plastic support.

Source: Picture CERN.

2.5   MSGC Performances

A peculiar feature of MSGCs, as well as of most other micro-pattern detectors, is that they can be operated at a gas gain close to 10^3, providing charge signals close in amplitude and shape to those of solid-state micro-strip detectors, despite the much lower ionization density; this has permitted the utilization of existing high-density ASIC integrated electronics for their readout. Both analogue charge recording and digital threshold sensors have been used, depending on the application (Sachdeva *et al.*, 1994; Clergeau *et al.*, 1997; Zeuner, 1997). Diodes are often added at the inputs to protect the sensitive electronics from the large pulses due to gas discharges. The multi-input readout PCB cards are connected to the anode strips by wire bonding or miniature connectors.

For localized ionization events, such as soft X-ray conversions or charged particles perpendicular to the plate, the (negative) amplified charge is shared between adjacent anodes, Figure 2.23 (Bouclier

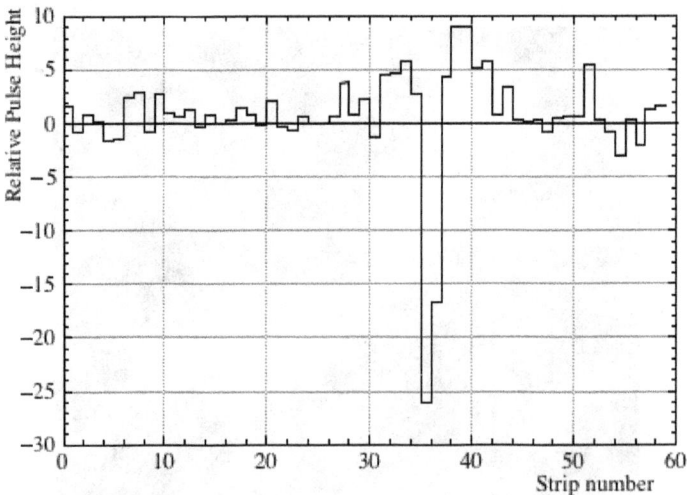

Figure 2.23:   Typical pulse height profile recorded on anode strips 200 μm apart; smaller signals of opposite polarity are induced on neighbouring strips by the motion of positive ions produced in the avalanche.

Source: Bouclier *et al.* (1995).

et al., 1995); smaller signals of opposite polarity are induced on neighbouring strips by the motion of positive ions produced in the avalanche process.

In their simpler conception, MSGCs provide one-dimensional coordinate localization; for charged particles, the space coordinates are obtained aligning several devices mounted at different angles to provide a stereoscopic view of the tracks. In the detection of neutral radiation (photons, neutrons) the second coordinate can be obtained placing a set of readout strips on the backside of the plates (Vellettaz *et al.*, 1995). In this case, the detected signal is due to the motion of positive ions, seen through the insulating plate; its amplitude depends on the ratio between the anode and cathode gap and on the support thickness, and it is therefore rather small for narrow-pitched devices. Special MSGC plate designs have been developed to increase the backplane signals (Fujita *et al.*, 2007).

A typical pulse height distribution recorded for minimum ionizing particles as well as the noise spectrum for a 200-μm pitch MSGC is shown in Figure 2.11.

The detection efficiency depends on the sensitivity of the front-end electronics employed. Figure 2.24 is a representative example of measured efficiency for minimum ionizing particles as a function of voltage and for a range of values of threshold, expressed in terms of the signal over noise ratios (Latronico, 2000).

While a simple digital signal-over-threshold detection provides a localization corresponding to the pitch, analogue recording of the pulse height profile permits, with a suitable averaging calculation (centre of gravity), to improve the localization accuracy.

Figure 2.25 is an example of measured distribution of the number strips providing a signal over threshold, or cluster size, for fast particles perpendicular to the substrate (Bouclier *et al.*, 1995). In the majority of events, the charge is shared between two adjacent strips, permitting the interpolation of the coordinate between strips. The position accuracy of the detectors for charged particles can be estimated from the distribution of the residuals, a difference of coordinates measured with a set of aligned MSGCs. An example is shown in Figure 2.26, recorded in a beam of fast charged particles

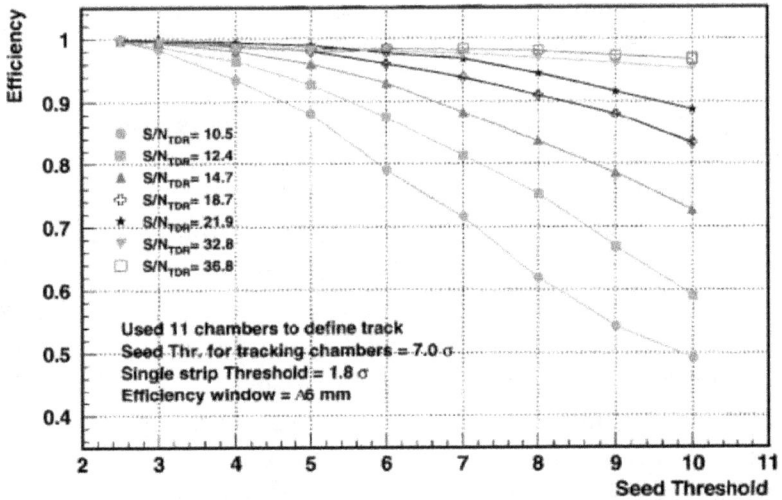

Figure 2.24:   Detection efficiency for minimum ionizing particles in a range of electronic signal over noise ratio and thresholds.

Source: Latronico (2000).

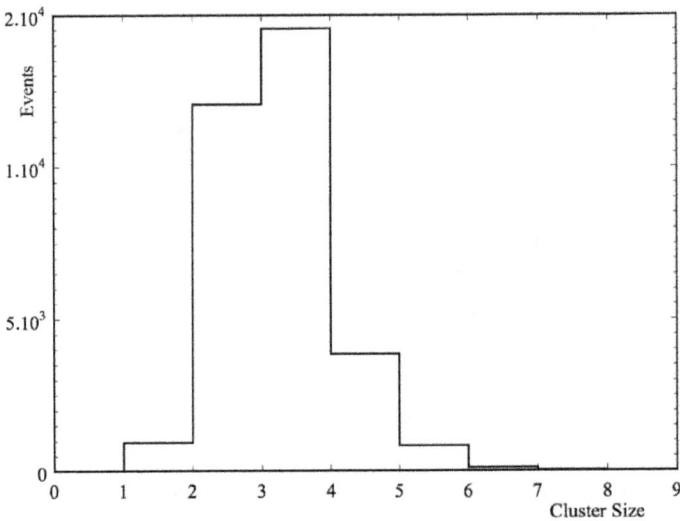

Figure 2.25:   Example of distribution of the number strips, at 200-μm pitch, providing a signal over threshold (cluster size) for fast particles perpendicular to the substrate.

Source: Bouclier *et al.* (1995).

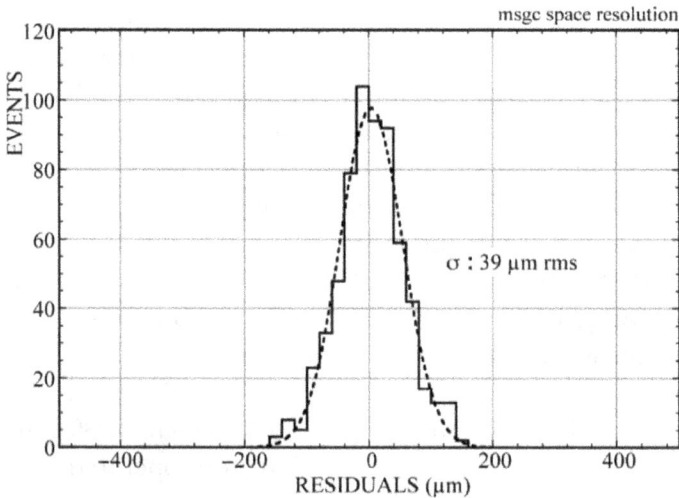

Figure 2.26: Position accuracy for fast particles perpendicular to the substrate. *Source*: Bouclier *et al.* (1995).

with three identical devices; the single-detector space accuracy for particles perpendicular to the plate is around 40-μm rms (Bouclier *et al.*, 1995).

2.6 Long-Term Operation, Ageing

The high counting rates that could be reached with MSGCs have made apparent a rate-related effect: ageing. This is a recurrent problem with gaseous devices and has been extensively studied (Kadyk, 1991; Va'vra, 2003; Sauli, 2003; Capeans, 2003). Permanent damage to the electrodes after protracted irradiation, with a severe degradation of performance, has been associated with the creation in the avalanches of polymeric compounds sticking to anodes and cathodes and perturbing the counting action or inducing discharges. Possibly because of the smaller effective area used for charge multiplication on the anode strips (a few microns as compared to a few hundred microns for wires), or a more effective polymerization process, MSGCs have been found to be particularly prone to fast radiation-induced deterioration.

Most ageing tests are made in the laboratory, using X-ray generators. To study in a reasonable time the survivability of the detectors after years of operation, the tests are usually performed at radiation fluxes several orders of magnitude larger than in real experiments; this can cast doubts on the interpretation of the results.

As for other gaseous devices, ageing of MSGCs is a rather controversial issue, rich in contradictory results; it seems clear, however, that only a proper choice of materials and gases can ensure a stable, long-term operation. Early observations led to the choice as quencher of dimethyl ether, CH_3OCH_3 (DME), a vapour well known from previous experience with wire chambers to induce small or negligible polymerization (Jibaly *et al.*, 1989). In similar operating conditions, the lifetime of detectors operated with argon-DME is orders of magnitude longer than using argon-methane (Alunni *et al.*, 1994). The presence of impurities in the gas flow or due to materials' outgassing strongly affects the result. Figures 2.27–2.29

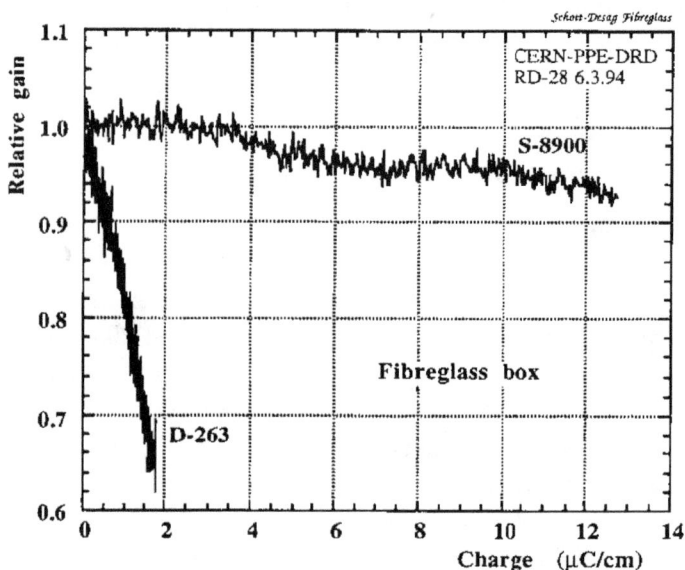

Figure 2.27:   Gain drop as a function of accumulated charge (ageing) for two supports in moderately clean conditions.

Source: Bouclier *et al.* (1994).

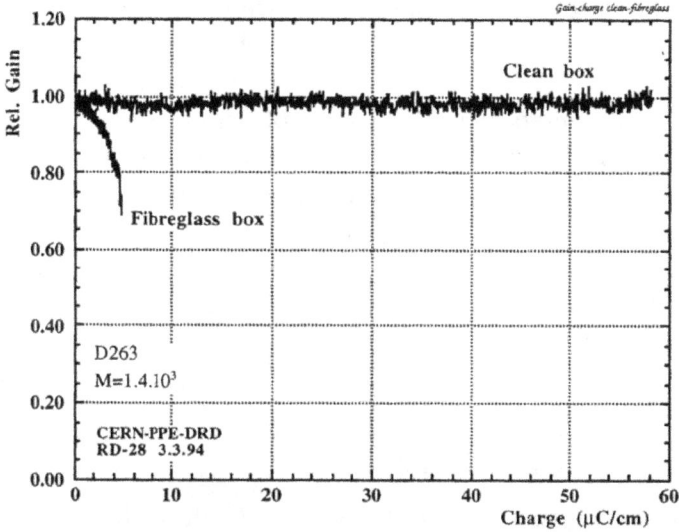

Figure 2.28: Ageing comparison for MSGCs operated in moderate (fibreglass box) and clean conditions (clean box).

Source: Bouclier *et al.* (1994).

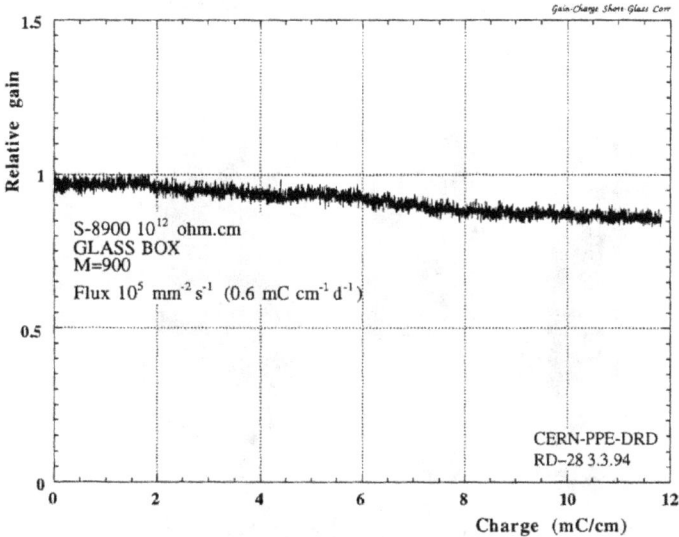

Figure 2.29: Relative gain as a function of accumulated charge for low-resistivity glass support and clean gas conditions.

Source: Bouclier *et al.* (1994).

show examples of the outcomes of a systematic study of the MSGC ageing processes (Alunni *et al.*, 1994). The relative gain, measured at intervals with a low-rate source, was recorded as a function of the charge collected per centimetre of anode strips under heavy irradiation for two detectors implemented on standard borosilicate and low-resistivity glass (Figure 2.27), and with two assembly frames, made either of standard fibreglass or using cleaner materials, glass and metals, Figure 2.28. With the "clean" construction, and in argon-DME, only a small drop in gain is observed after a total collected charge above $10\,\mathrm{mC}\,\mathrm{cm}^{-1}$, corresponding to an integrated minimum ionizing particle flux of $\sim 10^{11}\,\mathrm{mm}^{-2}$ (Figure 2.29).

2.7   Discharges and Breakdown

Well performing in the laboratory, MSGCs have been found to be prone to discharge when exposed to particle beams, and their fragile electrodes can be easily damaged (Bouclier *et al.*, 1995; Bellazzini *et al.*, 1998; Bressan *et al.*, 1999). Depending on the energy involved in the discharges, the outcome can vary from microscopic local damages (pitting), Figure 2.30, to complete local destruction of the strips (Figure 2.31).

Figure 2.30:   Moderate strips' damage (pitting) caused by low-energy discharges. *Source*: Picture CERN.

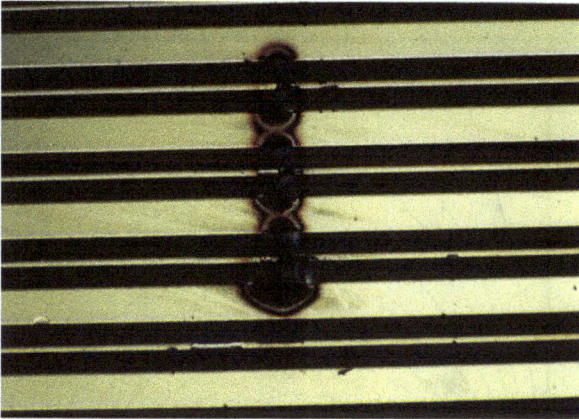

Figure 2.31: Local large damage due to a strong discharge.
Source: Picture CERN.

The processes leading to discharges have been extensively studied both experimentally and by model calculations. Aside from manufacturing defects, two sources of sudden breakdowns have been identified: spontaneous field emission of burst of electrons from the cathode strip edges and occurrence in the sensitive volume of heavily ionizing events during operation. Electrons ejected by the cathode edges are pre-amplified in the high local field, and are then further multiplied on reaching the anodes, thus resulting in an abnormally large avalanche size.

Figure 2.32 shows the electric field in the region close to the substrate, as well as the computed lines of equal gain for electrons multiplying all the way to the anode (Beckers *et al.*, 1994). For electrons emitted at the cathode, the overall charge can then approach the so-called Raether limit, $\sim 10^7$ electrons, in most cases causing a transition to a streamer and/or a discharge (Raether, 1964).

The second process leading to breakdown can be initiated by large radiation-induced ionization losses in the gas. In a 3-mm-thick gas gap operated at atmospheric pressure, minimum ionizing particles and soft X-rays liberate around 100 electron-ion pairs, while an electromagnetic shower or a neutron interaction can release three orders of magnitude more charge. Observed in most gaseous

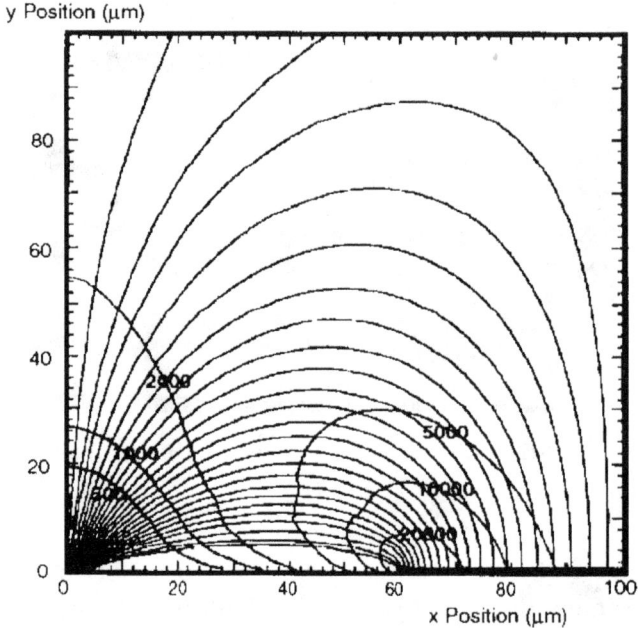

Figure 2.32:   Computed gain for electrons released near the cathode strip edges. *Source*: Beckers *et al.* (1994).

devices, the discharge process has been extensively studied, see, for example, Sauli (2014, Section 8.8). It can be investigated in the laboratory exposing the detectors to controlled sources of heavily ionizing radiation, ~MeV alpha particles from an external source as ^{241}Am, or from the decay of radioactive isotopes such as ^{220}Rn introduced with the gas flow.[4] In both cases, the primary charge released is around 10^5 electron-ion pairs, three orders of magnitude larger than for fast particles; at a typical gain of 10^3, the Raether limit is easily exceeded.

While the operating voltage of an MSGC depend on geometry and gases, it appears that the discharge limit is almost an invariant when expressed as a function of total created charge, consistent with the Raether conjecture. Figure 2.33, part of a systematic study

---

[4]Emitted by the decay of ^{232}Th isotopes from a natural thorium oxide cartridge.

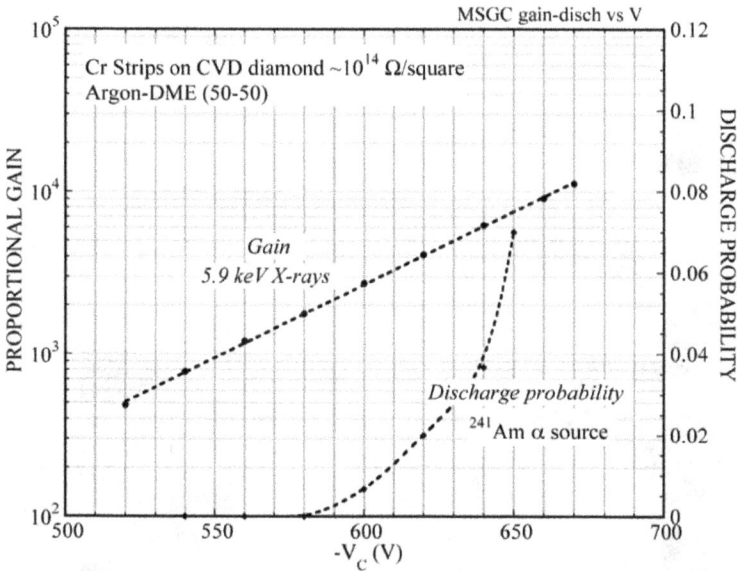

Figure 2.33: Gain for soft X-rays and discharge probability on exposure to heavily ionizing particles as a function of voltage.

Source: Bressan *et al.* (1999).

of discharge limits in micro-pattern devices, shows an example of experimental results obtained with a 200-μm pitch MSGC manufactured on a CVD-coated glass plate and operated in argon-DME (Bressan *et al.*, 1999). The gain is measured as a function of voltage with a low-intensity 5.9-keV ^{55}Fe X-ray radioactive source, and the discharge probability recorded exposing the detector to $\sim$5-MeV alpha particles. While the definition of the discharge onset is somewhat arbitrary, the fast rise of its probability sets a clear limit to the operation of the detector at gains of a few thousand.

2.8 Applications of MSGCs

As indicated in the introduction, micro-strip detectors were originally developed and successfully employed for applications in neutron spectrometry, using ^{3}He as sensitive gas or internal thin converter foils, at moderate counting rates; the large primary ionization charges produced by the interactions thus permit operation at safe values

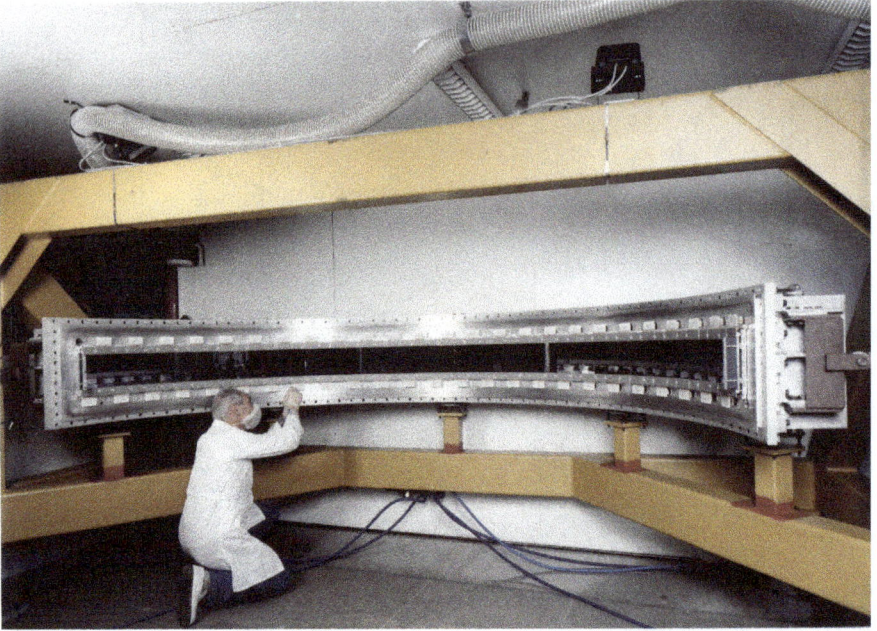

Figure 2.34:   The D20 powder diffractometer.
Source: Clergeau *et al.* (2001).

of gain. The D20 powder diffractometer, assembled at the ILL laboratory in Grenoble in late 1990 and still operational with various refurbishments and modifications, includes 48 medium-sized MSGCs, each with 32 counting cells, Figure 2.34 (Clergeau *et al.*, 2001). Mounted on a vacuum-tight vessel, the detector operates with a mixture of ^{3}He and CF_4, added to reduce diffusion and increase the electrons' drift velocity, at pressures between 2 and 4 bars. After experiencing serious instability and reliability problems, the original MSGC plates manufactured on borosilicate glass were replaced with detectors built on electron-conducting glass S8900 (Ortuiio-Prados and Budtz-Jørgensen, 1995), and operated satisfactorily for many years.

Two large-sized MSGCs were built to perform X-ray detection on board the ESA's International Gamma Ray Astrophysics Laboratory (INTEGRAL), launched in 2002. The JEM-X detector, Figure 2.35,

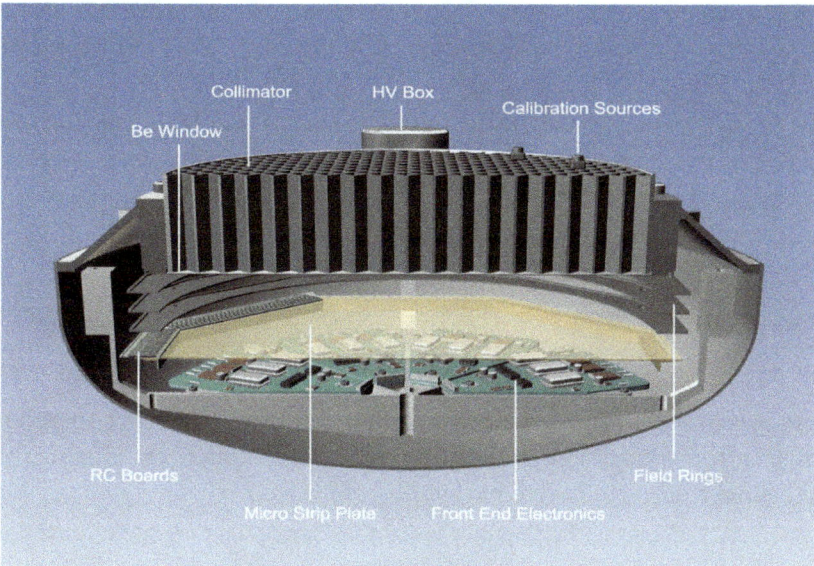

Figure 2.35:   The JEM-X detector.
Source: Courtesy C. Budtz-Jørgensen, DTU Space, Denmark.

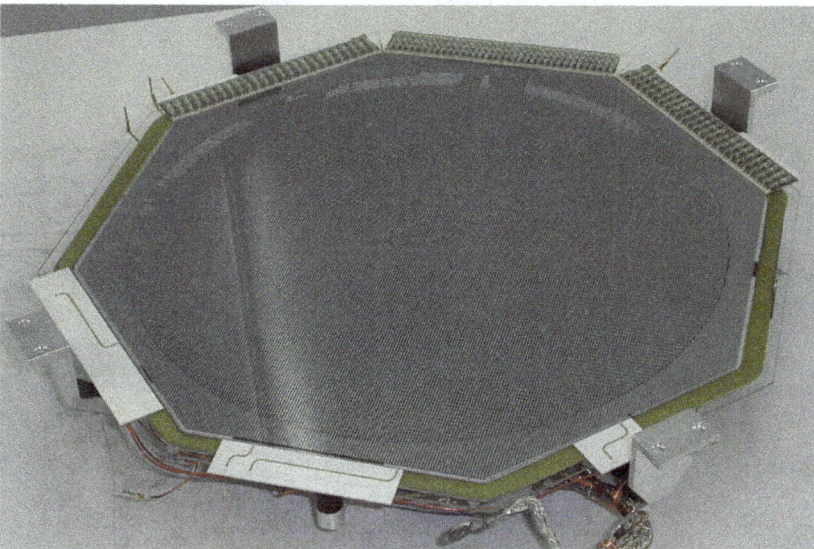

Figure 2.36:   The MSGC plate of the JEM-X telescope on board the INTEGRAL satellite.
Source: Courtesy C. Budtz-Jørgensen, DTU Space, Denmark.

is a xenon-filled coded-mask X-ray telescope, covering an energy range between 3 and 35 keV; the MSGC plate, 25 cm in diameter, is shown in Figure 2.36 (Lund *et al.*, 1999, 2003).

Initially suffering discharge and instability problems, the detector could be operated reducing the gain to the minimum required for operation, successfully mapping the galactic X-ray sources; completion of the mission is foreseen in 2029.

In one experiment, the Compact Muon Solenoid (CMS) at CERN's LHC, both the central and the end-cap trackers were designed making use of nearly 15,000 medium-sized MSGCs (Muller, 1998). Figure 2.37 schematically shows the barrel assembly. The project was discontinued, and replaced by solid state silicon microstrips, after long-term reliability problems appeared during the prototype studies.

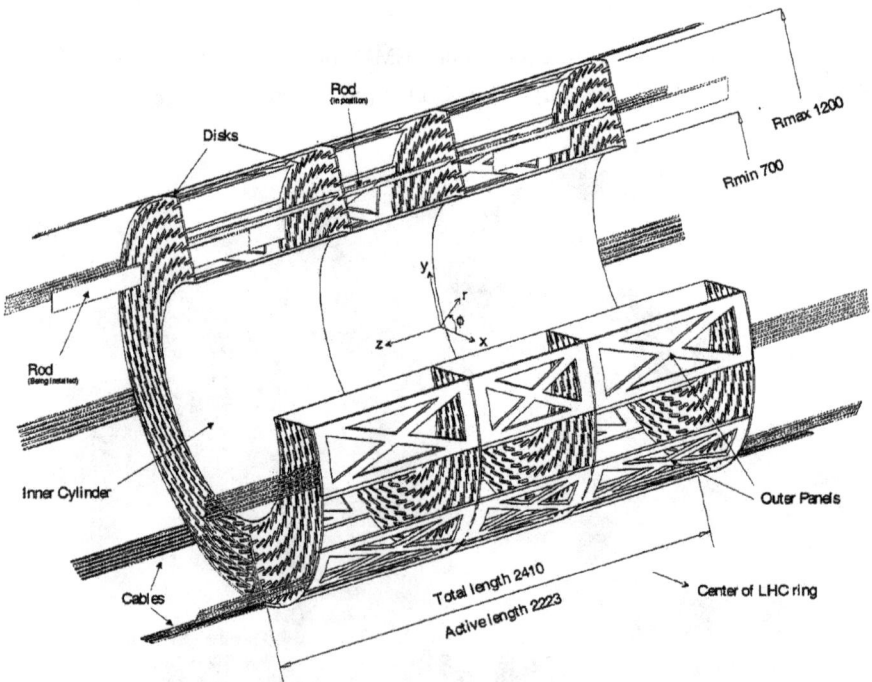

Figure 2.37:   The MSGC-based CMS barrel tracker project.
Source: Muller (1998).

2.9 Combined MSGC and GEM: The HERA-B Tracker

Built at the end of the nineties, the HERA-B inner detector system at DESY included 200 large-sized MSGCs with a total of 150,000 readout channels (Zeuner, 1997). Designed to operate at rates up to $10^4\,\mathrm{mm^{-2}\,s^{-1}}$, it soon suffered from severe degradation problems due to progressive destruction of the anodes caused by discharges. It was therefore decided to adopt the newly introduced Gas Electron Multiplier technology (see Chapter 5) as pre-amplifier of the ionization to reduce the operating voltage of the fragile microstrips. Figure 2.38 schematically shows the MSGC-GEM assembly (Zeuner, 2000), and Figure 2.39 is an example of combined gain curves measured as a function of voltage (Beirle *et al.*, 1999). To achieve the needed gain of around 5000, and operating the GEM at an effective pre-amplification around 50, the MSGC voltage could be decreased to safe values, resulting in a reduction in the discharge probability by a factor of 10^5 (Hott, 1998). The pre-amplifier foils were added progressively to the existing plates; the improvement required the development of the technology for the production and quality assessment of a large number of GEM foils, done at the CERN workshops. Figure 2.40 shows one of the GEM foils, $30 \times 30\,\mathrm{cm^2}$ held by the developers.

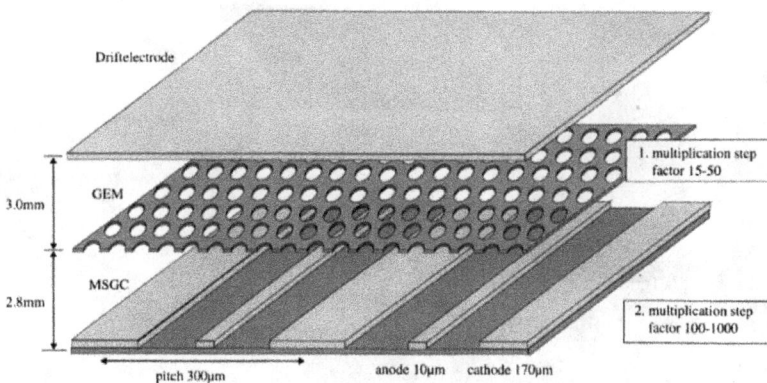

Figure 2.38:   HERA-B MSGC+GEM.

Source: Zeuner (2000).

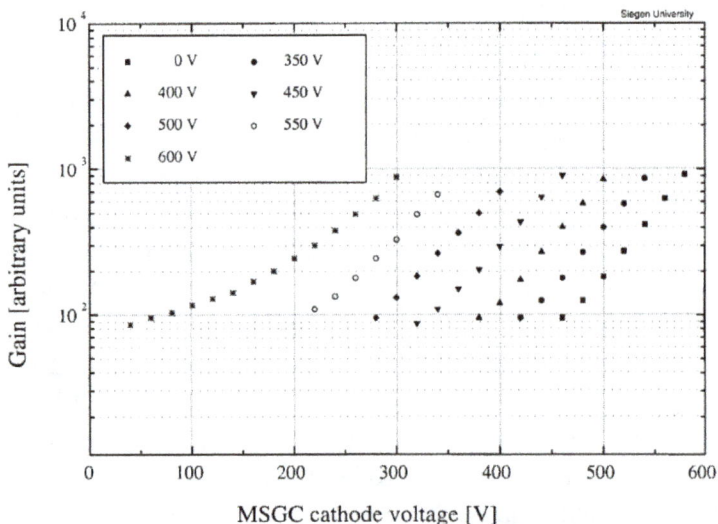

Figure 2.39:   Combined MSGC+GEM gain curves.
Source: Beirle *et al.* (1999).

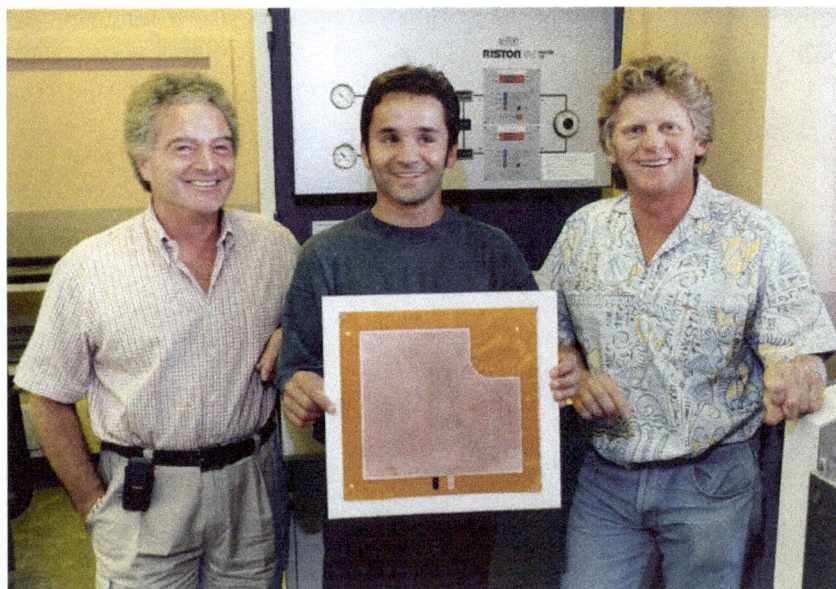

Figure 2.40:   One of the HERA-B GEMs produced at CERN. Left to right:
A. Gandi, R. De Oliveira and J-C. Labbé.
Source: Picture CERN.

The HERA-B inner tracker operated for several years, with a recorded discharge rate in running conditions of a few per day (Bagaturia *et al.*, 2002). A similar assembly of a cascaded GEM and an MSGC was used for the DIRAC high-resolution spectrometer at CERN (Adeva *et al.*, 2003), later replaced by a system of micro-drift chambers (Adeva *et al.*, 2016).

Chapter 3

Micro-Pattern Gas Detectors

3.1 Introduction

Fostered by the development of micro-strip counters and aiming
to improve performances and mitigate the long-term reliability
problems, numerous innovative detector geometries have been conceived and manufactured with photolithographic or silicon foundry
technologies both on rigid and plastic supports. Characterized by
narrow-pitch sensing electrodes, they are collectively named micro-
pattern gas detectors (MPGDs). Although these are found to be
interesting for some of their performances, the simpler structures
have been discontinued after the introduction, at the end of the
nineties, of the micro-mesh gaseous structure (MICROMEGAS)
and the gas electron multiplier (GEM), which are powerful and
easier to manufacture in large sizes, as described in the following
chapters. More sophisticated developments of some of these devices
are described in Chapter 8.

3.2 Micro-Gap, Small-Gap and More

Targeting a solution of the discharge problem, the so-called Coated
Cathode Conductive Layer (COCA COLA) chamber, depicted
in Figure 3.1, avoids breakdowns between anode and cathode
strips placing the two electrodes on opposite sides of a substrate
with sufficient conductivity to prevent the charging-up processes.
Manufactured on Tedlar,[1] with bulk resistivity $\sim 10^{12}\,\Omega\,\mathrm{cm}$, small

---

[1]Tradename Du Pont de Nemours.

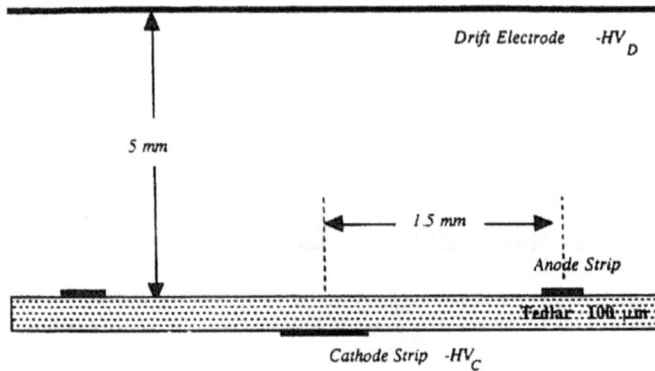

Figure 3.1:   Schematic diagram of the COCA COLA MSGC, with anodes and cathodes on opposite sides of a high resistivity substrate.

Source: Bouclier *et al.* (1991).

prototypes proved to operate satisfactorily for sources rates of few tens of Hz/cm^2 (Bouclier *et al.*, 1991). Due to the unavailability of suitable supports with lower bulk resistivity, this work was not pursued.

Introduced in 1993, the micro-gap chamber (MGC) aims at largely increasing the rate capability by reducing to a minimum the insulating gap between anode and cathode (Angelini *et al.*, 1993). Built on an insulating or high-resistivity substrate, the structure consists of thin metal anodes separated from the metallic cathode by silicon oxide or polyamide insulating strips, few μm thick, slightly wider than the anodes (Figure 3.2). An upper drift electrode defines the sensitive volume. In later designs, anode and insulating strip had the same width without noticeable difference in performance. On application of suitable potentials to the electrodes, an electric field is generated, as shown in Figure 3.3, with a high density of field lines connecting the anodes and the near cathode; this results in a fast development of the avalanches and a quick collection of positive ions (Bellazzini, 1996). Examples of gain measured as a function of anode voltage for different gas mixtures are given in Figure 3.4.

The energy resolution in the detection of soft X-rays is remarkably good: around 15% FWHM for 5.4 keV (Angelini *et al.*, 1993). Added to the reduced size of the insulating surfaces, minimizing

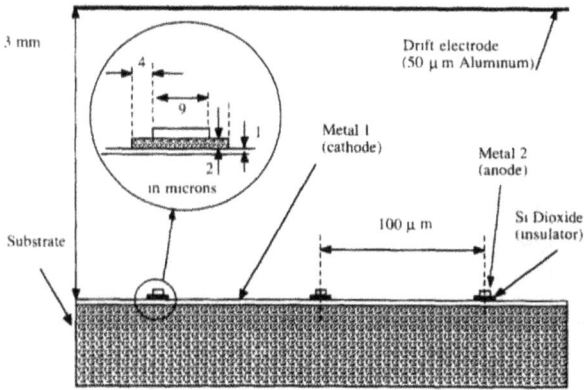

Figure 3.2:   Schematic diagram of the MGC.
Source: Angelini *et al.* (1993).

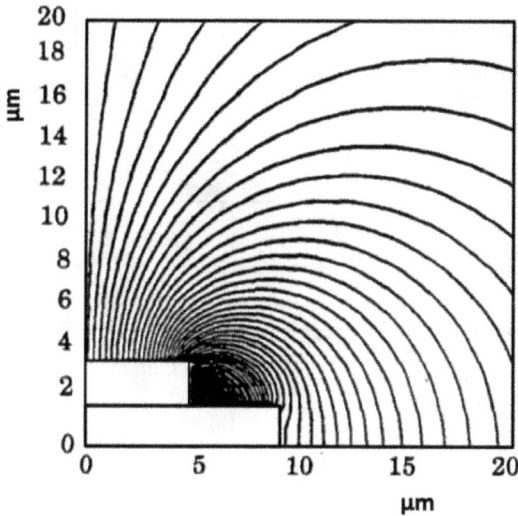

Figure 3.3:   Electric field lines close to the anode in the MGC.
Source: Bellazzini (1996).

charging-up, the fast clearing of the positive charge results in a rate capability reaching 10^7 Hz mm^2; see Figure 3.5 (Bellazzini, 1996).

Owing to the small distance between electrodes, a large positive signal is induced on the cathode, which can be exploited to obtain

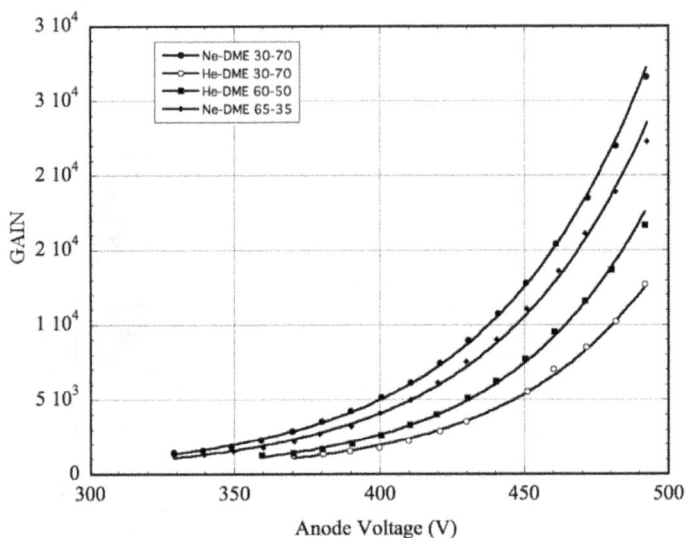

Figure 3.4:   Gain as a function of anode voltage of the MGC for different gas mixtures.

Source: Bellazzini (1996).

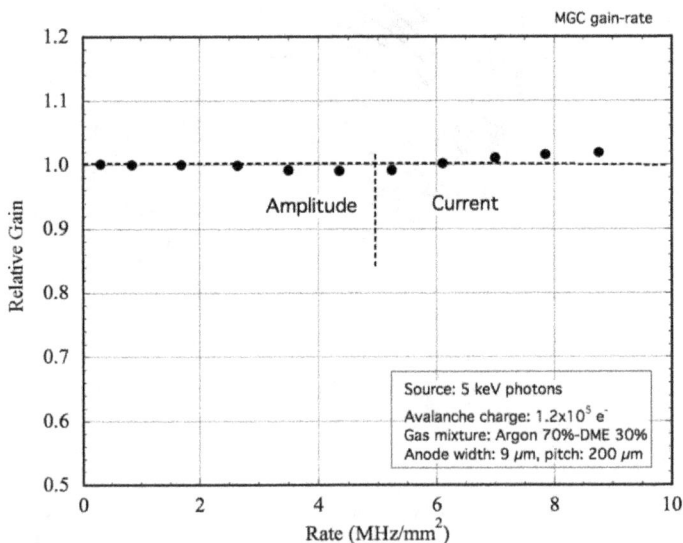

Figure 3.5:   Relative gain of the MGC as a function of irradiation rate.

Source: Angelini *et al.* (1993).

two-dimensional localization with a suitable patterning of the electrode (Bellazzini and Spandre, 1995).

The high rate performance, long-term stability and robustness against discharges have been studied for a wide range of metals, substrates and operating gases, with the rather predictable conclusion that a small detector capacitance and the use of metals with high melting point, such as titanium and tungsten, reduce the damages in the case of discharges (Cho *et al.*, 1998).

Since the essential parameter to ensure response uniformity is the accuracy in the artwork, the best results have been obtained with MGC manufactured on silicon using integrated circuit technologies and a self-alignment technique, therefore limiting the detector size to the availability of standard wafer production lines (5 or 8 inches diameter) (Clergeau *et al.*, 1997; van den Berg *et al.*, 1998; Barthe *et al.*, 1998).

Aimed also at alleviating the discharge problems in MSGCs, the small-gap chamber (SGC) alternates anode and cathode strips on a substrate, at a small distance, filling the gap between electrodes with an overlay of insulating polyimide; see Figure 3.6 (Chorowicz *et al.*, 1997). Manufactured also with silicon foundry technology, the device could reach gains above 10^3 with only a moderate (15%) drop at power on due to the initial charging up of the insulator.

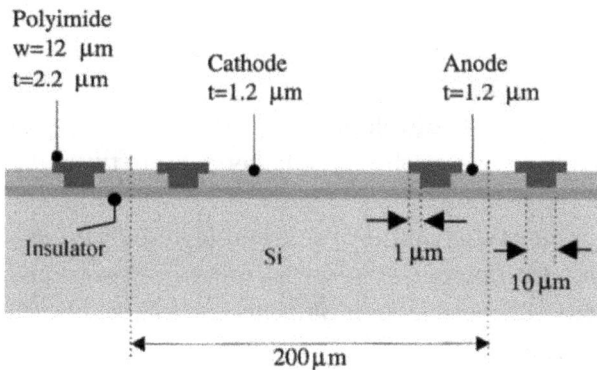

Figure 3.6: Schematic diagram of the SGC.
Source: Chorowicz *et al.* (1997).

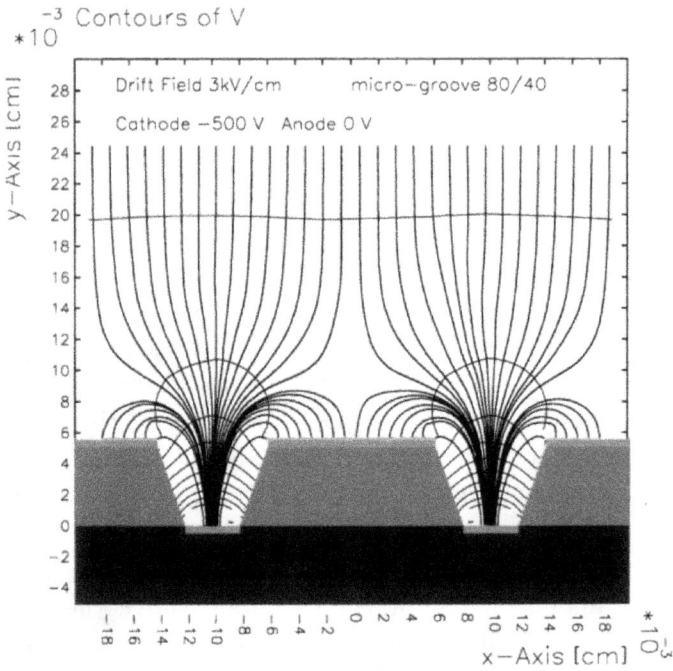

Figure 3.7:   The micro-groove detector.

Source: Bellazzini *et al.* (1999a).

Extensively tested in harsh beam conditions, SGCs were found
to suffer essentially from the same problems as the MSGCs, with
discharges often followed by irreversible damages to the structure
(Bouvet *et al.*, 2000).

Prefiguring later developments, the micro-groove detector
(MGD) has anode and cathode strips etched at two different levels,
separated by an insulator; see Figure 3.7 (Bellazzini *et al.*, 1999a).
To manufacture the structure, a metal-clad polyimide foil is first
patterned on the bottom side with strips or pads and glued to a
thin substrate. The upper side is then etched by photolithography
to form the anode strips, and buried channels are created by wet
etching of the insulator using the anode as mask. Compared with
the MGC, the new design has the advantage of a longer high-field
path for multiplication and therefore permits one to attain larger

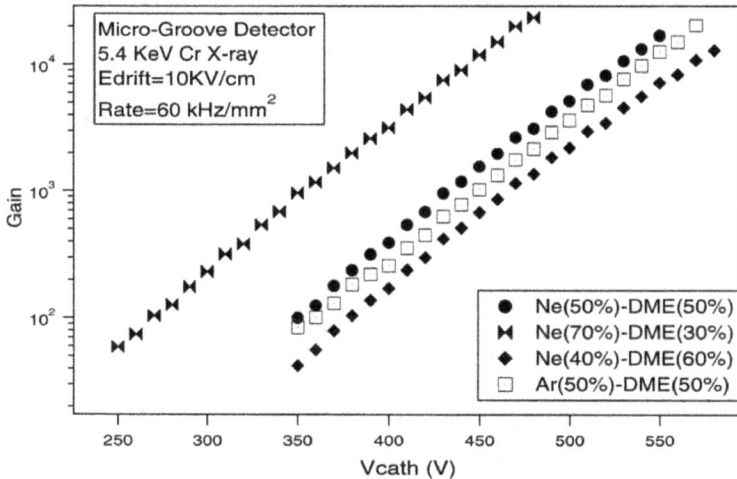

Figure 3.8: Gain of the MGD as a funnction of cathode voltage in several gas mixtures.

Source: Bellazzini *et al.* (1999a).

gains; see Figure 3.8. The use of thicker metals for the electrodes, typically 5 μm copper, increases the robustness of the detector in the case of discharges.

An outsized version of the MGD, built with conventional printed circuit board technology and named thick groove detector (TGD), can cover large detection areas at low cost. With 500 μm resolution and modest rate capability, it was found suitable to cope with the less demanding needs of cosmic ray tomography (Biglietti *et al.*, 2016).

3.3 Pixel Detectors

The micro-pattern devices described above rely on the charge amplification between thin anode strips and cathodes on the same substrate or buried in insulating trenches. Proposed in 1996, the "Compteur A Trous" (CAT) originally had a single hole, 1 mm in diameter mechanically pierced on a printed circuit board, facing the anode collecting and amplifying the electron charge released in the gas; see Figure 3.9 (Bartol *et al.*, 1996). Despite the large difference in the field line length and strength, the counter exhibited surprisingly

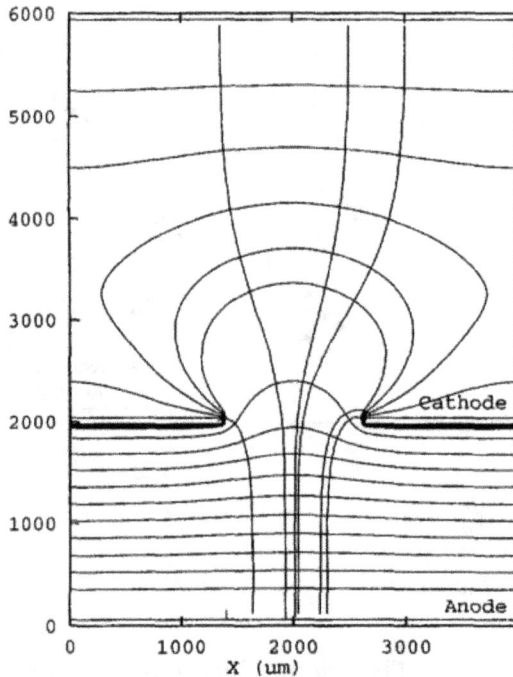

Figure 3.9:   Schematic diagram of the CAT.

Source: Reproduced from Bartol *et al.* (1996) with kind permission of AIP Publishing.

good energy resolution for soft X-rays; see Figure 3.10. A detailed analysis of detected charge as a function of position of a collimated source confirmed that, while the resolution is affected for the off-axis ionization, the gain is only slightly influenced (Chaplier *et al.*, 1999; Chaplier *et al.*, 2000).

Extended to a two-dimensional array of identical holes, the device permits the detection of ionization on individual anode pads or strips, prefiguring the later introduction of the gas electron multiplier (see Chapter 5).

A scaled down version of CAT, the WELL detector, consists in a two-dimensional matrix of sub-mm diameter holes, pierced with an advanced printed circuit technology on a 50-μm polyimide foil and pasted to a rigid readout board patterned with pads or strips;

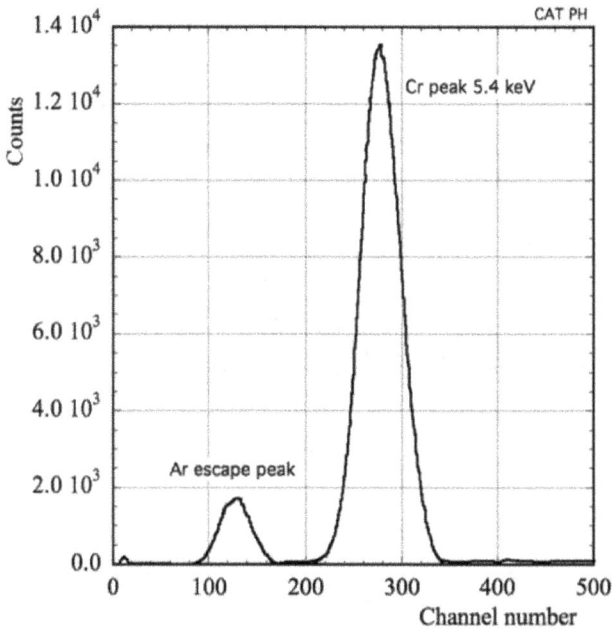

Figure 3.10: Energy resolution of CAT measured for a 5.4 keV X-ray source. *Source*: Reproduced from Bartol *et al.* (1996) with kind permission of AIP Publishing.

see Figure 3.11 (Bellazzini *et al.*, 1999b). With an energy resolution of around 20% for 5.9 keV X-rays, the counter can reach proportional gains above 10^4 in a wide choice of gases; see Figure 3.12. Named also micro-WELL, the detector finds applications in large area imaging detectors for astrophysics (Deines-Jones *et al.*, 2002).

While more robust than MSGC, the device suffered from similar breakdown problems when exposed to harsh particle beams; the problem could be solved after the development of the technology to add a high resistivity underlayer to separate the active electrodes from the readout. This is developed in order to solve similar problems in other micro-pattern devices at the cost of a reduced rate capability (Peskov *et al.*, 2012) (see also Section 4.4).

Manufactured with silicon foundry technology, the MICRODOT device consists in a matrix of hexagonal cells with anodic "dots"

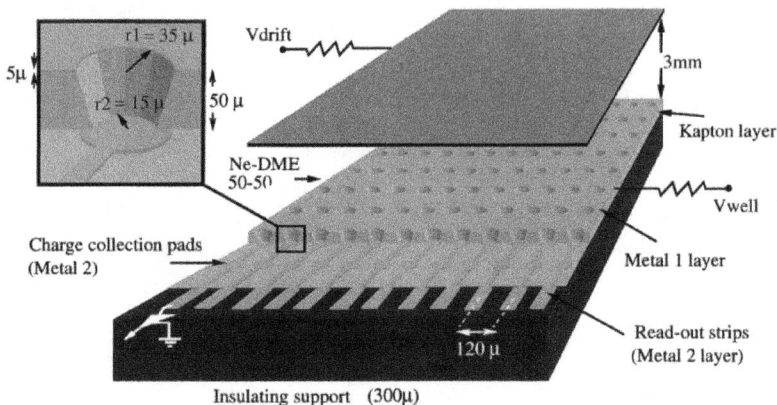

Figure 3.11:   Schematic diagram of the WELL detector.
Source: Bellazzini *et al.* (1999b).

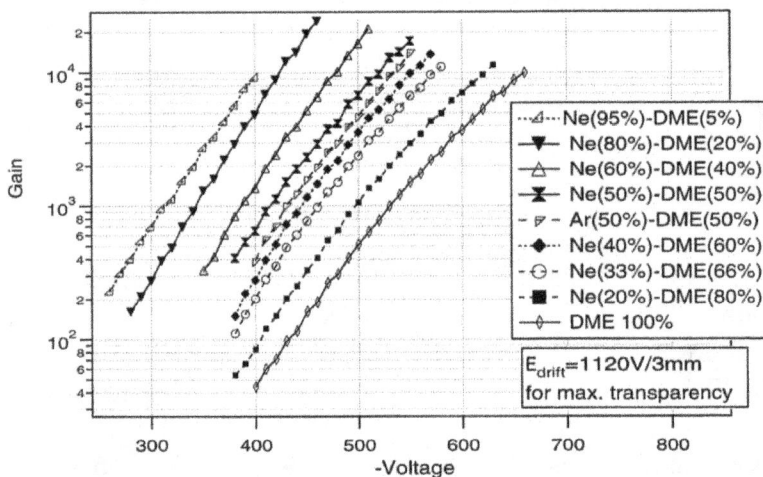

Figure 3.12:   Gain of the WELL detector measured in several gas mixtures.
Source: Bellazzini *et al.* (1999b).

surrounded by field rings built on a silicon oxide support; see
Figure 3.13 (Biagi and Jones, 1995). With the substrate ion-
implanted to prevent charging up of the insulator between electrodes, the detector provides gains above 10^4 at an incident X-ray

Figure 3.13: Schematic diagram of the MICRODOT detector.
Source: Biagi *et al.* (1997).

flux up to $1\,\mathrm{MHz\,mm^{-2}}$ (Biagi *et al.*, 1997). With a boron-doped amorphous silicon carbide semiconducting coating, the detector could withstand an accumulated charge up to 120 mC/cm with only a small gain loss (Biagi *et al.*, 1998). Possibly because of the cost and size limitations imposed by the manufacturing technology, the MICRODOT approach was discontinued in favour of the newly developed structures exploiting simpler photolithography processing.

A three-dimensional version of the device, the micro-pin array (MIPA), as shown in Figure 3.14 (Rehak *et al.*, 2000), avoids the surface-related charging and breakdown problems receding the anodes from the cathodes. Similar in design, but manufactured with conventional printed circuit board technologies, the micro-pixel (μ-PIC) detector is described in Section 8.3.

Reminiscent of the multi-pin array, developed in the eighties (Comby *et al.*, 1980), the pin pixel detector uses arrays of commercial multi-pin connectors inserted in a machined brass plate acting as cathode; see Figure 3.15 (Bateman *et al.*, 2002).

Similar in conception, the LEAK microstructure consists in a matrix of pins or wires assembled on an insulating plate and surrounded by a metallic sheet with holes serving as cathode; see Figure 3.16 (Lombardi and Lombardi, 1997). Conical nickel-plated needles with rounded tips around 10 μm in radius emerge about 1 mm from the structure and act as individual counters, permitting one to reach very high gains; Figure 3.17 shows an example of

Figure 3.14:   Microscope view of the MIPA detector.
Source: Courtesy G. Smith, BNL.

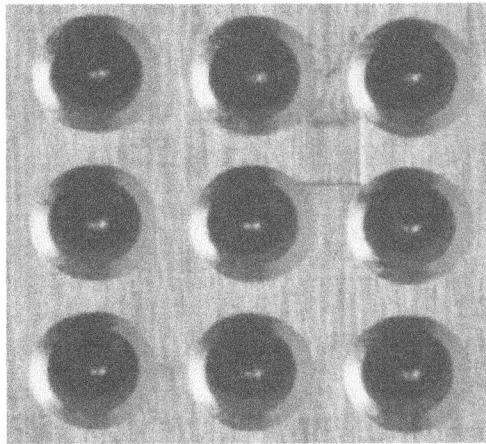

Figure 3.15:   Close view of the pin pixel detector.
Source: Bateman *et al.* (2002).

single-electron pulse-height distribution for a LEAK counter operated in low pressure propane. The detector was used for experiments
in nanodosimetry at the INFN Laboratory in Legnaro, Italy (Ferretti
et al., 2008).

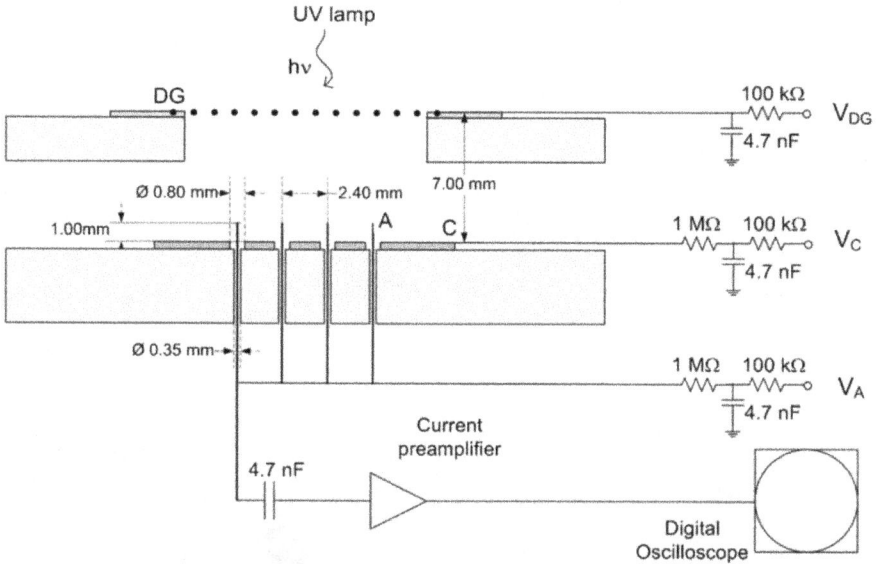

Figure 3.16: The LEAK microstructure.

Source: Ferretti *et al.* (2008).

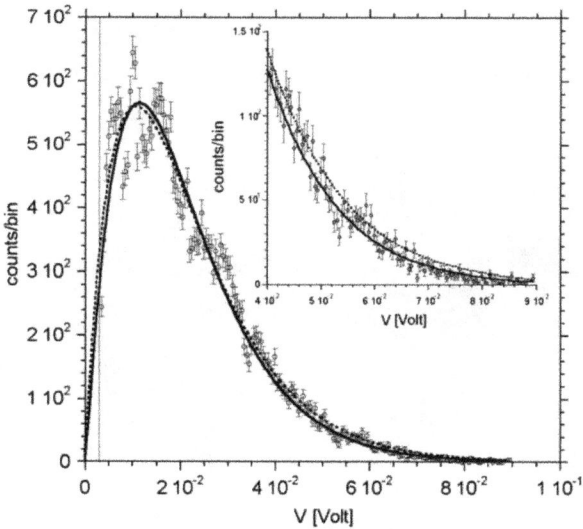

Figure 3.17: Single-electron pulse-height distribution recorded with the LEAK detector in 3 mbar propane.

Source: Ferretti *et al.* (2008).

Chapter 4

Micromegas

4.1 Introduction

Proportional counters, multiwire chambers and most micro-pattern detectors described in the previous chapters exploit the process of avalanche multiplication in the high electric field regions close to thin anodes to provide charge amplification. The progress of these devices sidestepped the development of an older gaseous device, the parallel plate avalanche counter (PPAC), having in principle several advantages over MPGDs: mechanically simpler and sturdier, owing to the absence of thin electrodes, PPACs have intrinsically better energy resolutions, since the charge multiplication process, occurring in uniform fields, suffers less from dispersive effects due to tolerances in the shape and size of thin anodes, see for example (Alkhazov, 1970). The rate capability is also inherently higher, due to the spread of the positive ions' space charge over the whole gap, as against the accumulation around the anodes that modify the multiplying electric field (Mathieson and Smith, 1988; Hendrix, 1984).

Parallel plate counters have been used successfully in the detection of highly ionizing tracks, heavy ions (Harrach and Specht, 1979) and X-rays (Bleeker *et al.*, 1980). However, at the higher gains needed for detection of fast particles, PPACs have the tendency to discharge, and had therefore a limited amount of applications. It was observed however that, for very narrow gaps, the saturation of the first Townsend coefficient at increasing electric fields, Figure 4.1 (Giomataris, 1998), results in a reduction of the dependence of the

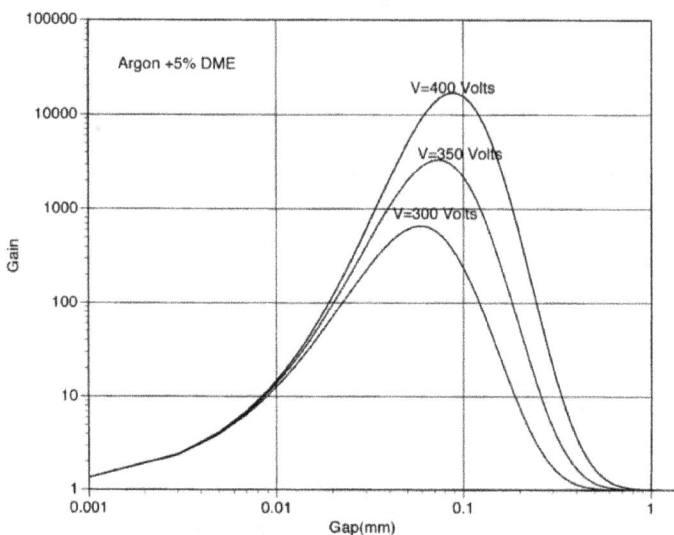

Figure 4.1:   Computed dependence of gain from the gap thickness in uniform fields for different applied voltages.
Source: Giomataris (1998).

multiplication factor from the gap thickness, thus permitting more stable operation at large gains.

Introduced in 1996 (Giomataris *et al.*, 1996), the micro-mesh gaseous chamber (MICROMEGAS) exploits this property, and led to a new family of powerful detectors used for many applications in particle physics and other fields.

Figure 4.2 shows a photo of Joanis Giomataris with examples of the new detectors.

4.2   MICROMEGAS Geometry and Operation

The basic MICROMEGAS design, optimized for the detection of minimum ionizing particles perpendicular to the detector, consists of a conversion volume delimited by a drift electrode, followed by a multiplication gap, typically $100\,\mu$m thick; the two regions are separated by a metallic mesh, Figure 4.3 (Giomataris *et al.*, 1996).

The gap uniformity, of paramount importance for the proper operation of the detector, is ensured by regularly spaced insulating

Figure 4.2:   Joanis Giomataris with some MICROMEGAS detectors.
Source: Courtesy J. Giomataris.

pillars manufactured with the techniques described in Section 4.3 (Figure 4.4). Figure 4.5 shows the equipotentials and field lines computed for typical MICROMEGAS geometry and operating potentials (Barouch *et al.*, 1999).

On application of suitable potentials to the electrodes, electrons released by ionization in the conversion gap drift through the mesh and multiply in the high electric field. Figure 4.6 is a model simulation of the electron transport and multiplication process in typical operating conditions (Colas *et al.*, 2004). Owing to diffusion, the charge spreads transversally; collected on sets of anodic electrodes, strips or pads, the charge permits to improve localization by interpolating the recorded signals using a centre-of-gravity (COG) algorithm.

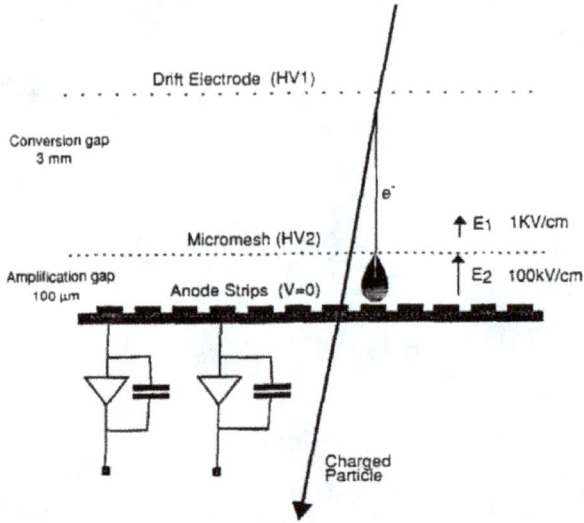

Figure 4.3:   Schematics of the MICROMEGAS detector.
Source: Giomataris *et al.* (1996).

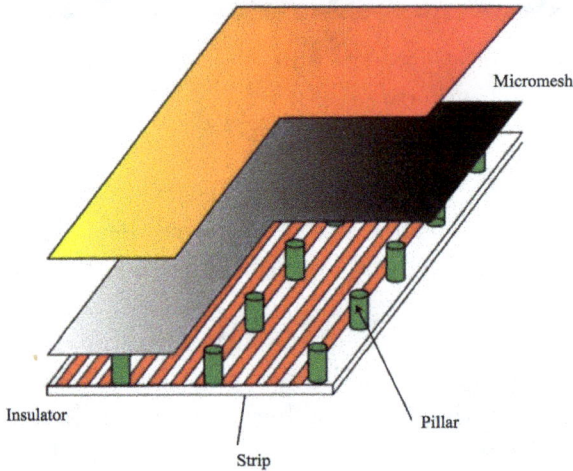

Figure 4.4:   Basic structure of MICROMEGAS, with insulating pillars ensuring the gap uniformity.
Source: Colas *et al.* (2004).

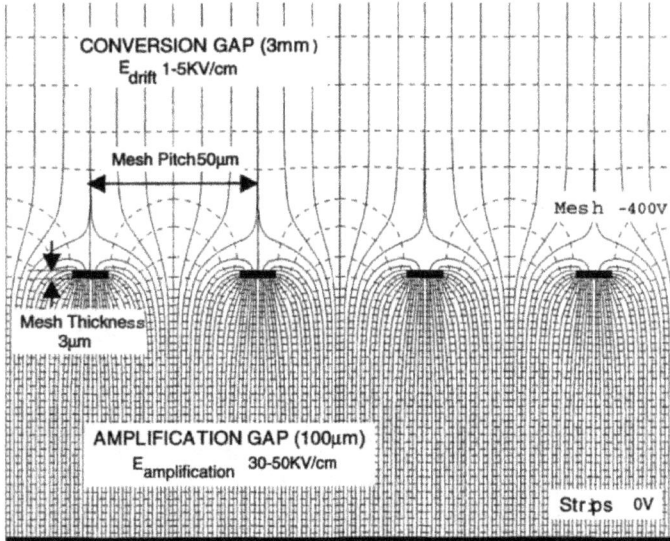

Figure 4.5: Equipotentials and field lines in the MICROMEGAS.
Source: Barouch *et al.* (1999).

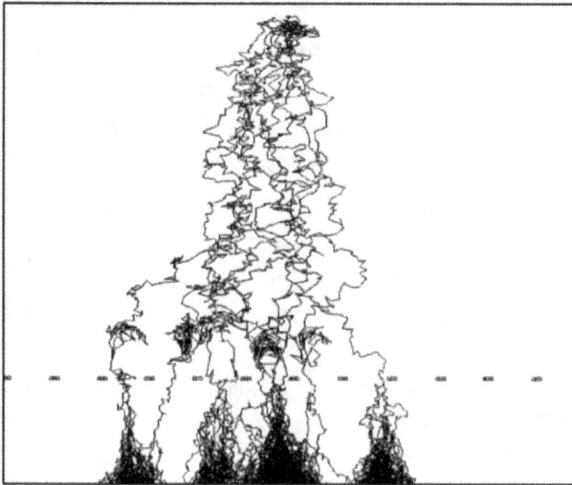

Figure 4.6: Model simulation of the electrons drift and multiplication in
MICROMEGAS.
Source: Colas *et al.* (2004).

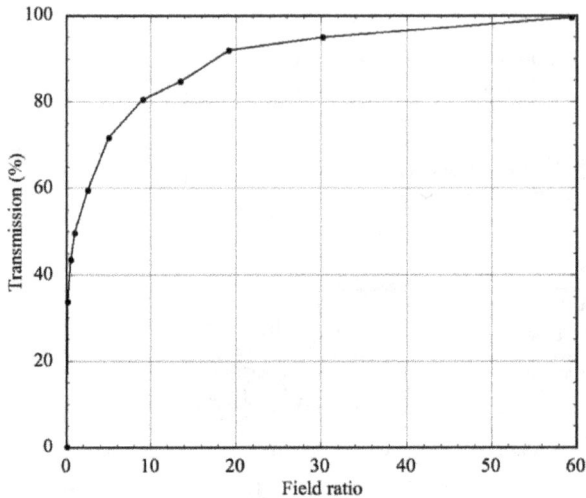

Figure 4.7:   Measured electron transparency as a function of the ratio between amplification and drift fields.

Source: Giomataris *et al.* (1996).

The fraction of ionization electrons transmitted into the multiplication gap depends on the mesh geometry and ratio of fields. Figure 4.7 is an example, measured with a MICROMEGAS with a 3-μm-thick crossed wire mesh at a 25-μm pitch. A good collection is obtained for drift fields exceeding 30–40 times the multiplication fields; this is easily achieved for short drift gaps, but has to be taken into account when designing detectors with long drift volumes, as time projection chambers, see Chapter 6.

The electron transparency has been extensively studied both experimentally and by model calculations. Figures 4.8 and 4.9 give the dependence from field ratio, computed for different models and choices of the detector geometrical structure (Bhattacharya *et al.*, 2013; Kuger *et al.*, 2016).

The largest fraction of the positive charge created in the multiplication gap is collected by the mesh proportionally to the field ratio. However, as mentioned above, the drifting electron cloud can be partly lost passing through the mesh; this results in a dependence of the fractional ion backflow from the mesh geometry and operating

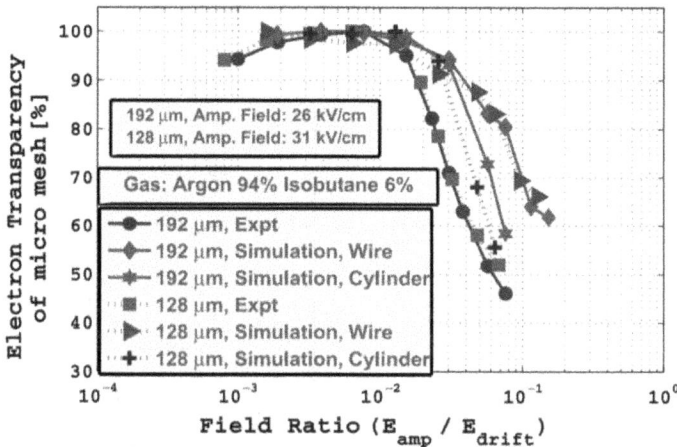

Figure 4.8: Micro-mesh electron transparency, measured and computed with different models and geometry.

Source: Bhattacharya *et al.* (2013).

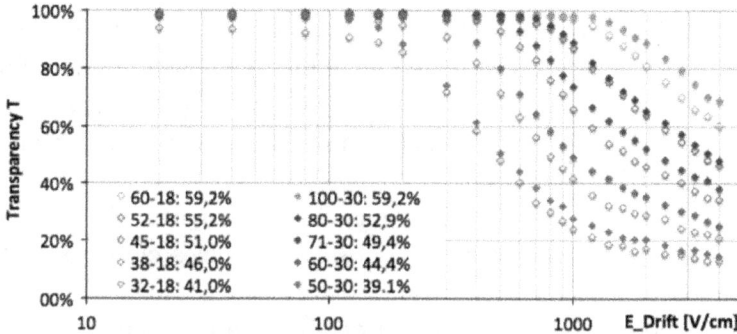

Figure 4.9: Computed electron transparency for different mesh geometries, labeled with the aperure width and wire diameter in μm, and percentage of the open area.

Source: Kuger *et al.* (2016).

gas, and is affected also by the presence of an external magnetic field (Section 6.5).

Owing to the narrow gap and high field, the development and collection of charges is very swift. As shown in Figure 4.10, the fast electron signal is accompanied by a slower ion signal, whose length

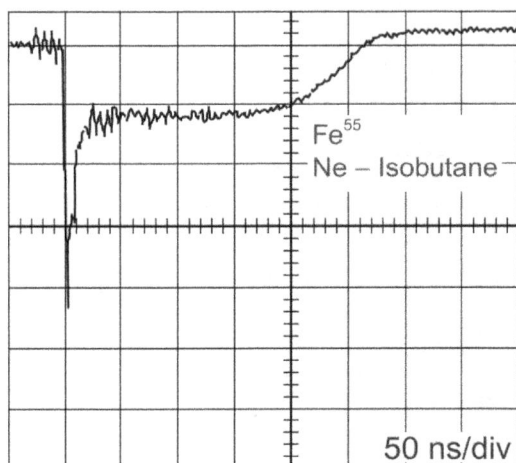

Figure 4.10:   Signal recorded on the anode with a current amplifier, with the fast electron pulse followed by a slower signal due to positive ions.
Source: Bay *et al.* (2002).

depends on the gas and gap thickness (Bay *et al.*, 2002). Figure 4.11 is an example of signals recorded with fast current amplifiers for two values of the gap, 50 and 100 μm; for the thicker gap (bottom track), a slower signal due to the positive ions adds to the faster electron signal (Barouch *et al.*, 1999).

Operated with an Ar-DME gas filling at gains above 10^4, the early prototypes could reach detection efficiencies close to 100% for fast particles.

Since the charge amplification occurs in a quasi-uniform electric field, the MICROMEGAS holds the excellent energy resolution properties of PPACs. Measured with a small-sized device, Figure 4.12 shows pulse height distributions recorded for X-rays in a range of energies; at 22 keV, the resolution is 5.4% FWHM, comparable to the one obtained with the best proportional counters (Charpak *et al.*, 2002).

Many detector geometries and gases have been tested in a systematic search for the best operating conditions to match the experimental requirements. In CF_4-isobutane mixtures, minimizing electron diffusion, the average cluster size, defined as the number of

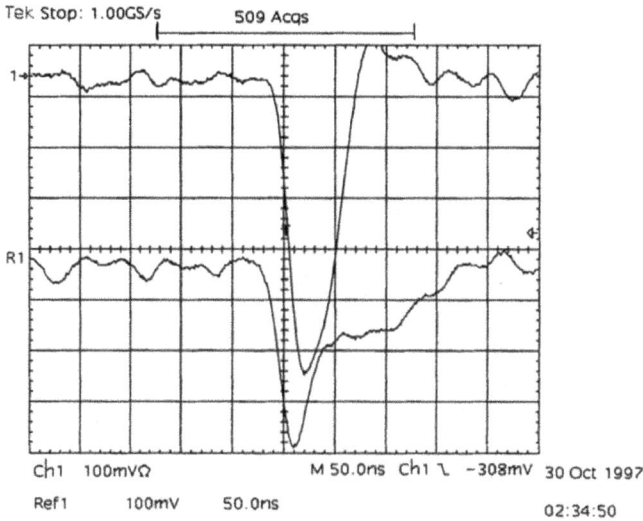

Figure 4.11: Fast anode signal in the MICROMEGAS. The width of the ion-induced component depends on the gap thickness, as shown for 50-μm (top track) and 100-μm gaps (bottom).

Source: Barouch *et al.* (1999).

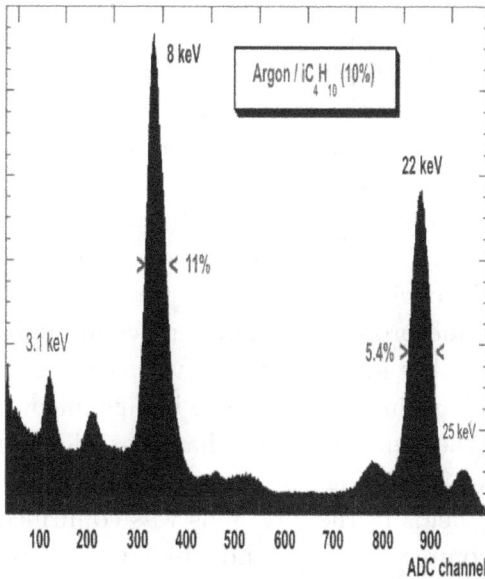

Figure 4.12: Energy resolution for X-rays.

Source: Charpak *et al.* (2002).

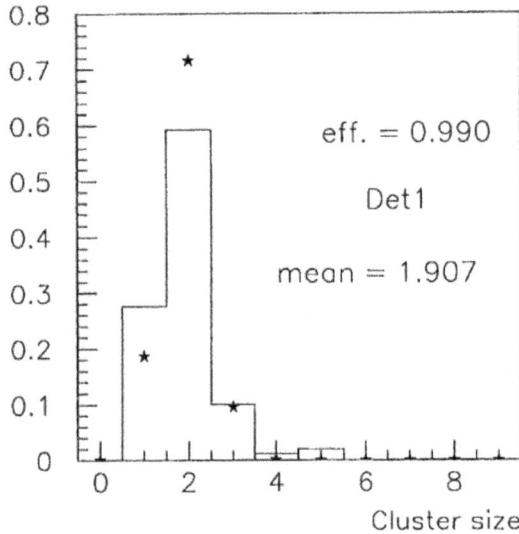

Figure 4.13:   Cluster size distribution in strips units (100 μm wide). *Source*: Derré *et al.* (2001).

strips at a 100-μm pitch with signals above threshold, is around 2, Figure 4.13 (Derre *et al.*, 2001); in these conditions, a position accuracy for fast particles perpendicular to the structure close to 17-μm rms has been obtained, Figure 4.14 (Derre and Giomataris, 2002).

Detection efficiency and localization accuracy depend on the fraction of ionization electrons transmitted through the mesh, a function itself of the ratio between drift and multiplication fields as shown in Figures 4.15 and 4.16 (Bortfeldt *et al.*, 2013). This can be a limitation in the choice of design parameters for thick drift volume devices, as the time projection chamber, where high drift field may lead to uncomfortably large operating voltages.

Exposure to high-intensity particle beams uncovered a drawback of the detector, the tendency to discharge at high gains, most likely a consequence of exceeding the Raether limit for events releasing large ionization yields in the gas. This was confirmed by systematic measurements comparing gains and discharge rates on exposure to heavily ionizing alpha particles, shown in Figure 4.17. The maximum gain before onset of the discharge is around 2,000, as observed in

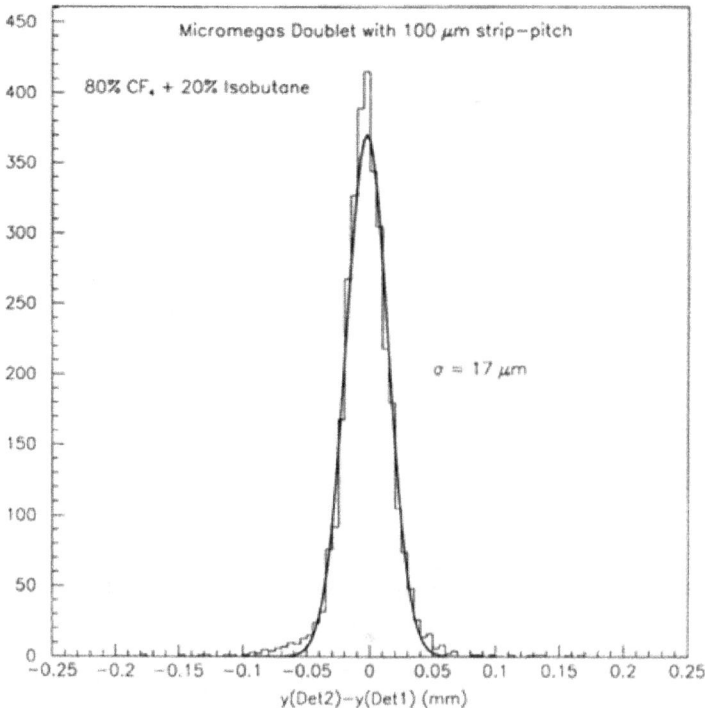

Figure 4.14: Position accuracy for fast particles measured with a MICRO-MEGAS pair.

Source: Derre and Giomataris (2002).

other devices that were tested (Bressan *et al.*, 1999). The problem is particularly severe in hadron beams, while the discharge rate is at least two orders of magnitude lower for muon beams of similar flux (Bay *et al.*, 2002). Although the structure has been demonstrated to withstand a large number of discharges unscathed, the unavoidable dead time following a breakdown poses serious limitations to the use of the detector in harsh operating conditions.

Figure 4.18 (Delbart *et al.*, 2002) is an example of measured discharge probability, normalized to the hadron beam intensity, as a function of gain for three gas fillings, favouring lighter mixtures. Studies in the laboratory with exposure to ionizing radiation confirm that the discharge probability has a steep increase when the total

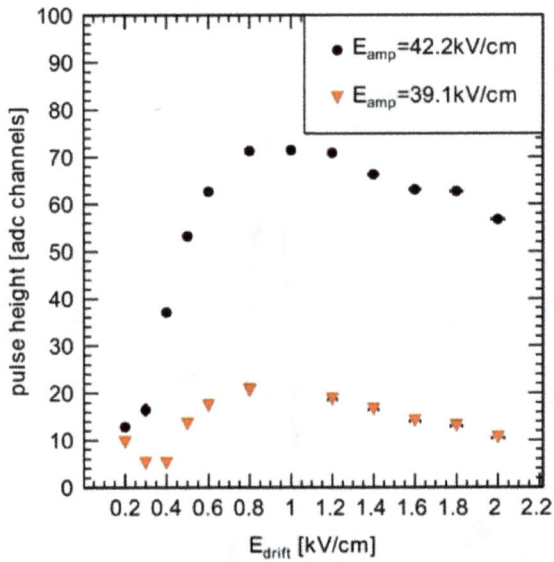

Figure 4.15:   Pulse height versus drift field for two values of the multiplication field.

Source: Bortfeldt *et al.* (2013).

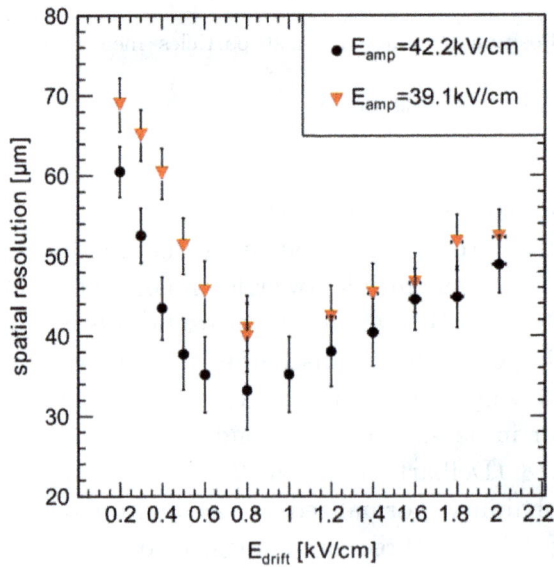

Figure 4.16:   Space accuracy as a function of drift field.

Source: Bortfeldt *et al.* (2013).

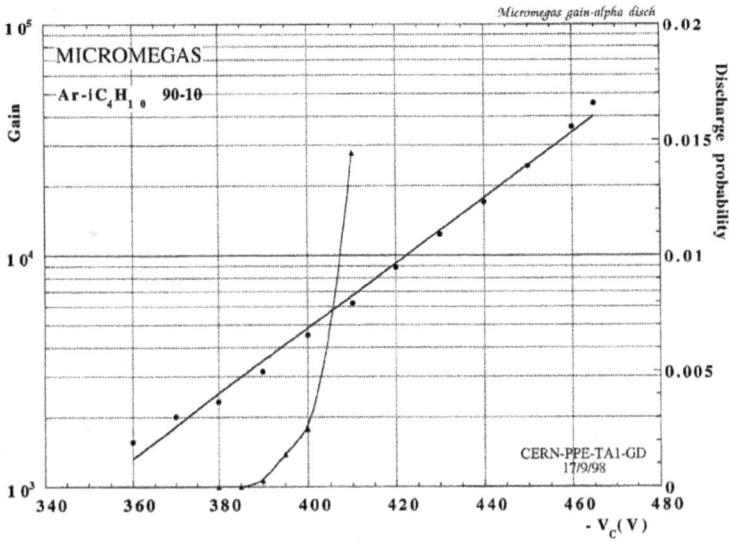

Figure 4.17: Gain and discharge rates on exposure to alpha particles.
Source: Bressan *et al.* (1999) and Bachmann *et al.* (1999).

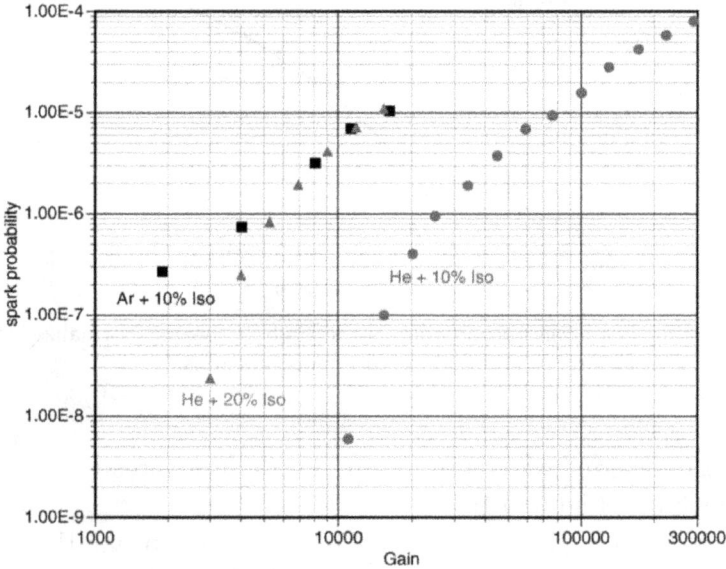

Figure 4.18: Spark probability, normalized to the hadron beam intensity, as a function of gain for two gas mixtures.
Source: Delbart *et al.* (2002).

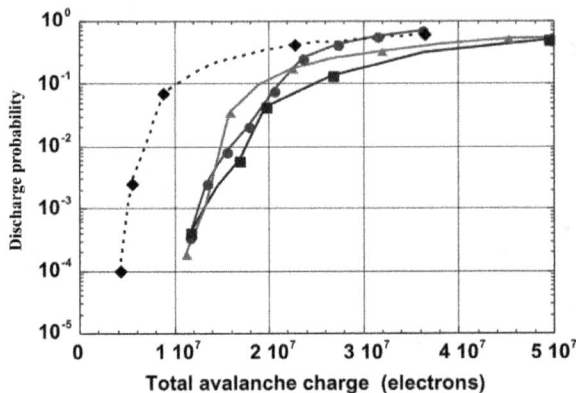

Figure 4.19: Discharge probability on soft X-rays as a function of total avalanche size in different gas mixtures.

Source: Bay *et al.* (2002).

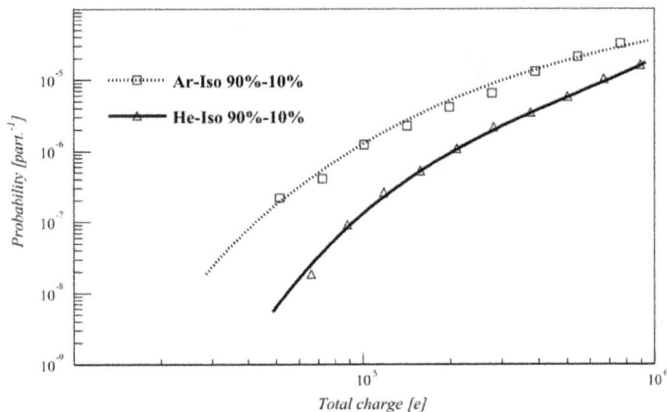

Figure 4.20: Discharge probability in a hadron beam, normalized to the particles' rate, for argon- and helium-based gas mixtures.

Source: Bay *et al.* (2002).

avalanche charge exceeds $\sim 10^7$ electrons, the already mentioned Raether limit (Figure 4.19).

Upon exposing the detector to a high-intensity beam, the reduced probability of interactions with large energy loss in light gas mixtures results in a reduction of the discharge probability for a given gain, as shown in Figure 4.20, and permits to ensure full detection efficiency

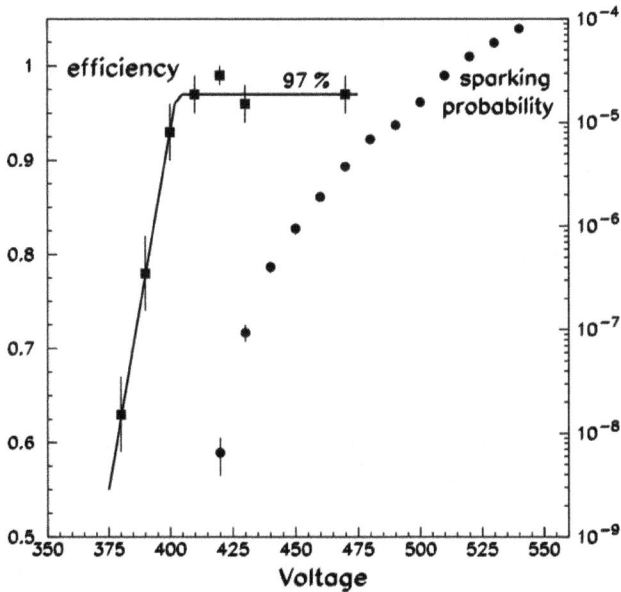

Figure 4.21:   Detection efficiency for fast particles (left) and discharge probability as a function of voltage (right) in He–isobutane (90–10).

Source: Bay *et al.* (2002).

at a tolerable discharge rate, Figure 4.21 (Bay *et al.*, 2002). These studies have been extended to a wide range of operating conditions (Thers *et al.*, 2001; Abbon *et al.*, 2001). A GEANT4 simulation of different energy loss processes in the detector exposed to hadron beams confirms the hypothesis that the observed discharges result from the release of highly ionizing particles, both within the gas and from the materials of the detector; amplified by the avalanche process, the total charge locally exceeds the Raether limit leading to a spark. In Figure 4.22, the observed spark rate in a pion beam is compared with the prediction of the simulation, for two assumptions on the modelling of gain (Procureur *et al.*, 2010).

Substantial progress in the reduction of the discharge rate has been accomplished with the development of resistive protection layers, Section 4.4, or adding a GEM pre-amplification stage, Section 4.8.

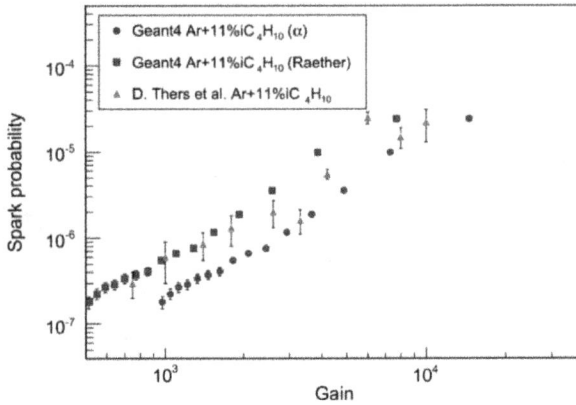

Figure 4.22:   Comparison of measured (triangles) and computed spark probabilities in Ar–isobutane.

Source: Procureur *et al.* (2010).

4.3   Manufacturing Techniques and Materials

Crucial for the correct operation of the detector is the uniformity of the multiplication gap, achieved with the insertion of insulating spacers. Initially realized with thin nylon wires or machined fibreglass grids, in the more advanced versions of the detector, the gap uniformity is ensured with built-in, regularly spaced insulating "pillars" between the anode and the mesh, manufactured with various techniques. For narrow gaps, the flatness of the anode is an essential requirement, imposing for patterned electrodes the use of adequate manufacturing techniques or fillers to smooth the gaps between strips. The development of resistive electrodes naturally solves this issue (Section 4.4).

One of the methods used for the construction exploits the high-accuracy etching techniques developed by the CERN workshops for the fabrication of flexible printed circuits. Figure 4.23 schematically shows (Delbart *et al.*, 2001) that the process makes use of a thin, metal-clad polymer foil, chemically etched through different steps to result in a structure with a finely patterned metal grid on the top side and regularly spaced, insulating pillars in contact with the printed circuit board constituting the anode.

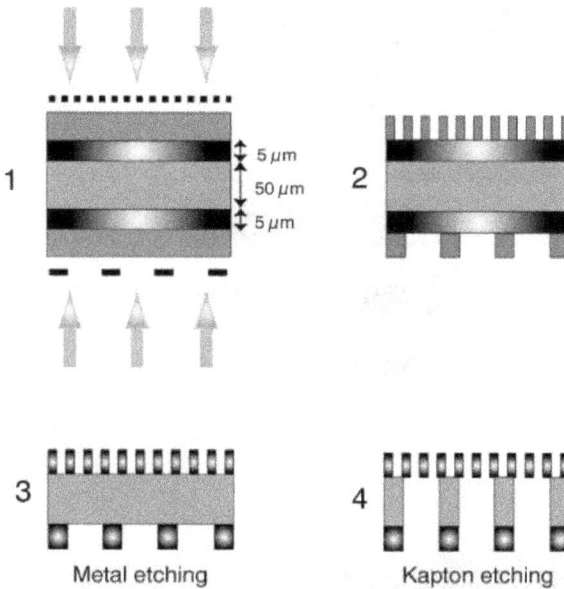

Figure 4.23: The early MICROMEGAS manufacturing process, with chemically etched insulating pillars between anode and cathode mesh.

Source: Delbart *et al.* (2001).

Better suited for the fabrication of large-area detectors, and also developed at CERN, an improved manufacturing method named bulk MICROMEGAS makes use of metallic micro-meshes, available commercially in a wide range of geometries and metals, as shown schematically in Figure 4.24 (Giomataris *et al.*, 2006). The process consists of laminating at high temperature a mesh between two layers of photoresist resin; after exposure to UV light through a mask, the resin is chemically removed, leaving regularly spaced insulating pillars encapsulating the mesh and ensuring the gap uniformity and stability. Figure 4.25 shows a close view of the structure, with pillars 400μm in diameter at 2-mm spacing. Preventing the collection of ionization electrons locally, the pillars result in a small local decrease of efficiency. As shown in Figure 4.26, a plot of the number of reconstructed tracks perpendicular to the detector shows a loss of detection in the positions corresponding to the pillars, accounting to about 1% of the average, and corresponding to the geometrical

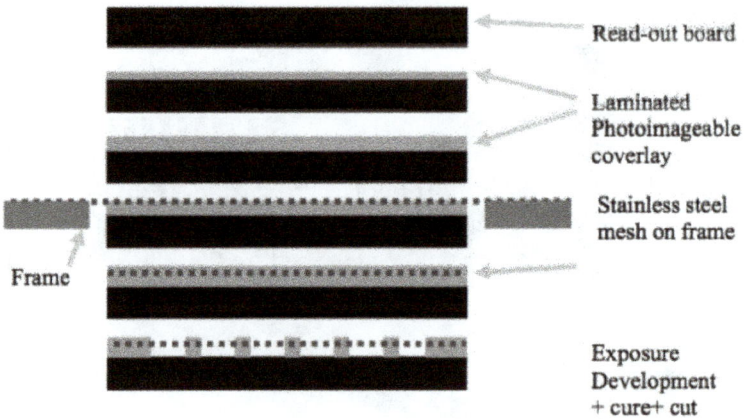

Read-out board

Laminated
Photoimageable
coverlay

Stainless steel
mesh on frame

Frame

Exposure
Development
+ cure+ cut

Figure 4.24:   Bulk MICROMEGAS manufacturing scheme.
Source: Giomataris *et al.* (2006).

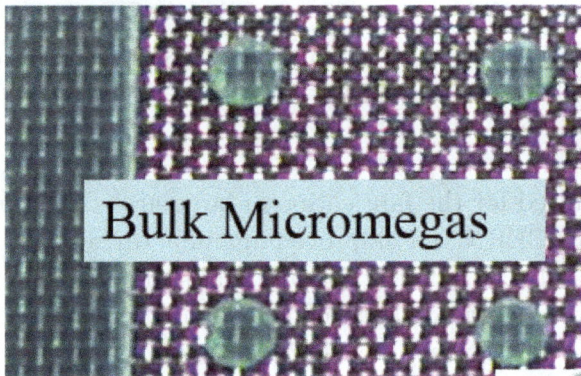

Bulk Micromegas

Figure 4.25:   Close view of a bulk MICROMEGAS.
Source: Giomataris *et al.* (2006).

dead area (300 μm diameter pillars at 2.5-mm pitch) (Alexopoulos *et al.*, 2010).

The effect of the insulating pillars on efficiency and local track distortions has been studied by model calculations (Bhattacharya *et al.*, 2013, 2015). Figure 4.27 is an example of the simulation result, showing the distorted drift trajectories of electrons in the region close to a pillar.

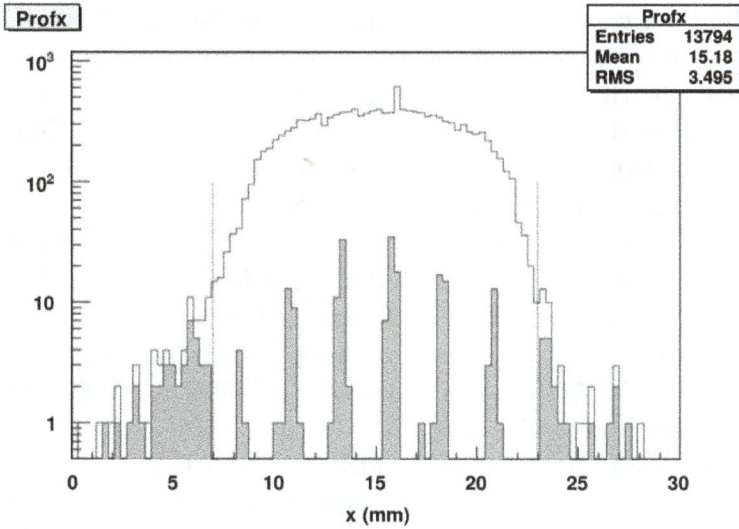

Figure 4.26: Beam profile in the X-projection; the grey histograms correspond to the inefficiency in the region of the pillars.

Source: Alexopoulos *et al.* (2010).

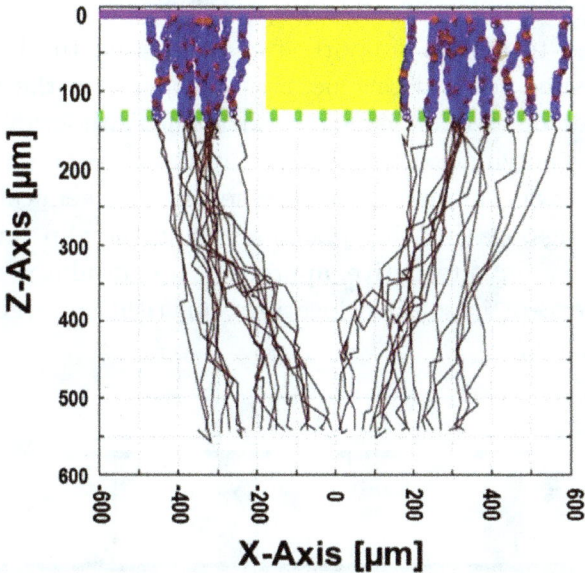

Figure 4.27: Computed distortion of the electrons' trajectory near a pillar.

Source: Bhattacharya *et al.* (2015).

The bulk construction technique permits to manufacture large-sized devices, as those used for the readout of time projection chambers for the T2K experiment (Anvar *et al.*, 2009) and for the ATLAS muon detector upgrade (Alexopoulos *et al.*, 2010). Combined with the development of spark-protected resistive devices, it has been adopted in many other experimental setups.

Realized on thin printed circuit boards, the bulk technology permits the design of cylindrical devices, adapted to central tracker systems (Aune *et al.*, 2009). An example is the cylindrical ASACUSA detector, designed to reconstruct antiproton–nucleon interactions (Radics *et al.*, 2015). Another example of cylindrical MICROMEGAS device is described in Section 4.5.

Reminiscent of the micro-groove detector, but realized with a technology inspired by the development of GEM devices (Section 5.3 of Chapter 5), the micro-bulk MICROMEGAS structure (Andriamonje *et al.*, 2010) improves the detector stability and response uniformity, and has been used to implement several experiments: CAST, NEXT, nTOF and others (Section 4.5). As shown schematically in Figure 4.28, a flexible Kapton foil, metal coated on each side, is glued to a rigid support after patterning the bottom layer with the desired readout scheme, strips or pads. In the micro-bulk, the top layer of metal is etched with a photolithographic process, exposing the polyimide in a matrix of round holes, typically 30 μm in diameter at 80-μm pitch. The polymer is then chemically removed in the openings, leaving tiny pillars ensuring the gap thickness and uniformity. In an alternative approach, the insulator is removed completely except in selected locations constituting larger pillars at millimetre pitch.

Figure 4.28:   Micro-bulk MICROMEGAS.

4.4 Resistive MICROMEGAS

The vulnerability of MICROMEGAS detectors to breakdown when operated in harsh conditions has stimulated the elaboration of methods to mitigate the appearance and growth of radiation-induced discharges. Inspired by the works on resistive plate chambers (RPCs), coating one or more electrodes with controlled resistivity layers permits to locally drop the voltage in case of an excess current, effectively damping the growth of a spark at the cost of a reduced rate capability (de Oliveira *et al.*, 2010; Peskov *et al.*, 2012). The design of a resistive bulk MICROMEGAS is shown schematically in Figure 4.29 (Alexopoulos *et al.*, 2011; Wotschack, 2012). The protection consists in a thin layer of photoimageable coverlay with high-resistivity strips

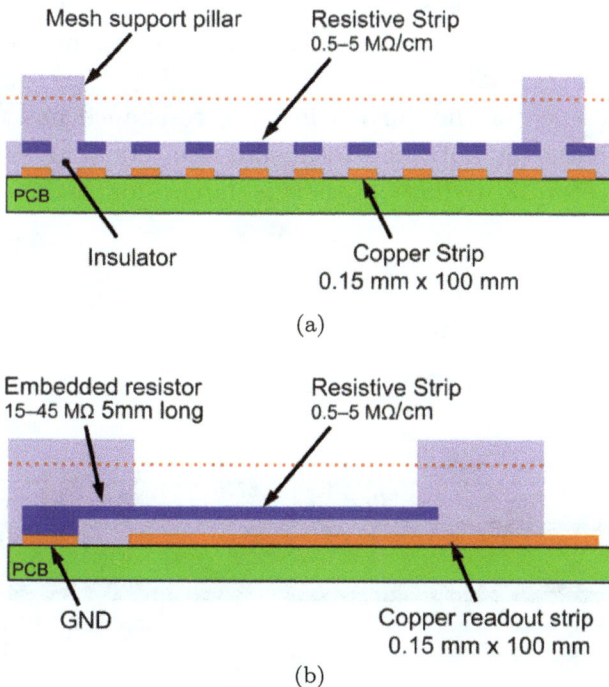

(a)

(b)

Figure 4.29: (a) and (b) Schematic diagram of the resistive bulk MICROMEGAS.

Source: Alexopoulos *et al.* (2011).

deposited on the active side, with anodic readout strips on the outside. Realized with various designs, the strips are grounded at one or both ends with high-value protection resistors.

Various materials have been investigated to realize the resistive electrodes: continuous carbon-loaded Kapton and resistive paste with surface resistivity from a few to a few tens of MΩ /square (Manjarres *et al.*, 2012; Galan *et al.*, 2013). Industrialized using sputtering technologies on thin polyimide foils, the process is used to manufacture large-area detectors (Bianco *et al.*, 2016). The bulk MICROMEGAS anode structure adopted for the ATLAS muon detector upgrade (see Section 4.5) is shown in Figure 4.30.

The efficacy of the resistive spark protection has been verified with extensive exposures of the detectors to high-intensity beams. Figure 4.31 shows the spark rates observed in a beam of $\sim 10^6$ neutrons $cm^{-2}s^{-1}$ as a function of gain for two argon-CO_2 mixtures (Alexopoulos *et al.*, 2011).

The presence of the current-limiting resistance has the consequence of reducing both the local and the global rate capability of the detector, a reduction that is larger for higher values of

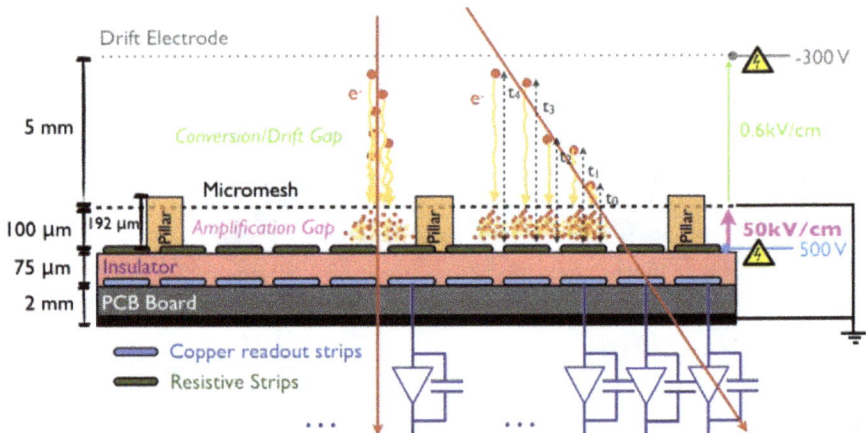

Figure 4.30:   Schematic diagram of the resistive MICROMEGAS for the ATAS muon detector upgrade.

Source: Alexopoulos *et al.* (2019).

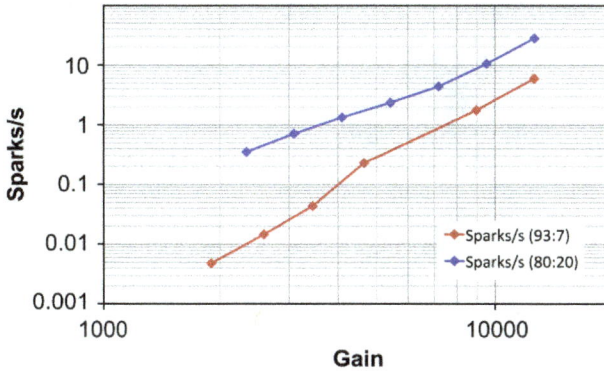

Figure 4.31: Spark rate normalized to the neutron flux in Ar–CO$_2$ 80–20 (higher points) and 97–3 (lower points) as a function of gain of the resistive MICROMEGAS.

Source: Alexopoulos *et al.* (2011).

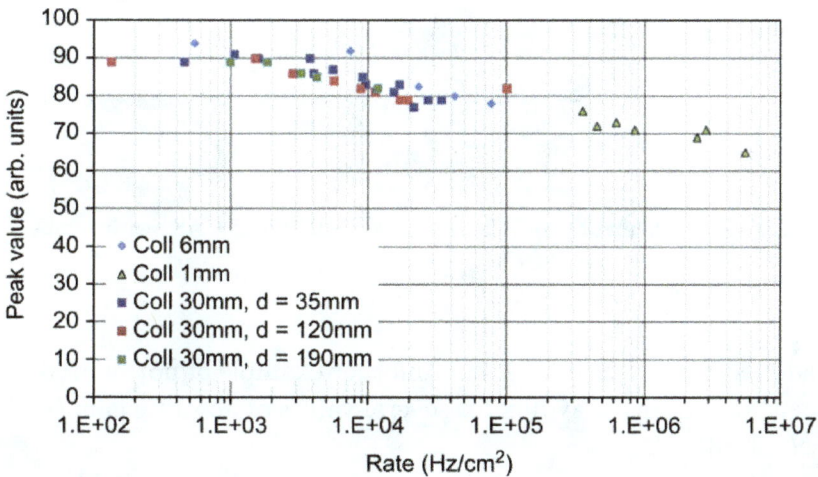

Figure 4.32: Relative gain of a resistive MICROMEGAS as a function of 8-keV X-ray flux and different collimations.

Source: Alexopoulos *et al.* (2011).

resistivity. Figure 4.32 is an example, measured with an 8-keV X-ray generator as a function of rate on a MICROMEGAS with strips having a resistance of $2\,M\Omega\,cm$ and $15\,M\Omega$ to ground. A gain drop is observed above $\sim\!10^4\,Hz\,cm^{-2}$, independent from the

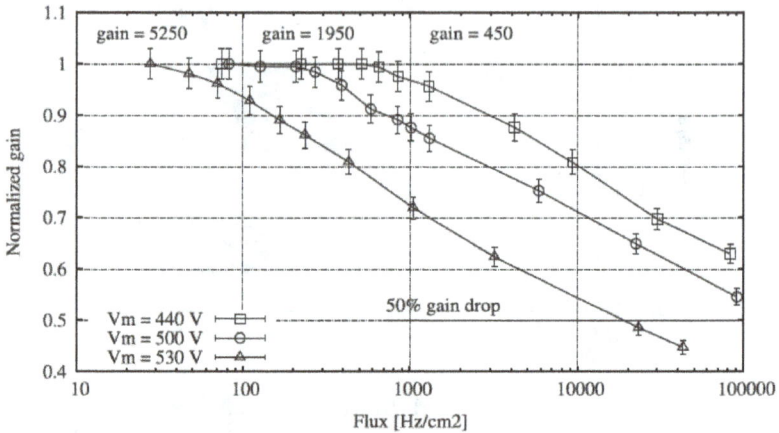

Figure 4.33:   Normalized gain as a function of flux at different values of the operating voltage.

Source: Galán *et al.* (2013).

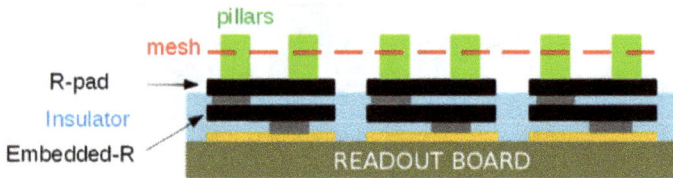

Figure 4.34:   Schematics of the resistive MICROMEGAS with embedded resistors.

Source: Chefdeville *et al.* (2016).

irradiated surface. Figure 4.33 is another example of normalized gain dependence from X-ray flux for several values of absolute gain (Galan *et al.*, 2013).

Systematic measurements of rate capability in a wide range of strips' resistance and sizes have been reported in the initial development of the ATLAS detector upgrade at CERN (Wotschack, 2012). The rate performance can be considerably improved using more complex protection structures, with resistive pads connected to the readout electrode with controlled value patterned resistors, Figure 4.34. Figure 4.35 shows the anode current as a function

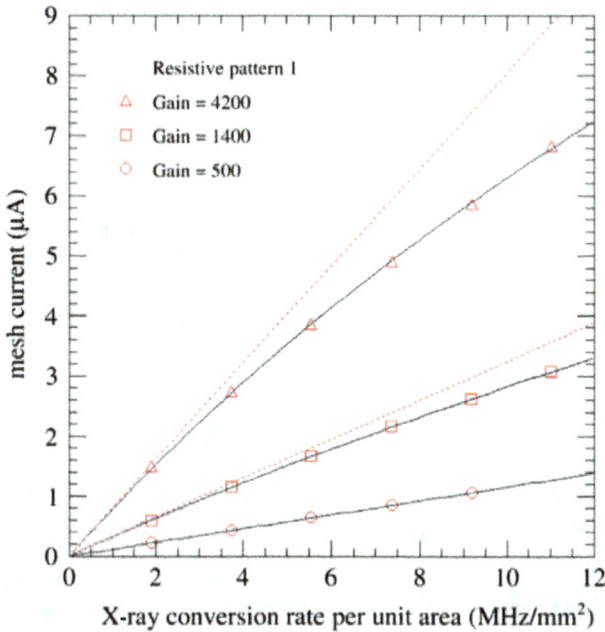

Figure 4.35:   Rate dependence of the cathode current for different values of gain. *Source*: Chefdeville *et al.* (2016).

of X-ray flux for several values of gain; the deviation from a linear dependence, dashed lines, indicates the onset of charging-up processes (Chefdeville *et al.*, 2016).

For applications needing a low material budget, for example in X-ray transmission radiography and heavy ions' detection, the described resistive assemblies, requiring a relatively thick printed circuit board support, are not adequate. An alternative scheme, named floating strips MICROMEGAS, permits to reduce the thickness of the detector to a few percent of radiation length, Figure 4.36. The anode consists of narrow copper strips, individually connected to the voltage through high-value resistors; underlying perpendicular readout strips, separated from the anode by an insulating foil, are used to read out the signals induced by the multiplying charges (Bortfeldt *et al.*, 2017).

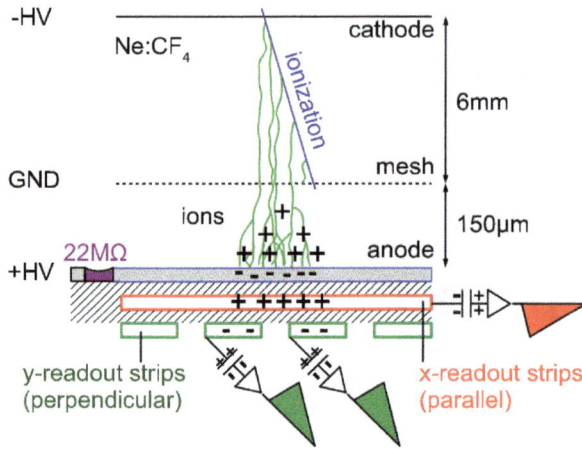

Figure 4.36:   Low material budget floating strips MICROMEGAS. *Source*: Bortfeldt *et al.* (2017).

4.5   Applications in Particle Physics

The first medium-scale system of MICROMEGAS detectors has been built and operated in the COMPASS spectrometer at CERN (Abbon *et al.*, 2007). The tracker included twelve identical detectors with an active area of $40 \times 40\,\text{cm}^2$ each, assembled in doublets of identical chambers mounted back-to-back and rotated by $90°$ to provide orthogonal coordinates. Signals are read out on anode strips, with 360-μm and 420-μm pitch for the central and outer parts of the setup, respectively. Figure 4.37 shows a fully assembled detector installed in the beam line, with the readout electronics mounted at a distance from the active area to keep it outside the acceptance of the spectrometer (Bernet *et al.*, 2005).

Using a gas mixture of Ne–C_2H_6–CF_4 (80–10–10), optimized to reduce the discharge rates when exposed to the high-flux hadron beam, the chambers delivered a detection efficiency close to 99% and a localization accuracy of 70-μm rms when operated at moderate beam intensities (Figures 4.38 and 4.39) (Bernet *et al.*, 2005). At high particle flux, aside from space charge and voltage drop limitations, the detection efficiency is affected by occupancy, the probability of hitting the same strip within the resolution time of the electronics;

Figure 4.37: A MICROMEGAS fully equipped detector in the COMPASS experiment at CERN.

Source: Bernet *et al.* (2005).

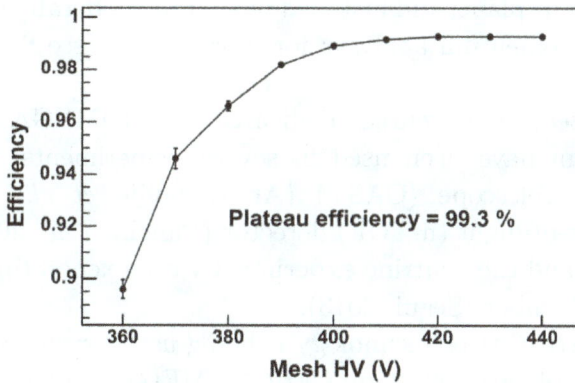

Figure 4.38: Efficiency as a function of voltage for fast particles.

Source: Bernet *et al.* (2005).

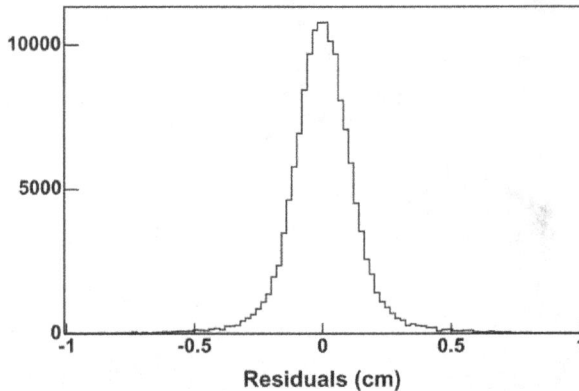

Figure 4.39:   Localization accuracy of the COMPASS MICROMEGAS, having 70-μm rms.

Source: Bernet *et al.* (2005).

this has been largely improved using a pixelated readout on narrow pads in the high-flux area of the detector (Neyret *et al.*, 2012).

A large system of modular detectors has been operated in the Tokay-to-Kamiokande (T2K) experiment to study the neutrino oscillations. The system includes three medium-sized time projection chambers, each equipped with a matrix of 12 bulk MICROMEGAS on each readout plane, and installed in a large, moderate field magnet (0,2 T) for momentum analysis; for more details, see Section 6.3 of Chapter 6.

Small-sized, high-accuracy dedicated MICROMEGAS devices of similar design have been used in several experiments: the CERN Axial Solar Telescope (CAST) (Andriamonje *et al.*, 2004), the neutron time-of-flight (n-TOF) detector (Pancin *et al.*, 2004; Belloni *et al.*, 2012) and the neutrino experiment with a xenon time (NEXT) projection chamber (Segui, 2013).

The resistive strips technology is being used for the construction of a system of large (1–2 m^2) MICROMEGAS chambers for the planned Small Wheel upgrade of CERN's ATLAS muon spectrometer. The images in Figures 4.40 and 4.41 show a prototype under construction (Wotschack, 2012). Assembled in quadruplets of four layers per module, the detectors have been extensively tested in

Figure 4.40:   The ATLAS MICROMEGAS detector during assembly.
Source: Picture CERN.

Figure 4.41:   An assembled ATLAS MICROMEGAS chamber.
Source: Picture CERN.

realistic beam conditions, at particle fluxes up to $15\,\text{kHz}\,\text{cm}^{-2}$, demonstrating their high detection efficiency. A detailed description of the construction, tests and event reconstruction algorithms is given in Bianco *et al.* (2016 and Alexopoulos *et al.* (2020). The avalanche charge is recorded on each layer on sets of parallel readout strips, $300\,\mu\text{m}$ wide at 415-μm pitch; the main coordinate in the radial direction (X) is determined with a COG algorithm on signal on adjacent strips (or cluster). A small (1.5°) stereo angle between layers permits a coarser reconstruction of the second (Y) coordinate.

The tight requirements on planarity and tolerance on the narrow gaps are ensured by a system of external tension adjustment screws; the details are shown in Figure 4.42 (Alexopoulos *et al.*, 2020).

Detector prototypes have been installed in the high-intensity gamma source facility GIF at CERN to assess their long-term endurance under irradiation, Figure 4.43 (see also Section 4.7).

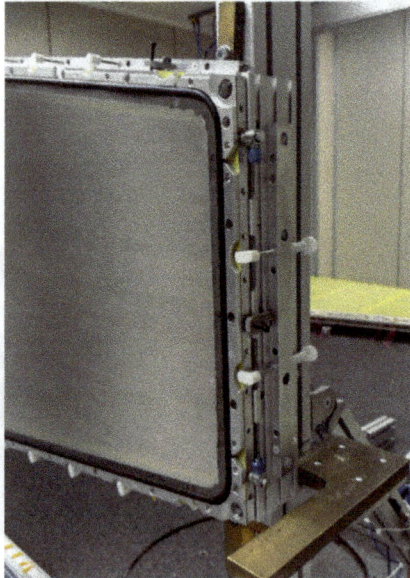

Figure 4.42:   Details of the tensioning screws of the ATLAS MICROMEGAS module.

Source: Alexopoulos *et al.* (2020).

Figure 4.43: ATLAS prototypes installed in the high-intensity gamma source GIF at CERN.
Source: Picture source CERN.

Figure 4.44 shows the detection efficiency for minimum ionizing particles measured with the four layers in a module, and Figure 4.45 provides the cluster reconstruction efficiency as a function of readout strip number. Figure 4.46 shows the position resolutions, measured on a test beam in the X and Y directions (Alexopoulos *et al.*, 2020).

By combining the centroid determination and a time of drift measurement, in the so-called micro-TPC operating mode (Hattori *et al.*, 2007), the detector can achieve $\sim$100-μm position resolution for large incidence angles in the presence of an external magnetic field up to 0.3 T, Figure 4.47 (Kuger, 2017). When fully implemented, the system will include more than 500 chambers, covering an area of $1{,}200\,\mathrm{m}^2$ and will be capable to withstand the increased event rates generated by the luminosity of the LHC collider upgrade (Iakovidis, 2013).

The use of MICROMEGAS for the readout of time projection chambers is discussed in Section 6.3 of Chapter 6.

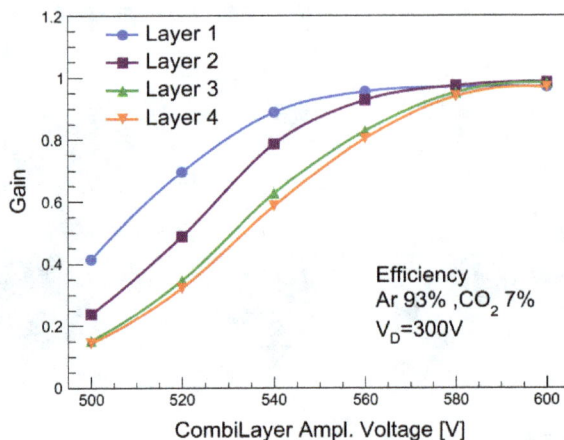

Figure 4.44:   Detection efficiency for minimum ionizing particles measured with several ATLAS MICROMEGAS modules.

Source: Bianco *et al.* (2016).

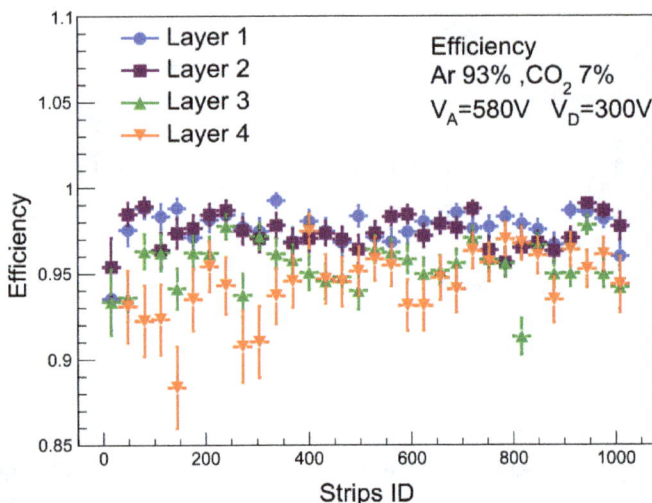

Figure 4.45:   Cluster reconstruction efficiency along the detector.

Source: Bianco *et al.* (2016).

While the most widely used MICROMEGAS detectors are planar, a system of cylindrical devices has been built and operated in the CLAS12 experiment at the Jefferson Laboratory. As seen in Figure 4.48, the detector includes a barrel tracker with six

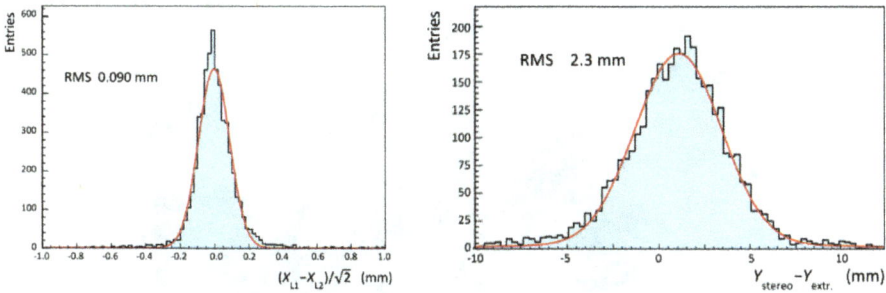

Figure 4.46:   Cluster localization accuracy in the X and Y directions.
Source: Alexopoulos *et al.* (2020).

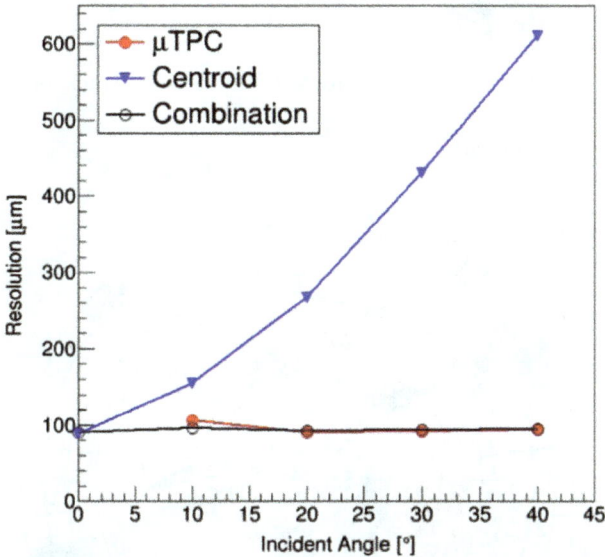

Figure 4.47:   Space accuracy of the ATLAS MICROMEGAS achieved with a COG algorithm (triangles) and in the micro-TPC operating mode.
Source: Kuger (2017).

layers of cylindrical MICROMEGAS followed by six circular forward chambers; the system is installed in a 5-T solenoidal magnetic field (Acker *et al.*, 2020). The detectors' construction is based on the resistive MICROMEGAS technology described in Section 4.4; a five-prong recorded event is shown in Figure 4.49.

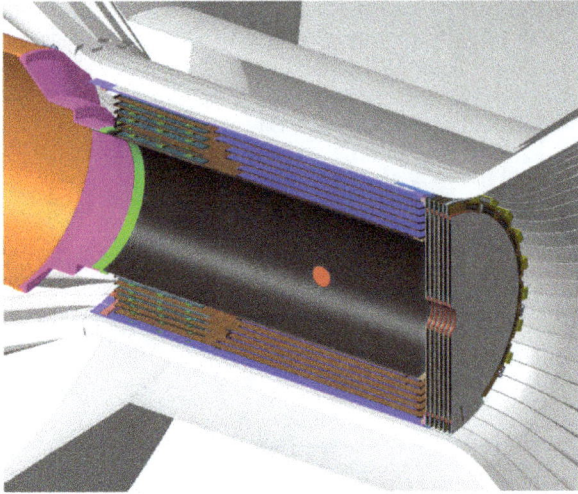

Figure 4.48:   The Barrel and Forward MICROMEGAS Trackers for the experiment CLAS12 at JLab.

Source: Acker *et al.* (2020).

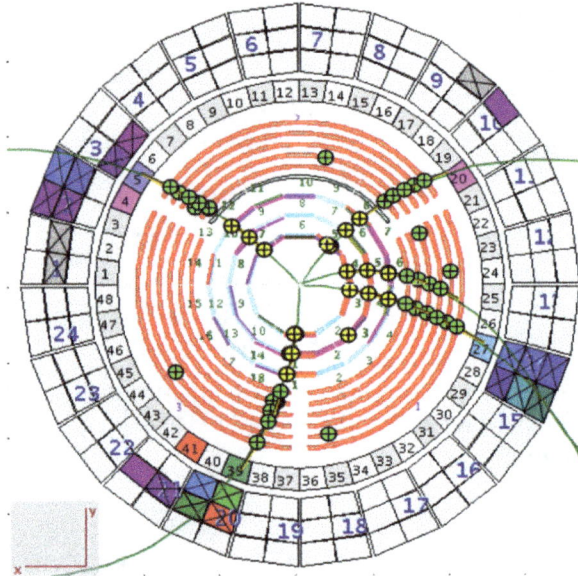

Figure 4.49:   A five-prong event recorded with the CLAS12 MICROMEGAS detector.

Source: Acker *et al.* (2020).

4.6 Applications in Other Fields

In particle physics, to handle the high particle rates and multiplicities, signals are recorded on individual pickup electrodes, strips or pads, using highly integrated, custom-made electronics circuitry. Various systems of encoding have been developed to reduce cost and complexity of the readout for low-rate applications. With the so-called genetic multiplexing, several strips are grouped and connected to a single electronic channel (Procureur *et al.*, 2013). The interconnection pattern exploits as a fundamental assumption the fact that at least two neighbouring strips register a signal from an incoming particle, thus creating an information redundancy; combinatorial considerations lead then to an interconnection scheme in which the information lost by multiplexing is compensated by the redundancy. For a medium-sized detector with two-dimensional readout strips at $\sim$500-μm pitch, the multiplexing scheme can reduce the number of readout channels from 1,024 to 61 for each plane.

The genetic multiplexing scheme has been used to encode the information of MICROMEGAS detectors developed for imaging thick absorbers using cosmic rays, with a technology named transmission muon radiography or muography. The method consists of recording the integrated density of cosmic muon tracks as a function of position with a telescope of position-sensitive chambers. Figure 4.50 shows an example of muon radiography of the Saclay water tower with (left) and without water filling, recorded with a four-day continuous exposure (Bouteille *et al.*, 2016).

The muon-based radiography technique has been used for the investigation of hidden structures in volcanology, archaeology, homeland security and other fields; for a review, see Procureur (2018). A particularly interesting application is the search of hidden voids in Khufu's pyramid (Morishima *et al.*, 2017). Proposed by Luis Alvarez in the 1960s, the method consists of installing large position-sensitive detectors inside the pyramid or near a face, recording the directions and intensity of muons crossing the structure. Originally

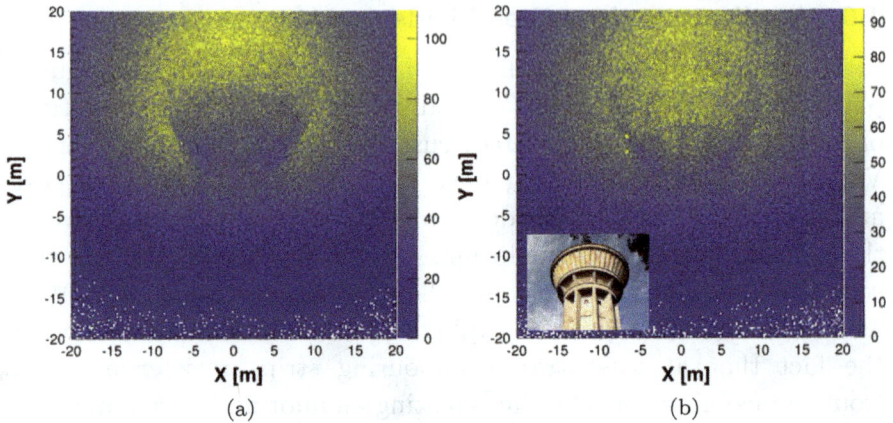

Figure 4.50:   Muon transmission radiography of the Saclay water tower, filled (a) and empty (b).

Source: Bouteille *et al.* (2016).

implemented using spark chambers, the use of large MICROMEGAS telescopes largely increases the sensitivity and resolution of the system. The long exposure time, counted in years, in a rather hostile environment with large excursions in temperature and humidity has required detailed studies to ensure the operating stability of the detectors.

Figure 4.51 is an example of measurement of the muon excess, revealing the presence of two internal cavities, the (known) Grand Gallery and of the so-called Big Void, confirming the observations of other detectors placed inside the structure (Procureur and Attie, 2019).

4.7   Long-Term Operation, Ageing

The issue of permanent damage, or ageing, has been studied extensively with long-term exposure to high radiation fluxes. While polymerization processes of organic compounds or residual pollutants are expected to occur in all gaseous detectors, their effect on the MICROMEGAS and similar devices is predictably minor due to the detectors' conception, with the multiplying field provided between

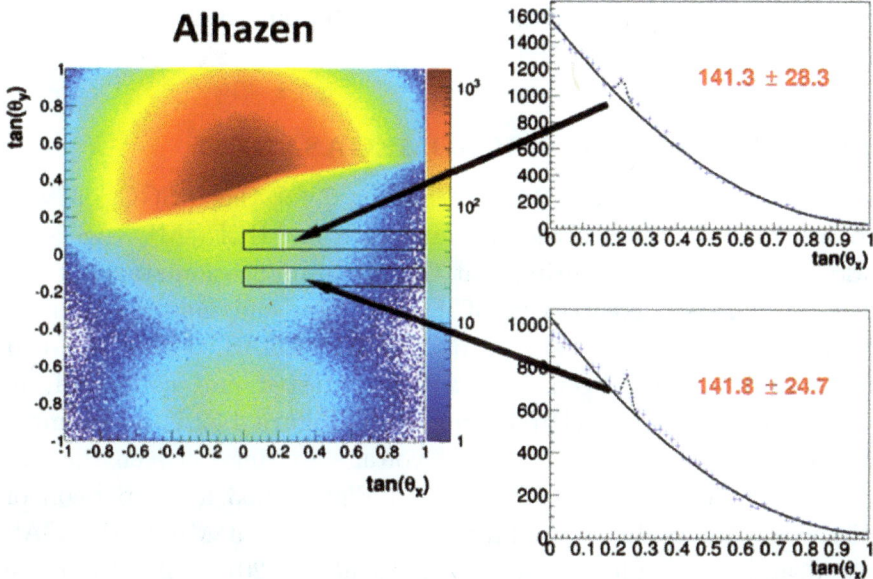

Figure 4.51: Presence of two voids inside the Khufu pyramid, recorded with the CEA-Saclay muography system.

Source: Reproduced from Procureur and Attie (2019) with kind permission of Elsevier Masson SAS.

electrodes and not, as in wire and micro-strip chambers, near the surface of thin electrodes that can be easily coated by insulating deposits as described in Section 2.6 of Chapter 2. Measurements during the early development demonstrated that no change in performance occurred with long-term exposures to high-flux X-rays up to a collected charge of about $20\,\mathrm{mC\,mm^{-2}}$. This has been confirmed with systematic exposures to gamma rays and neutrons of resistive MICROMEGAS prototypes of the ATLAS muon upgrade (Manjarres *et al.*, 2012). Measured in the laboratory with a high-intensity X-ray beam, no degradation is observed after a total collected charge up to $250\,\mathrm{mC\,cm^{-2}}$, corresponding to five years of operation at CERN's high-luminosity Large Hadron Collider (Galan *et al.*, 2012). Exposure to a very-high-flux gamma ray at CERN's gamma irradiation facility (GIF) has confirmed the long-term performance up to an accumulated charge of $100\,\mathrm{mC\,cm^{-2}}$, corresponding to an integrated detected photon flux of $20\,\mathrm{MHz\,cm^{-2}}$ (Sidiropoulou *et al.*, 2017).

4.8   Hybrid Systems: MICROMEGAS+GEM

As discussed in Section 2.9 of Chapter 2 for the MSGCs, a way to substantially reduce the sparking rate in MICROMEGAS is to add a Gas Electron Multiplier at the pre-amplification stage, at the cost of increased complexity of construction and material budget. Performances of several models of MICROMEGAS coupled to GEM pre-amplifiers have been systematically tested and compared, in particular for their discharge properties. Figure 4.52 is an example of measured gain as a function of voltage sharing between the two devices (Procureur *et al.*, 2011), and Figure 4.53 a comparison of discharge probability as a function of gain for a MICROMEGAS alone and a cascaded device (Moreno *et al.*, 2011). A systematic study of discharge rate for different MICROMEGAS models coupled to GEMs is given in the reference (Procureur *et al.*, 2011).

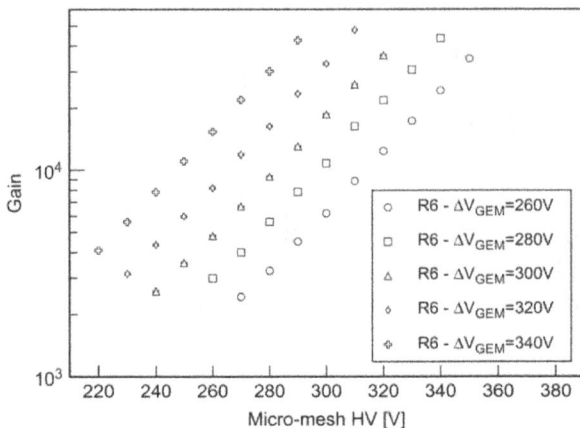

Figure 4.52:   Gain versus voltage of a MICROMEGAS+GEM structure. *Source*: Procureur *et al.* (2011).

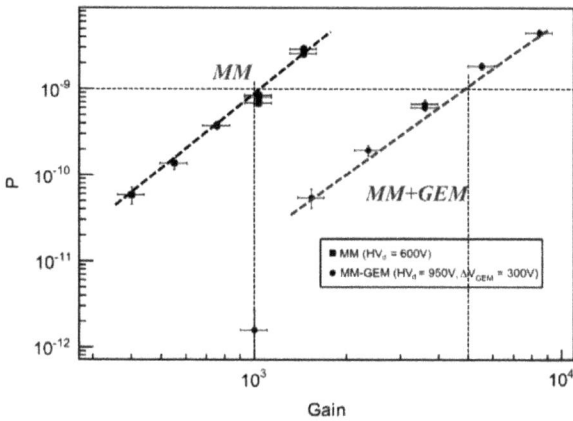

Figure 4.53: Discharge probability as a function of overall gain for MICROMEGAS and MICROMEGAS+GEM.

Source: Moreno *et al.* (2011).

The combination of two MICROMEGAS or a MICROMEGAS with one or more GEM pre-amplifiers has been also studied aiming at the reduction of the ions' backflow for time projection chambers (Conceicao *et al.*, 2010; Ratza *et al.*, 2018) and in Cherenkov Ring Imaging devices (Alexeev *et al.*, 2017), see Sections 6.5 of Chapter 6 and 7.4 of Chapter 7.

Chapter 5

Gas Electron Multiplier

5.1 Introduction

The reliability problems encountered with micro-strip counters are mainly caused by the use of fragile electrodes in the presence of the high electric fields needed to achieve gains, typically around 10^4, for detection of small ionization yields. Under these conditions, the occurrence in the gas of infrequent but highly ionizing events due to nuclear interactions or electromagnetic showers may lead to the creation of a local charge density exceeding the Raether limit ($\sim 10^7$ electron–ion pairs), resulting in the transition of the avalanche to a streamer and ultimately a discharge.

This difficulty was met long ago with wire chambers, operated at the high gains required for the detection of single electrons in photosensitive gases. An innovative device, the Multi-step Avalanche Chamber (MSAC) solved the problem, introducing the concept of pre-amplification: a dual structure, with a region of high field between two meshes imparting to the ionization electrons a first boost of gain, followed by a second amplifying structure (Charpak and Sauli, 1978). The combined amplification of the cascaded assembly, each operated below the critical gain for discharges, added to the suppression of photon-mediated feedback processes due to self-absorption in the gas, permitted to successfully perform detection and localization of single photoelectrons in Cherenkov ring imaging (RICH) applications (McCarty *et al.*, 1986).

Inspired by the same basic concept, the gas electron multiplier (GEM), introduced by the author in 1997, is a thin polymer foil,

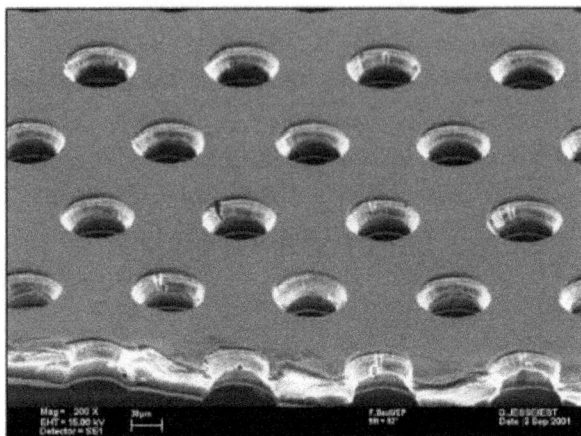

Figure 5.1:   Electron microscopic image of a section of a GEM electrode, 50 μm thick. The holes' pitch and diameter are 140 and 70 μm, respectively.
Source: Picture CERN.

metal coated on both sides and pierced with a high density of holes, typically 50–100 mm^{-2}, Figure 5.1 (Sauli, 1997). Inserted between a drift and a charge collection anode, and with the application of appropriate potentials, the GEM electrode develops the field lines and equipotential shown in Figure 5.2. The large difference of potential applied between the two faces of the foil creates a high field in the holes; electrons released in the upper region drift towards and into the holes, acquiring sufficient energy to cause ionizing collisions with the molecules of the gas filling the structure. A sizeable fraction of the electrons produced in the avalanche leaves the multiplication region and transfers into the lower gap of the structure, where it can be collected by an electrode or injected into a second multiplying section. This gives a great flexibility of use to the detector, exploited to achieve higher gains and better performances. The thin polymer foils can be shaped to match the experimental requirements; their flexibility permits to build non-planar devices, as discussed later. Figure 5.3 is a close-up picture of a standard GEM foil, with 70-μm holes at 140-μm pitch, in a hexagonal pattern; the quality of the photolithographic artwork can be seen in the electron microscope picture of a single hole in Figure 5.4.

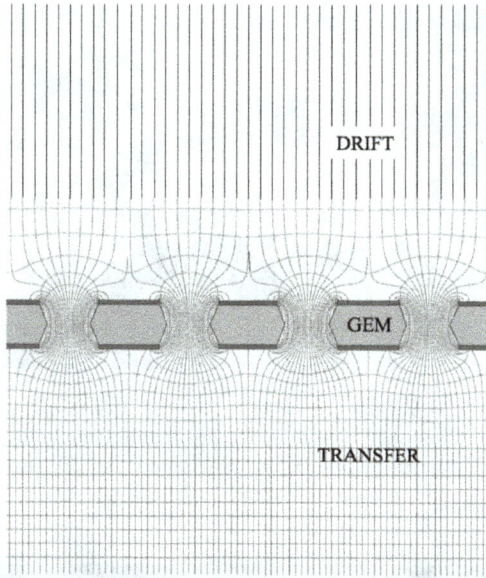

Figure 5.2:   Electric field near a GEM electrode.

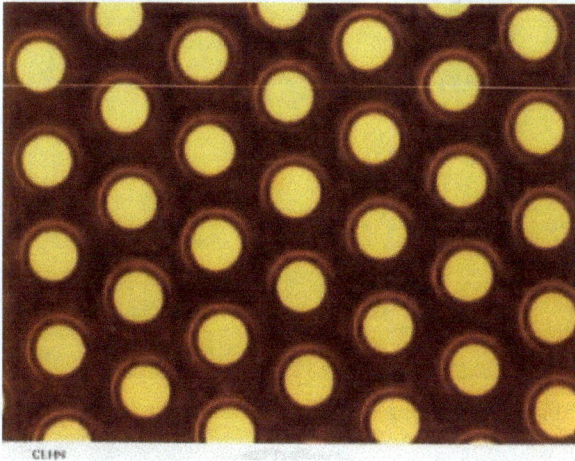

Figure 5.3:   Close-up of a GEM electrode.
Source: Picture CERN.

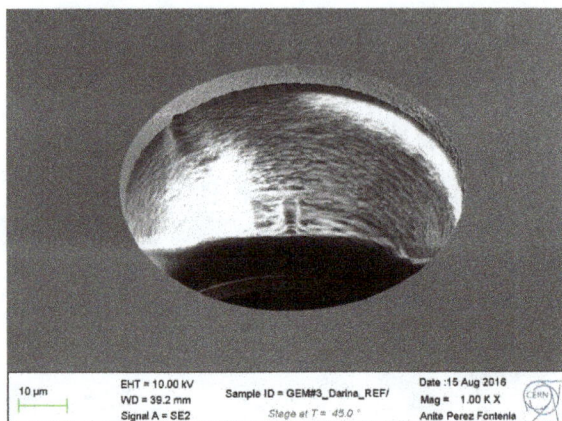

Figure 5.4:   Electron microscopic view of a hole, 70 μm in diameter.
Source: Picture CERN.

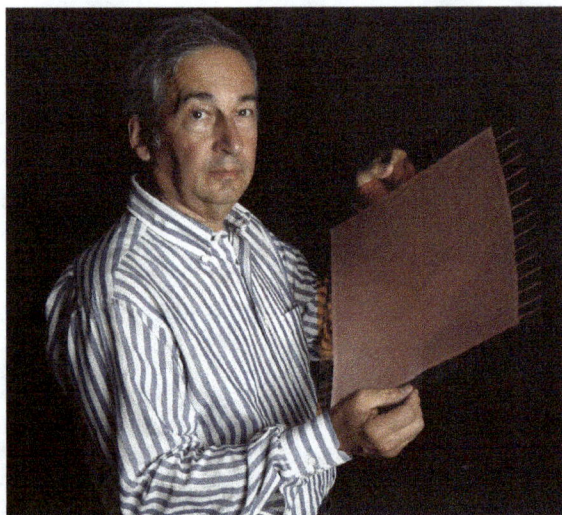

Figure 5.5:   Fabio Sauli holding one of the GEMs built for the COMPASS tracker.
Source: Picture CERN.

Figure 5.5 shows the author holding a medium-sized GEM foil built for the COMPASS experiment at CERN. A detailed review of GEM performances and applications can be found in the reference (Sauli, 2016).

5.2 Basic GEM Operation

Different from other gaseous counters, the (negative) signal on the GEM anode is induced only by the collection of electrons, without a contribution from the slow positive ions, making the device potentially very fast and minimizing space charge problems. Moreover, the moderate field between multiplying and sensing electrodes, named as transfer or induction gap, reduces the probability of the propagation of a discharge to the fragile front-end readout electronics (see Section 5.10).

To perform localization, the anode can be patterned with strips or pads; an example of a two-dimensional readout is shown in Figure 5.6 (Bressan *et al.*, 1999). The charges equal and opposite in sign induced on the bottom GEM electrode can be used as energy trigger, permitting the detection and localization of ionization produced by neutral radiation.

Replicated in a cascade of GEM foils, the pre-amplification and transfer process permits to attain very high proportional gains without the occurrence of discharges, as discussed in the next sections.

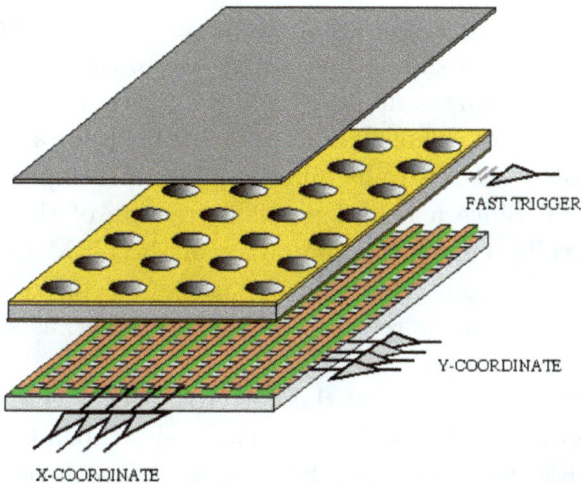

Figure 5.6: Schematics of a single-GEM detector with two-dimensional strip readout.

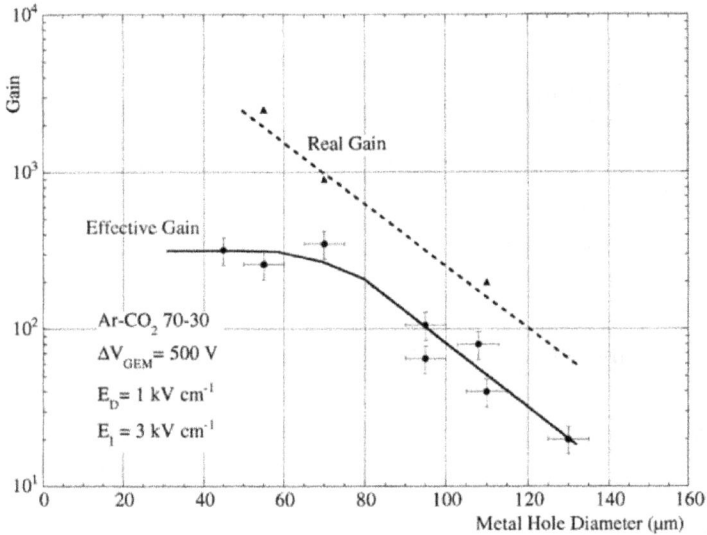

Figure 5.7:   Effective and real gain at fixed GEM voltage as a function of the holes' diameter.

Source: Bachmann *et al.* (1999).

The diameter and shape of the holes have a direct influence on the performance and long-term operation stability of a detector. It was found already in early studies that to ensure high gains, the optimum diameter of the holes should be comparable to the foil thickness, as shown by the measurements in Figure 5.7 (Bachmann *et al.*, 1999). Indeed, while narrower holes result in larger fields for a given voltage difference, losses on the walls and charging up compensate for the increased gain. Since a field-dependent fraction of the multiplying electrons is collected on the lower face of the GEM electrode, the useful or effective gain, defined as a ratio of detected to primary charge, is lower than the real gain of the multiplier, as shown in the figure.

Owing to the structure of the detector, the sharing of collected charges (electrons and ions) between electrodes depends on the value of fields, GEM geometry and filling gas; it has been extensively studied both with measurements and simulations. Figure 5.8 provides an example of currents measured on all electrodes as a function of

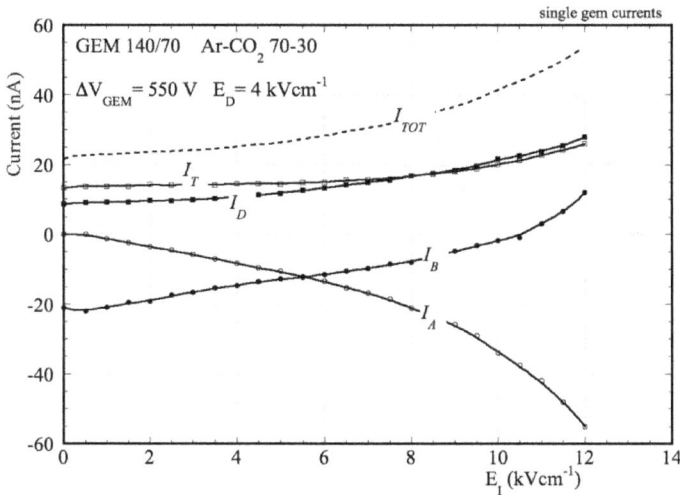

Figure 5.8: Currents under irradiation measured on the four electrodes of a single-GEM detector, at increasing induction fields. I_D, I_T, I_B and I_A are the currents on the drift, top and bottom GEM and anode, respectively; $I_{TOT} = I_A + I_B$ corresponds to the total charge gain of the structure.

Source: Bachmann *et al.* (1999).

the induction field, with the other fields kept constant (Bachmann *et al.*, 1999). Above ~8 kVcm^{-1}, avalanche multiplication begins in the induction gap; exploitable to attain higher gains, this is not a desirable feature since it might help in propagating a discharge through the structure.

The fraction of ionization electrons transferred though the GEM foil, or transparency, depends on the drift field, decreasing at high values due to losses to the top GEM electrode, as seen in Figure 5.9 for a "standard" geometry (70-μm holes at 140-μm pitch) at increasing values of the GEM voltage (Bachmann *et al.*, 1999); the initial loss at low fields is due to electron capture and diffusion. Transparency losses are important in determining the energy resolution, and directly affect the detection efficiency.

The maximum proportional amplification that can be attained before discharge depends on the GEM manufacturing quality, since a single local defect can affect the whole electrode. For small sizes and with a thorough selection of the electrodes, an effective gain well

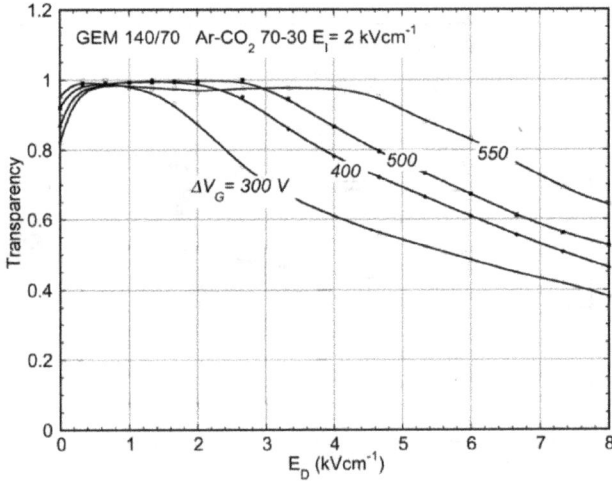

Figure 5.9: Electron transparency of a standard GEM electrode as a function of drift field for fixed induction field and several values of GEM voltage.

Source: Bachmann *et al.* (1999).

above 10^3 can be obtained, as shown in Figure 5.10, measured in argon–carbon dioxide mixtures at atmospheric pressures (Benlloch *et al.*, 1998). For larger areas, these conditions are seldom realized, and it is preferable to adopt a multi-GEM structure to safely reach large amplification factors (Section 5.5).

The energy resolution achievable with GEM detectors compares with the one of standard proportional counters, ~17% FWHM for 5.9-keV X-rays, Figure 5.11 (Bouclier *et al.*, 1997). However, gain non-uniformities and shifts due to charging up of the insulating surface of the hole may degrade the resolution.

The influence of the holes' shape, resulting from the manufacturing process, has been studied extensively both experimentally and by model calculations, and is discussed in Section 5.8.

GEM detectors have been operated successfully in a variety of gas mixtures and in a range of pressures, from a few torr to several atmospheres; examples are given in the following sections. The device has also been operated at cryogenic conditions in dual-phase detectors (see Chapter 10).

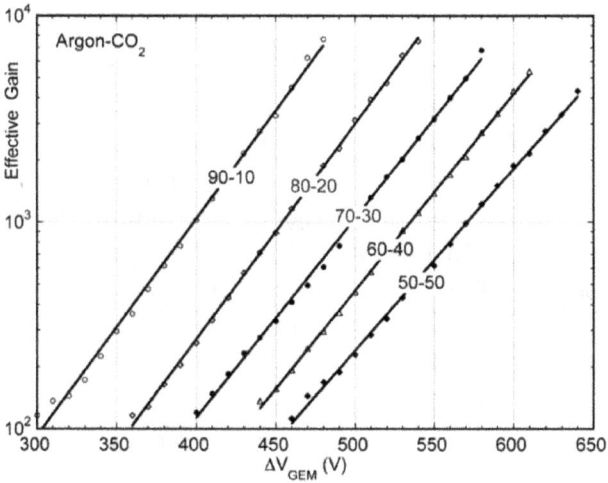

Figure 5.10: Single-GEM gain as a function of voltage in Ar–CO_2 mixtures at atmospheric pressure.

Source: Benlloch *et al.* (1998).

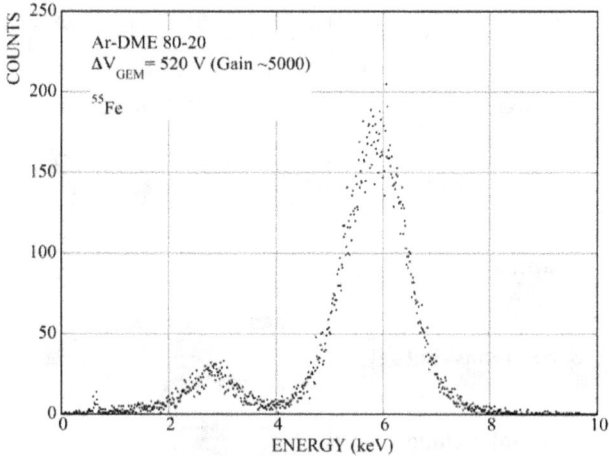

Figure 5.11: Pulse height spectrum on 5.9-keV X-rays recorded with a single GEM. The energy resolution is ~17% FWHM.

Source: Bouclier *et al.* (1997).

5.3   GEM Manufacturing

Small and medium-sized GEM electrodes are fabricated using a double-mask, high-quality wet etching technique. The production starts with the procurement of a high-grade polyimide foil, coated on each side with a thin metal layer (in most cases, 50-μm-thick Kapton with 5-μm copper claddings[1]). The manufacturing method[2] shown schematically in Figure 5.12 begins with the exposure of the metallized polymer foils, coated with a photosensitive resin layer, to ultra-violet light through masks from both side of the sheet. The exposed resin is chemically removed with a standard printed circuits technology. The metal is then etched in an acid bath reproducing the hole's pattern, and the foil is immersed in a solvent for the polymer[3] until holes dig in from the two sides, resulting in a characteristic double-conical shape, also named hourglass (Figure 5.13(a)).

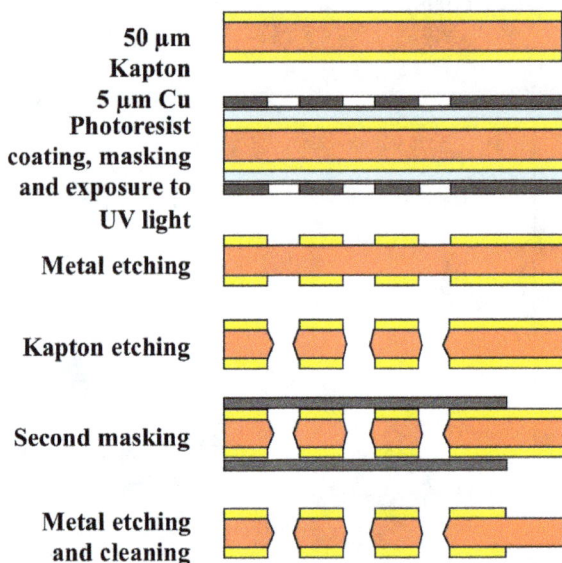

Figure 5.12:   Schematics of the double-mask GEM manufacturing process.

---

[1]Novaclad^{TM}, produced by Sheldahl Corporation.
[2]Processes developed by R. de Oliveira and collaborators, CERN Detector Technologies.
[3]Potassium hydrochloride, ethylene diamide or a combination of the two.

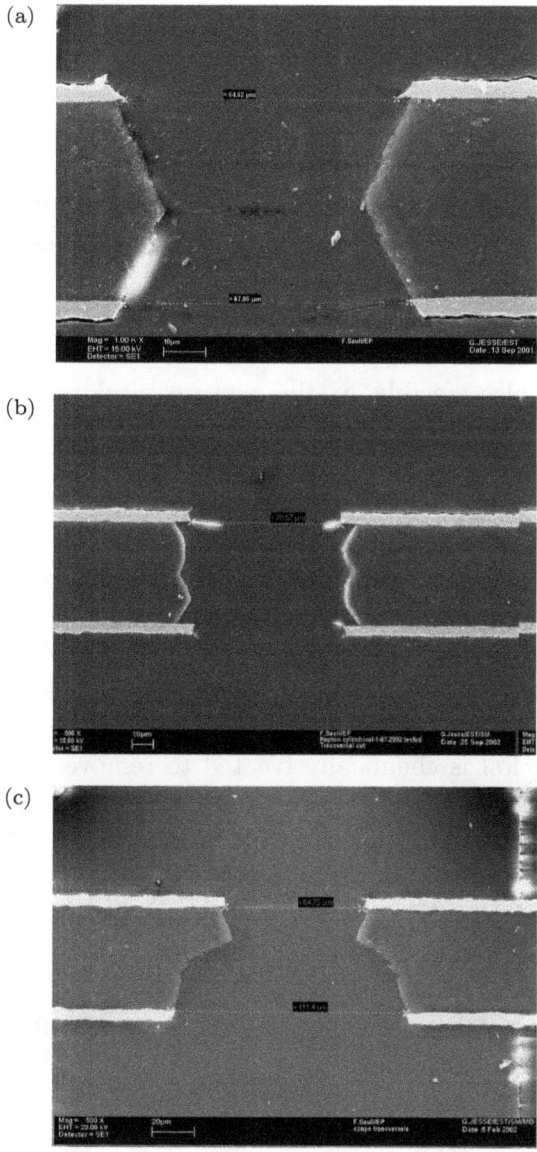

Figure 5.13:   Double-conical (a), cylindrical (b) and conical holes (c) in GEMs manufactured with double-mask technology.

Source: Pictures CERN.

A longer dissolution of the polymer results in quasi-cylindrical holes (Figure 5.13(b)), while the use of two masks with different hole diameters permits to realize conical shapes (Figure 5.13(c)).

The detailed shape of the holes plays an important role in determining the detector performances, see Section 5.8. A double-conical shape permits to reach high gains, but exhibits a slow gain increase at startup due to charging up of the insulating surfaces during operation. In contrast, a cylindrical shape results in a more stable operation, but is prone to discharges at high gains due to the strong electric fields created by the thin residual metal protrusions around the holes, visible in the picture, and resulting from the prolonged polymer dissolution process. Conical holes show a larger gain increase with time when irradiated from the wide side; they have some interest in the reduction of positive ions backflow (Section 6.5).

To satisfy the stringent requirements on diameter and pitch of the holes, the two masks have to be aligned with a tolerance of a few microns, an increasingly difficult requisite for large sizes. A single-mask process has been developed to permit the realization of larger areas, up to and above a square metre, Figure 5.14 (Duarte Pinto *et al.*, 2008). Following the masking, metal and polyimide etching processes, the foil is chemically treated to remove about half of the metal, opening the holes on the bottom side; a second polyimide etching permits to realize quasi-cylindrical holes, Figure 5.15 (Alfonsi *et al.*, 2010). The manufacturing process is described in detail in the reference (Villa *et al.*, 2011). A splicing technology has also been developed to join together several GEM foils with minimal efficiency loss (Alfonsi *et al.*, 2010).

The single-mask technology has been adopted for the construction of the large GEM foils used for the CMS muon detector and the ALICE TPC upgrades at CERN (Abbaneo *et al.*, 2013; Gasik, 2017), see Section 5.6. Shah *et al.* (2019) provide a comparison of performances between single- and double-mask processing.

Other groups have developed technologies to produce the holes' pattern on a variety of different supports (Tamagawa *et al.*, 2009; Shalem *et al.*, 2006; Martinengo *et al.*, 2011; Takahashi *et al.*, 2013;

Cu-clad Kapton

Single mask
Photoresist

Cu etching

Kapton etching

Second Cu etching

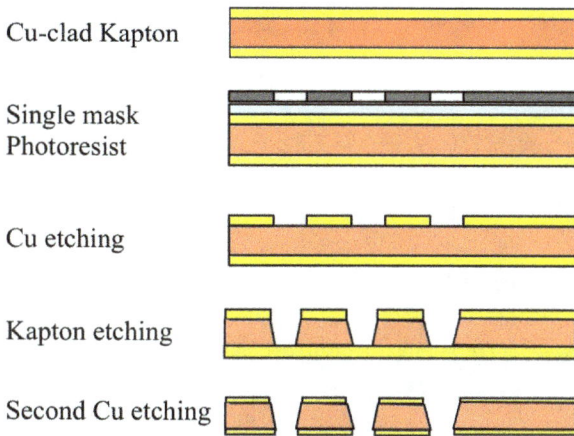

Figure 5.14:   Schematics of the single-mask GEM manufacturing.

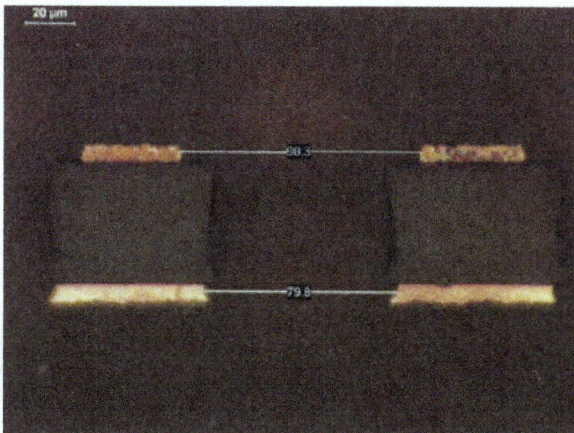

Figure 5.15:   Quasi-cylindrical hole shape obtained with the single-mask process.
Source: Pictures CERN.

Xie *et al.*, 2013). Figure 5.16 shows schematically the process used
to manufacture GEM electrodes on glass (G-GEM) (Fujiwara *et al.*,
2018) and Figure 5.17 shows a close-up of the hole's pattern and a
microscopic view of a section through a hole of the glass plate. The
quality of the manufacturing and the cylindrical shape of the holes

1. Photosensitive Glass Substrate

UV exposure

Photo Mask

2. UV exposure (1st_exp)

Crystal portion
(Li$_2$O SiO$_2$)

3. Crystal formation
   (heat treatment)

Via

4. Via etching
   (Hydrogen Fluoride wet etching)

HF(Spray etching)

Cr sputter

5. Metallization process I (Cr)

Grinder

Remove Cr in surface
(remains in via)

6. Remove metal except in via

Cu sputter
(inside via also)

7. Metallization process II (Cu)

Chemically etch
Cr in via
(Cu are also
removed with Cr)

8. Selectively etch metal in via

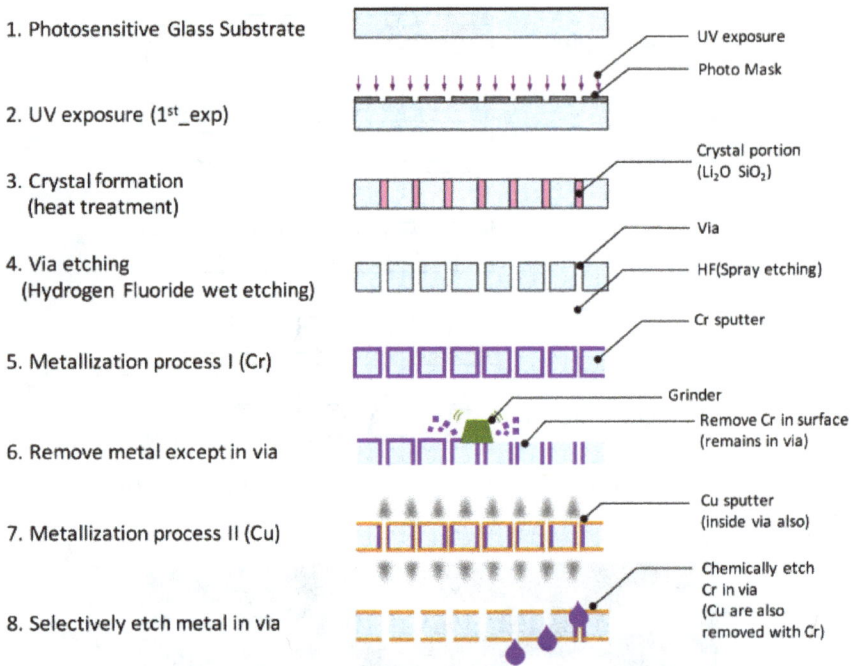

Figure 5.16:   Manufacturing process of a glass GEM.
Source: Fujiwara *et al.* (2018).

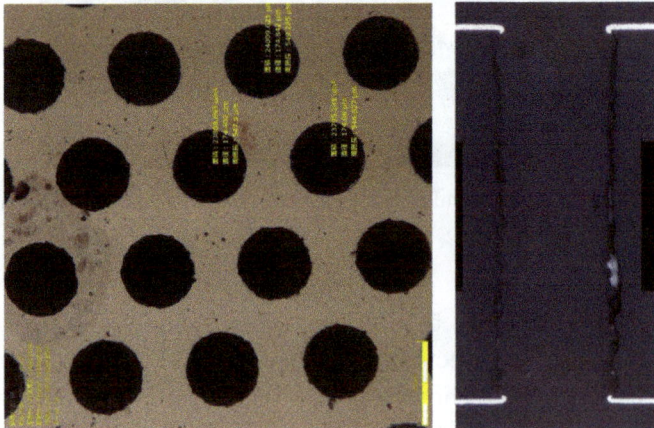

Figure 5.17:   Top view and section through a hole of the Glass GEM.
Source: Fujiwara *et al.* (2018).

result in moderate gain shifts due to charging up and excellent gain uniformity and energy resolution.

Alternative production methods have been developed, such as plasma etching and laser drilling (Inuzuka *et al.*, 2004; Tamagawa *et al.*, 2006); they have limited use due to high manufacturing costs. Dedicated hardware and software tools have also been developed to assess and certify the quality and uniformity of the manufacturing artwork over large areas and are employed for the construction of large detector systems, see Section 5.6 (Kalliokoski *et al.*, 2012; Hildén *et al.*, 2014). Figure 5.18 is an example of the measured

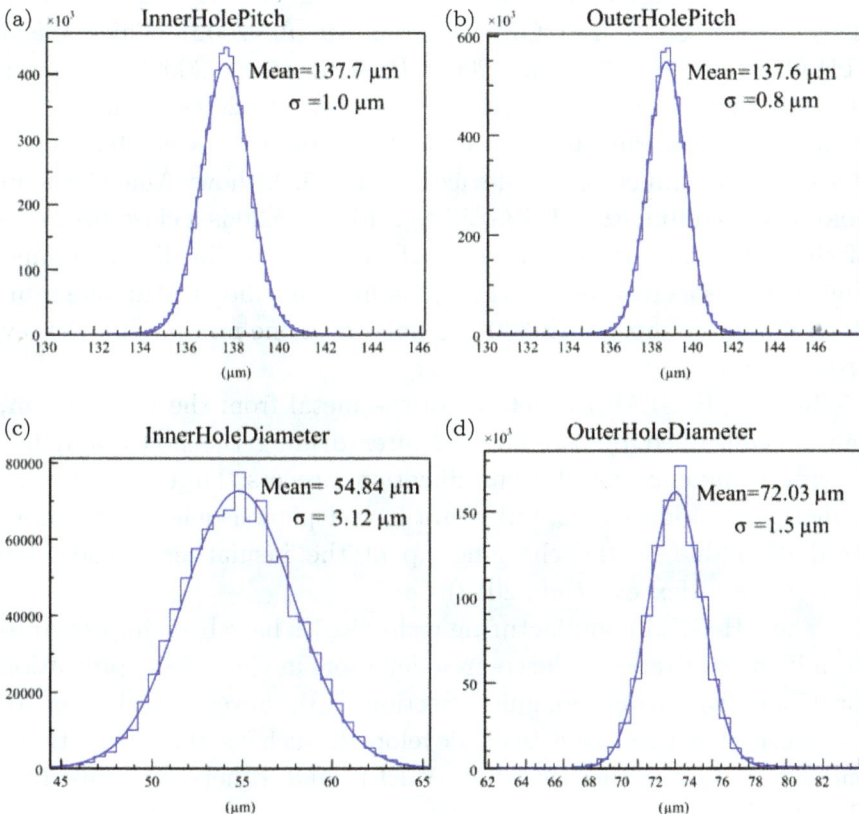

Figure 5.18: Distributions of the inner ((a) and (c)) and outer ((b) and (d)) GEM holes' pitch and diameter.

Source: Posik and Surrow (2015).

distribution of the holes' pitch and diameter over a 10×10-cm^2 GEM foil,[4] measured with an automated optical scanner (Posik and Surrow, 2015).

5.4 Thick GEM

The high-precision photolithographic processes described above are used to manufacture GEM foils with holes' diameter from few hundred down to few tens of microns, needed to ensure the best position accuracy. Mechanical drilling on thicker supports has been employed to manufacture devices entailing coarser performances: the so-called Optimized GEM (Periale *et al.*, 2002), Thick GEM (TH-GEM) (Chechik *et al.*, 2004; Breskin *et al.*, 2009) and large electron multiplier (LEM) (Badertscher *et al.*, 2009), suitable for applications requiring a large area and rigid electrodes, as photosensitive detectors and cryogenic devices. Figure 5.19 shows Amos Breskin holding a medium-sized TH-GEM foil; Figure 5.20 is a close-up views of the holes' pattern, one mm apart and 400 μm in diameter, and Figure 5.21 is a cross section through a mechanically drilled, one-mm-thick fibreglass sheet with 300-μm holes at a one mm pitch (Alexeev *et al.*, 2008).

In the TH-GEM, the retreat of the metal from the hole, or rim, can be controlled by wet etching after drilling the holes, and has a strong influence on the amplification process: large rims favour higher gains, Figure 5.22, but result in conspicuous gain shifts under irradiation due to the charging up of the insulating surface, see Section 5.8 (Alexeev *et al.*, 2012).

The TH-GEM manufacturing technologies have been improved to permit the realization of large-area detectors in view of an application for Cherenkov Ring Imaging (Section 7.3). Several variations of the basic structure have been developed, such as the Micro-Holes and Strip Plate, the Resistive WELL and others, described in Chapter 8.

---

[4]Manufactured by Tech-Etch Inc., 45 Aldrin road, Plymouth, MA 02360 (USA).

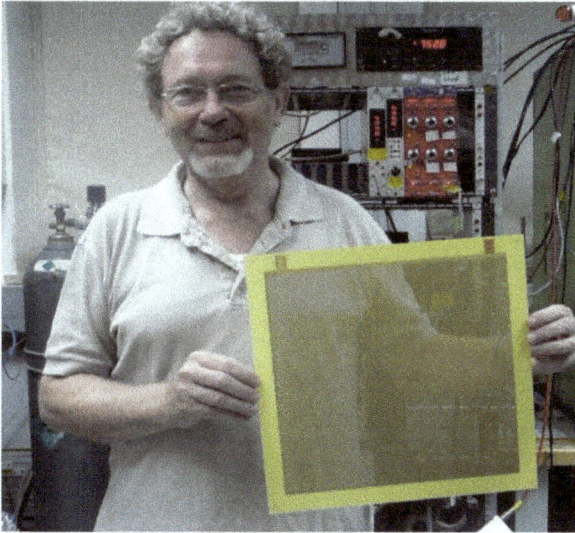

Figure 5.19:   Amos Breskin with a TH-GEM plate.
Source: Courtesy A. Breskin.

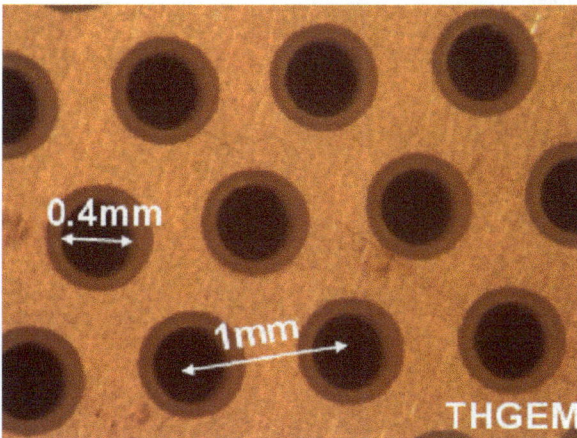

Figure 5.20:   Close-up view of TH-GEM holes' pattern.
Source: Breskin *et al.* (2009).

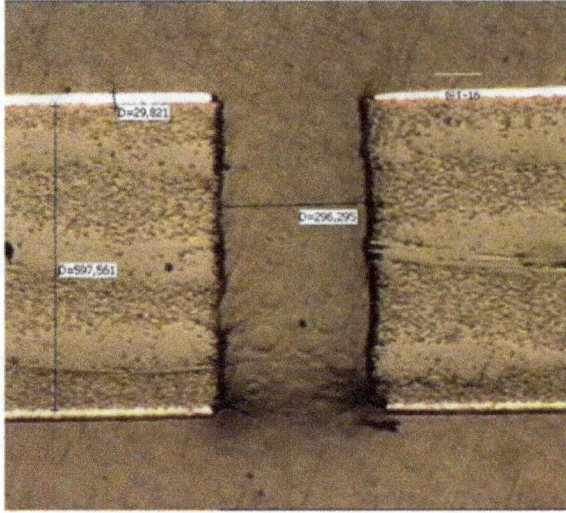

Figure 5.21:   Cross section through a 300-μm diameter hole of a TH-GEM.
Source: Alexeev *et al.* (2010).

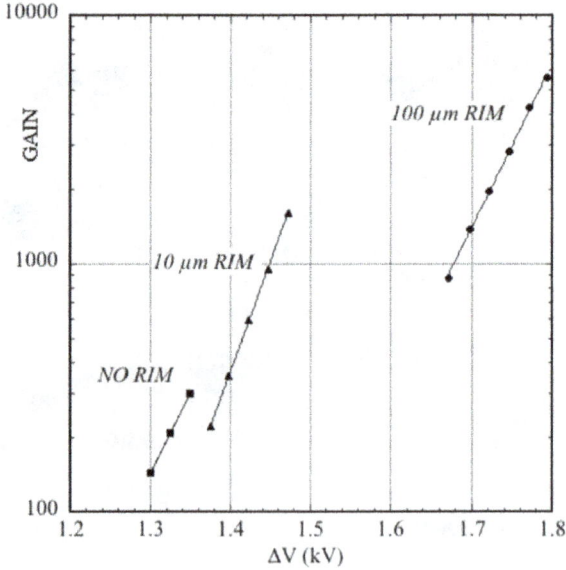

Figure 5.22:   Gain as a function of voltage for TH-GEMs with different insulating rims around the holes.

Source: Data from Alexeev *et al.* (2012).

5.5 Multi-GEM Assemblies

Single-GEM electrodes can reach gains well above 10^3 when tested in favourable laboratory conditions (Figure 5.10). However, similarly to that observed with other micro-pattern devices, when exposed to high rates and/or harsh beam environments, they tend to discharge at lower values of gain.

As shown in Figure 5.23, exposure to heavily ionizing alpha particles reduces the operating voltage, limiting the safe gain to a value below 10^3, comparable with similar results obtained with other MPGDs, MSGC and MICROMEGAS (Bressan *et al.*, 1999). Although the mechanical sturdier structures are not easily damaged as in MSGCs, high discharge rates affect the detection efficiency and may harm the front-end readout electronics. Different from the previously described MPGDs, however, GEM electrodes can be assembled in cascade, achieving large amplifications at moderate operating voltages for each element; this results in a substantial reduction of the discharge rates for equal gains. While double and

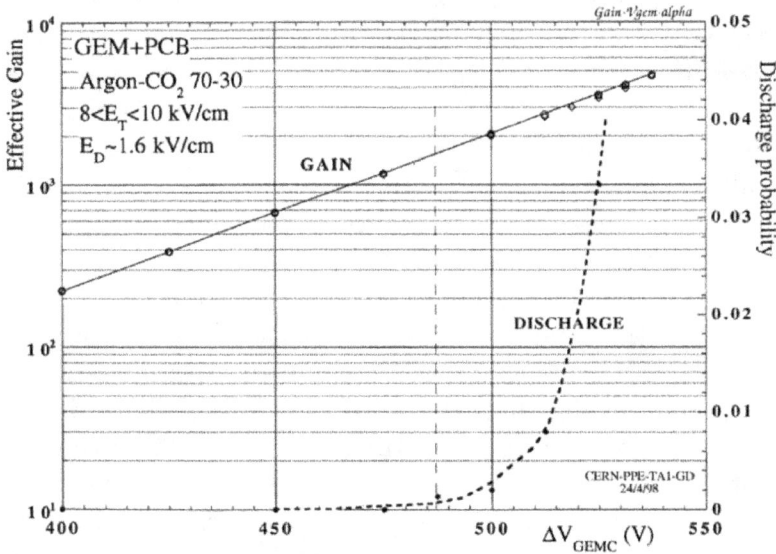

Figure 5.23: Single-GEM gain and discharge rate as a function of voltage. *Source*: Bressan *et al.* (1999).

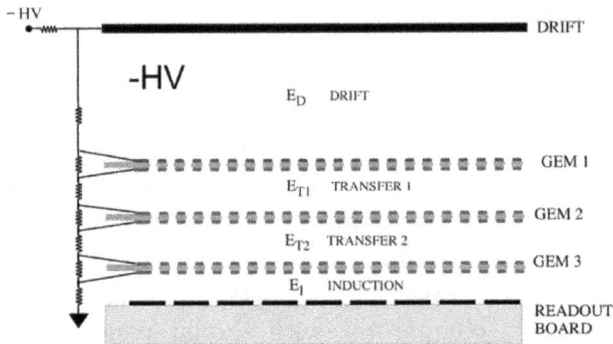

Figure 5.24:   Schematics of a TGEM.

triple GEM devices are the most widely used, cascades up to five electrodes have been tested (Bondar *et al.*, 2003; Dehmelt, 2015). The reference (Nath Patra *et al.*, 2018) describes a comparison of performances between a triple and quadruple GEM detector.

Figure 5.24 shows schematically a TGEM assembly, with three identical electrodes mounted at short distances, typically 2–3 mm; the drift gap is a few millimetres thick for the detection of fast particles, but can be increased to centimetres for detection of X-rays or to metres in drift and Time Projection Chambers. The resistive high-voltage distribution is the simplest and most widely used, although individual cascaded power supply systems have been developed that allow more flexibility in the choice of the voltage sharing (Corradi *et al.*, 2007).

The dependence of effective gain from voltage is shown in Figure 5.25 for single-, double- and triple-GEM assemblies. For convenience of operation, the voltage applied to each foil is usually identical, but a small unbalance between the foils, with the first GEM operated at higher gain, favours the reduction of the discharge rate (Section 5.10). The dashed lines in the figure denote the measured discharge probability on exposure to an external alpha particles source (Bachmann *et al.*, 2002); for a TGEM, the onset of the discharge occurs at a gain above 10^4, well above the value commonly used for the detection of fast particles.

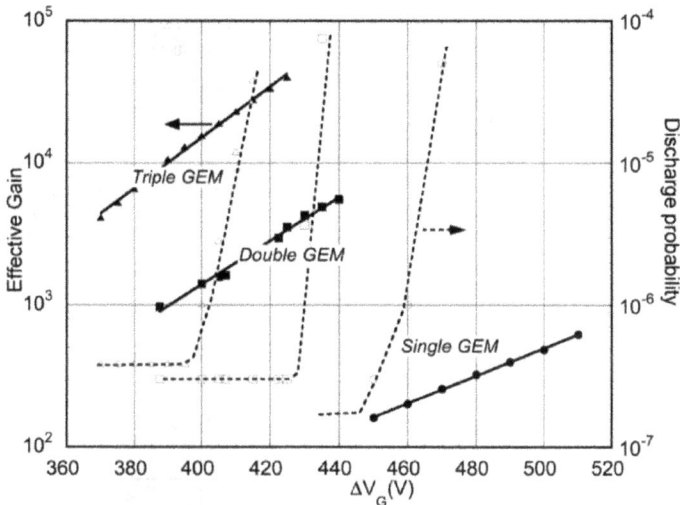

Figure 5.25: Effective gain of a single-, double- and triple-GEM detector measured on soft X-rays. The dashed lines are discharge probabilities on exposure to an internal alpha particles source.

Source: Bachmann *et al.* (2002).

It should be noted that the determination of the discharge threshold is somewhat arbitrary, and the result depends on many factors (detector geometry, gas filling and purity, moisture), making a comparison of results from different groups rather difficult. For a more detailed discussion on discharges, see Section 5.10.

In multiple structures, while the GEM voltages determine the gain of device, the electric fields in the transfer and induction gaps control the flow, spread and collection of the electron and ions charges. Increasing the field below a GEM electrode increases the electron current extracted from the foil, but decreases the current injected into the following foil; at each transfer, part of the charge is lost to the electrodes. A thorough optimization of the various field strengths permits to achieve the optimum operating performances in terms of gain, ion backflow reduction and discharge.

An example of measured currents on all electrodes of a double-GEM detector under uniform irradiation is given in Figure 5.26 as a function of transfer field between the two GEM foils' protection.

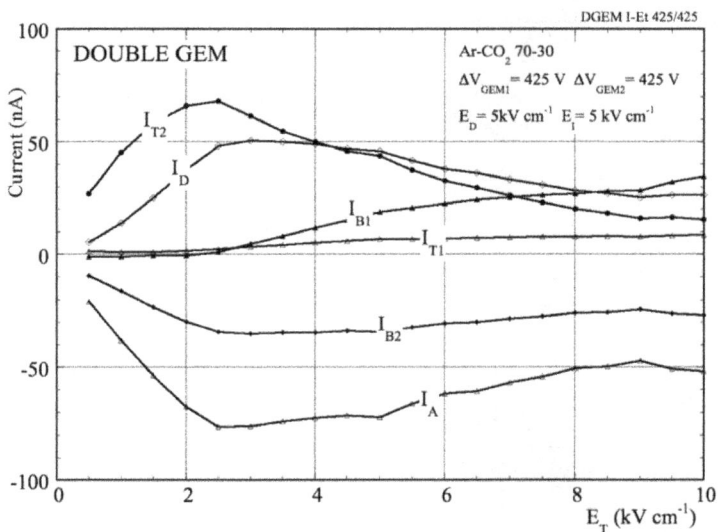

Figure 5.26: Measured currents on all electrodes of a double-GEM detector under uniform irradiation (D: drift; T1, B1: top and bottom of GEM1; T2, B2: top and bottom of GEM2; A: anode).

Source: Bachmann *et al.* (1999).

Dedicated programs have been used to simulate the multiplication and transfer processes in a variety of conditions, geometry, gas fillings and the presence of an external magnetic field: MAGBOLTZ to compute the electron and ion drift velocity and avalanche multiplication (Biagi, 1999), GARFIELD to map the electric fields and motion of the charges (Veenhof, 1998; Tikhonov and Veenhof, 2002). Two commercial programs provide also field calculations and drift properties, as well as three-dimensional representations of the detector geometry and charge flows: MAXWELL (Ansoft, 2019) and COMSOL (Multiphysics, 2019). As an example, Figure 5.27 shows the simulated propagation of the avalanching electrons through a four-GEM structure, part of a study aimed at the reduction of the positive ions' backflow in the ALICE TPC upgrade at CERN (Bhattacharya *et al.*, 2017). More detailed results of simulations are given in Sections 5.8 and 5.9.

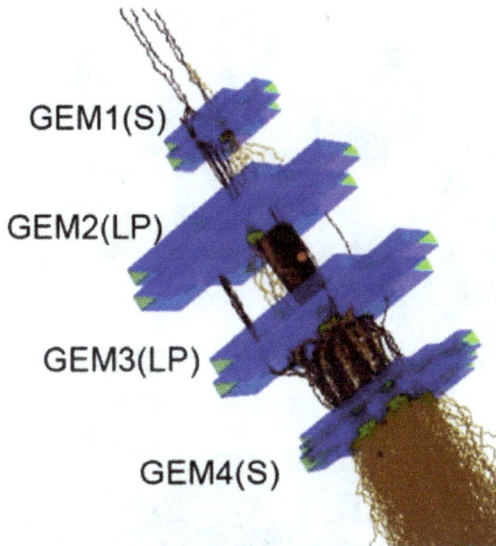

GEM1(S)

GEM2(LP)

GEM3(LP)

GEM4(S)

Figure 5.27: Simulated electron avalanche propagation through a quadruple-GEM structure.

Source: Bhattacharya *et al.* (2017).

5.6 GEM Detectors Construction

After manufacturing, GEM foils are thoroughly inspected, cleaned and HV tested, preferably in dry air or inert gas, with the procedures described, for example, in Altunbas *et al.* (2002). For moderate sizes, a common detector construction method (named "box" assembly) makes use of a set of frames serving the different functions and assembled with bolts and O-rings; drift and GEM electrodes are stretched on tensioning frames and glued on the final fibreglass frames. The anode with the readout pattern is commonly manufactured on a thin printed circuit board and serves also as a gas window.

An example of a detector with an active area of 10×10 cm^2, open during assembly, is shown in Figure 5.28 (Bressan *et al.*, 1999). A typical assembly includes, in sequence, a thin polymer window for gas containment, a drift electrode, one or more GEM foils and a patterned printed circuit board as the anode. A pictorial

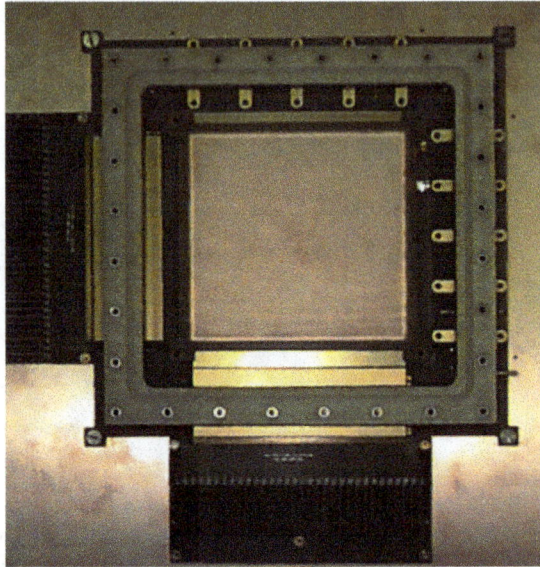

Figure 5.28:   A small-sized GEM detector during assembly.
Source: Picture CERN.

sequence of the box construction can be found in the web pages of the Gas Detectors Development group at CERN under "GEM framing and assemblage" (Sauli, 2010). Once finished, the detector is completed with high-voltage connections, gas inlet–outlet and readout electronics.

Normally operated at ground potential, the anode can be structured with arbitrary readout patterns, depending on the application: pixels, pad rows, one-, two- and even three-dimensional projective strips. Figures 5.29 and 5.30 are examples: two-dimensional Cartesian projective strips (Bachmann *et al.*, 2002), and pad rows read out individually or interconnected along three directions on the back side (Bachmann *et al.*, 2002). The latter scheme has been used for an ambiguity-free reconstruction of coordinates in case of multiple simultaneous tracks, as in the case of photon detection for Cherenkov Ring Imaging, see Section 7.4 (Meinschad *et al.*, 2004).

The multi-layer readout circuits are realized on thin-metal-coated polyimide foils, with a technology parent of the one used

Figure 5.29: Cartesian readout pattern, with two sets of metal strips at 400-μm pitch, separated by 50-μm polyimide ridges.

Source: Bressan *et al.* (1999).

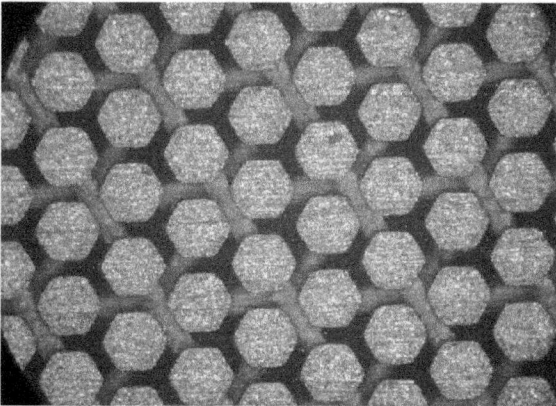

Figure 5.30: The hexaboard: individual readout pads, 300 μm in diameter, are interconnected along three directions on the back side of the board.

Source: Bachmann *et al.* (2002).

for manufacturing GEM foils, using the appropriate set on masks; holes for interconnections are metal plated as for conventional printed circuits. The strip width for the top and bottom layer in the circuit of Figure 5.29, 80 and 350 μm, respectively, has been optimized for equal sharing of the collected charge (Altunbas *et al.*, 2002). The box assembly has been used for the construction of larger detectors, as CERN's LHCb muon trigger (Bencivenni *et al.*, 2002).

Versatile and permitting the replacements of the components in case of failures, the box construction is expensive and has an unfavourable aspect ratio between active and total surface. A cheaper, fully glued construction has been developed for the production of the first large GEM beam tracker system of the COMPASS experiment at CERN (Capeans *et al.*, 2003; Altunbas *et al.*, 2002) and adopted since for the TOTEM setup (Antchev *et al.*, 2010). An exploded view of the assembly is shown in Figure 5.31. The structure makes use of thin mechanically cut fibreglass frames with narrow spacer inserts at regular intervals, shown in Figure 5.32; pre-stretched GEM foils, 30 cm on the side, are glued onto the frames serving both as support and spacer to the next electrode.

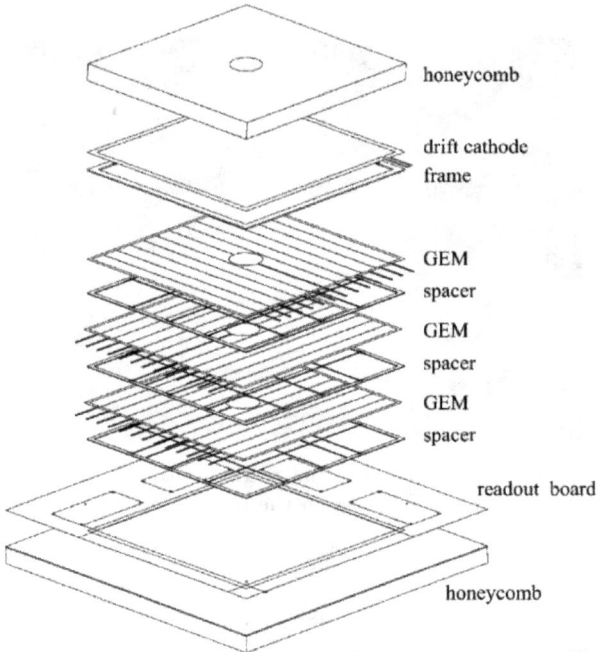

Figure 5.31:   Schematics of the COMPASS TGEM assembly.
Source: Altunbas *et al.* (2002).

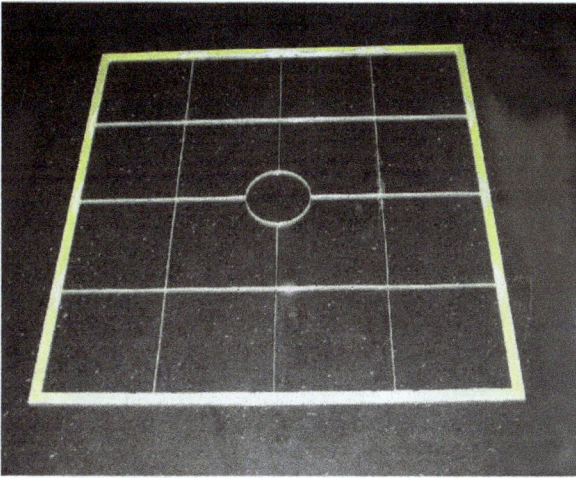

Figure 5.32:   Thin support frame with the gap-restoring spacers. The central round structure corresponds to the "beam killer".
Source: Picture CERN.

Following the outcomes of a systematic study on discharges, one side of the large GEM foils is divided in sectors, individually connected to the voltage through a high-value resistor to limit the energy available in case of a spark (see Section 5.10). The separation between sectors, typically 200 μm wide, is created with a second masking process, removing the copper cladding of the foil; for a 30×30-cm^2 foil, separating the active area in 12 sectors appears to be a safe option. As a rule of thumb, the surface of each sector should not exceed around 100 cm^2, corresponding for a 50-μm Kapton foil to a capacitance of about 5 nF (Bachmann *et al.*, 2002). The effect of the stored energy on the propagation of a discharge is discussed in Section 5.10. A central round sector, independently powered (the beam killer), allows to activate or inhibit the region of the beam. This permits to run the full detector at low rates for alignment; when inhibited, it prevents saturation of the electronics at the highest rates.[5]

---

[5]The detector itself is capable of operating at very high beam rates, see Section 5.9.

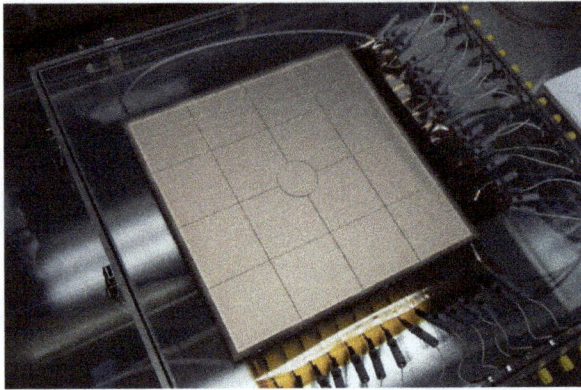

Figure 5.33:   A framed GEM foil inserted in the HV test box. The frame's spacers and the separations between sectors are visible in the picture.
Source: Picture CERN, Altunbas *et al.* (2002).

The framed foils are then assembled and glued in sequence, with the help of pins to ensure proper alignment; light honeycomb plates on each side ensure rigidity to the structure. All along the assembly process, the foils are high-voltage tested in a nitrogen-filled box, shown in Figure 5.33 (Altunbas *et al.*, 2002; Capeans *et al.*, 2003). Close to 30 TGEM detectors with two-dimensional readout have been built for the COMPASS experiment, and operated for many years (Ketzer *et al.*, 2004; Abbon *et al.*, 2007). Figure 5.34 shows one of the detectors, equipped with readout electronics and installed in the M2 muon and hadron beam at CERN. A similar construction has been used for the forward tracker of the TOTEM experiment at CERN; in this case, the TGEM detectors have a semi-circular shape to match the experimental requirement, Figure 5.35, illustrating the great flexibility of the technology. Figure 5.36 is a view of one half of the assembled TOTEM telescope, mounting ten GEM chambers in sequence (Croci *et al.*, 2013).

Optimized for operation in the high-flux area near the beam pipe, 12 double TGEM detectors with an active area of 20×24 cm^2 each have been built and operted in the LHCb experiment at CERN (Alfonsi *et al.*, 2006). The chambers, assembled in pairs, have an individual pad readout, OR-ed between the two detectors in the

Figure 5.34:   One of the COMPASS TGEM chambers in the beam.
Source: Picture CERN.

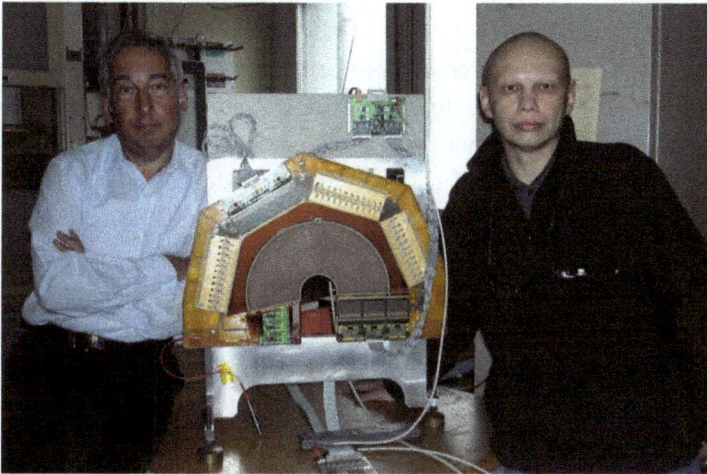

Figure 5.35:   The author and Leszek Ropelewski with a prototype of the semi-circular TGEM TOTEM module.
Source: Picture CERN.

Figure 5.36:   One half of the forward TOTEM telescope. The second half is inserted from the left, capturing the beam vacuum tube.

Source: Bagliesi *et al.* (2010).

pair. Designed mainly to ensure efficient triggering at particle rates up to 300 kHz cm^{-2}, the chambers are operated with with a fast gas mixture (Ar–CF4–CO$_2$) achieving 96% detection efficiency in a 20-ns coincidence window, see Section 5.7 (Cardini *et al.*, 2012). Figure 5.37 is a close-up view of part of the LHCb muon detector at CERN; three of the six TGEM chambers are visible, assembled close to the beam pipe.

The use of carbon tetrafluoride in high-rate gaseous detectors implies a thorough control of the water contamination in the mixture, as discussed in Section 5.11.

Larger size detectors, using a variant of the manufacturing technique used for COMPASS, have been developed to instrument the Super Bigbite Spectrometer (SBS) at the Thomas Jefferson National Laboratory (JLAB) in the US (Gnanvo *et al.*, 2015). Each chamber is an assembly, within the same gas volume, of four GEM modules, 60×50 cm^2 each, individually mounted on fibreglass frames with thin spacer supports. Similar in design and developed in view of

Figure 5.37:   The LHCb double-GEM chambers installed close to the beam pipe. *Source*: Picture courtesy G. Bencivenni.

applications at the planned electron–ion colliders, with the composite GEM structure of a trapezoidal shape, one metre in length, they have been extensively tested in a charged particles beam at Fermilab (Gnanvo *et al.*, 2016).

Built with thin windows to reduce self-absorption, GEM detectors of similar conception aim at particle identification, exploiting the difference in transition radiation emission by multi-layer radiators. Xenon-based gas filling optimizes the X-ray conversion efficiency in thin drift gaps; ionization charge is amplified by a TGEM stack and recorded. The first results with a prototype show that an e/p rejection factor of 5 can be achieved with a single-GEM-TRD module and a 10-cm composite radiator (Barbosa *et al.*, 2019).

Suitable for medium-sized GEM detectors, the COMPASS assembly technique is less appropriate for the manufacturing of larger devices, since the failure of any internal component cannot be repaired. A new assembly technique has been developed for the construction of the upgrade of the CMS high-η, muon detector at CERN; named "self-stretching", it relies on the use of thick frames

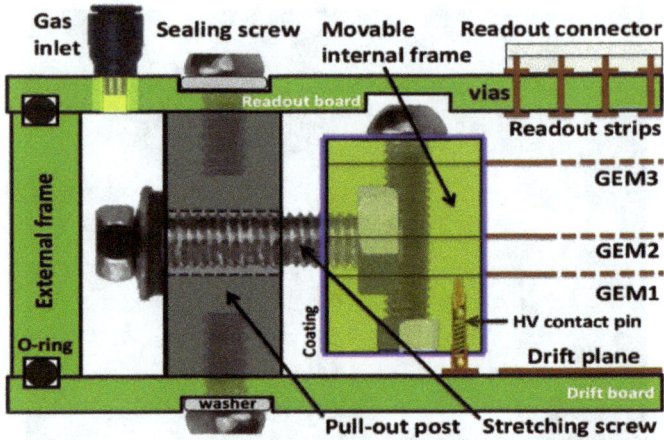

Figure 5.38:   Self-stretching technique for large-area GEM detectors' construction.

Source: Abbaneo *et al.* (2013).

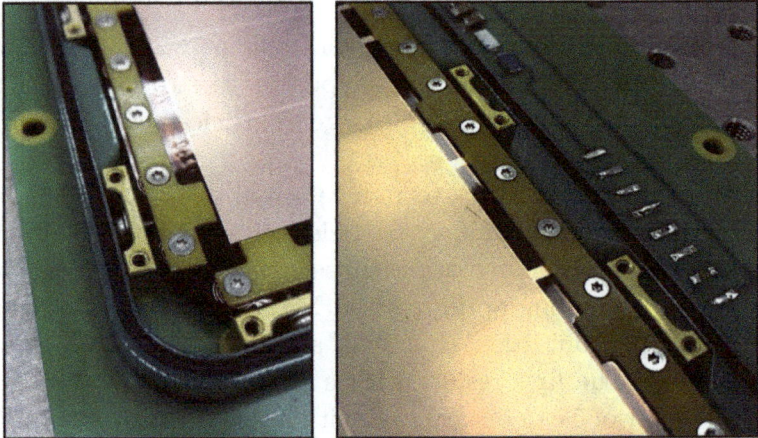

Figure 5.39:   Detailed views of the GEM stack stretching pins and frames.

Source: Abbaneo *et al.* (2017a).

and of a system of screws to tension the GEM foils all around their edges, Figure 5.38. Figure 5.39 shows detailed views of a chamber edge during manufacturing, and Figure 5.40 the mounting of a framed foil on the stretching table. The frames with stretched

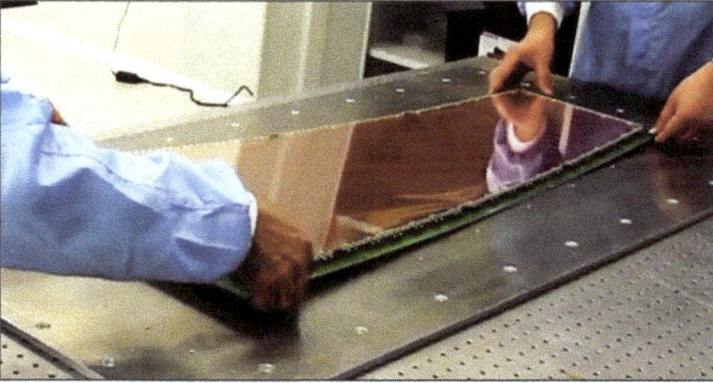

Figure 5.40: Moving the stack to the stretching table.
Source: Abbaneo *et al.* (2017a).

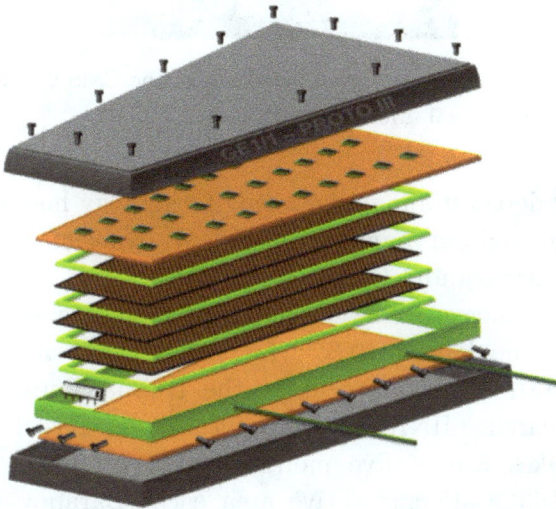

Figure 5.41: Exploded view of a CMS GEM module for the forward muon detector.
Source: Abbaneo *et al.* (2013).

foils are successively assembled with the help of centring pins. Figure 5.41 shows schematically a TGEM CMS module (Abbaneo *et al.*, 2013), and Figure 5.42 shows a fully assembled detector installed for systematic beam tests (Abbaneo *et al.*, 2017a).

Figure 5.42:   A fully assembled prototype of the CMS GEM module.
Source: Picture Courtesy J. Merlin, CERN.

A similar detector's manufacturing technology has been adopted for the compressed baryonic matter (CBM) experiment at the FAIR facility (GSI Darmstadt) (Adak *et al.*, 2017).

Designed to measure charged particles' momenta and multiplicity in a magnetic field up to 0.8 T for the Barionic Matter and the Nuclotron (BM&N) experiment at the Dubna Joint Institute for Nuclear Research (JINR), a tracking system based on T-GEM detectors makes use of five modules of 66×41-cm^2 and seven modules of 163×45-cm^2 active area each (Baranov *et al.*, 2017; Galavanov *et al.*, 2019).

The flexibility of the GEM electrodes permits the construction of non-planar devices; as charge multiplication occurs in the holes, bending of the foil does not modify the main detector properties. A four-layer cylindrical GEM detector, with an active length of 70 cm, built and operated for the KLEO-2 experiment in Frascati is shown in Figure 5.43 (Balla *et al.*, 2013). The readout pattern on the anode, with longitudinal and interleaved strips at 25°, provides

Figure 5.43:   Cylindrical GEM detector for KLOE-2.
Source: Balla *et al.* (2013).

Figure 5.44:   The BoNus radial GEM TPC chamber.
Source: Picture courtesy H. Fenker, JLAB.

a spatial resolution of 200- and 400-μm rms for the azimuthal and longitudinal coordinates, respectively.

Developed to serve as both target and tracker at Jefferson Lab, the BoNus detector is a radial time projection chamber with three cascaded curved GEM amplifiers; the readout is performed on rows of pads on the outer cylinder, Figure 5.44 (Fenker *et al.*, 2008).

A cylindrical TGEM device has been built also for the inner tracker upgrade of the BESIII experiment at IHEP in Beijing (Amoroso *et al.*, 2016).

5.7   Operating Characteristics, Efficiency and Localization Accuracy

Since their introduction, GEM-based detectors have been employed in a large number of experimental setups and applications. The initial developments aimed at the use of the technology for detection and localization of high-rate charged particle beams, but extended to other applied fields: astrophysics, biomedicine, material analysis and plasma physics. Detectors have been operated successfully with a variety of gas fillings and in a wide range of pressures, from few torr to several atmospheres; examples will be given in the following sections. The devices have also been operated at cryogenic conditions in dual-phase detectors (Chapter 10).

In a multi-GEM structure, the various electrodes collect the amplified charges (electrons and ions) resulting from the multiplication and drift processes in field-dependent relative proportions: Figures 5.8 and 5.26 provided representative examples. The anode, final electrode in the structure, only gathers the electrons leaving the last GEM in a cascade; the induced negative signal is therefore very fast, as its width corresponds to the drift of the high-mobility electrons over a few millimetres.

Figure 5.45 is an example of pulses detected with a fast current amplifier on the anode (Ziegler, 1998). Two tracks separated by less than ∼20 ns are well resolved; this corresponds to about a one-mm multi-track resolution. As shown in Figure 5.6, the anode can be patterned with one- or two-dimensional projective readout strips to perform localization. For a single GEM and localized ionizing events (soft X-ray conversions), Figure 5.46 shows the measured width of the charge induced on the anode as a function of induction field. For the narrower gap, the distribution fwhm is about 400 μm.

As far as the charge is shared between several adjacent strips, a centre-of-gravity algorithm permits to estimate the space coordinate

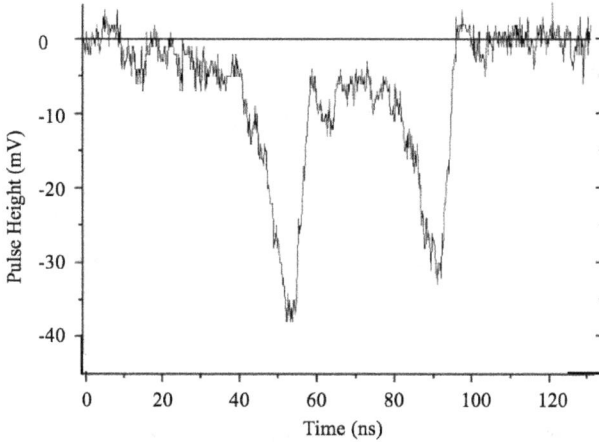

Figure 5.45:   Fast signals recorded on the GEM anode.
Source: Ziegler (1998).

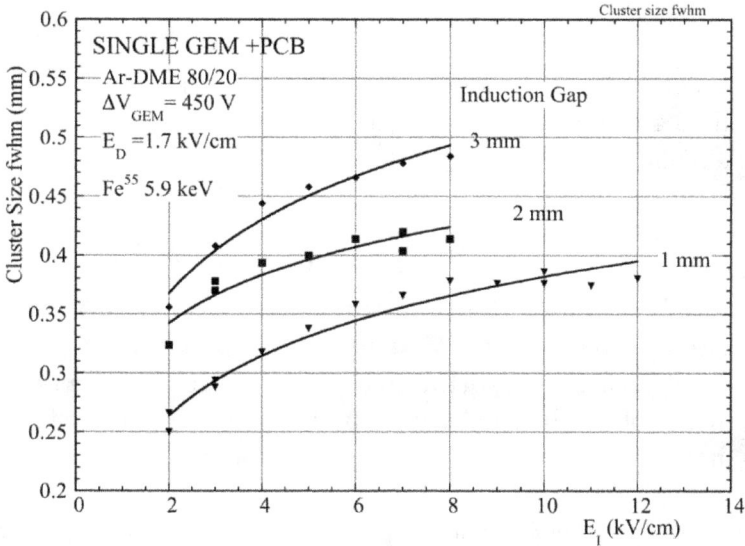

Figure 5.46:   Single-GEM cluster size for 5.9 keV X-rays in a range of induction gaps.
Source: Bressan *et al.* (1999).

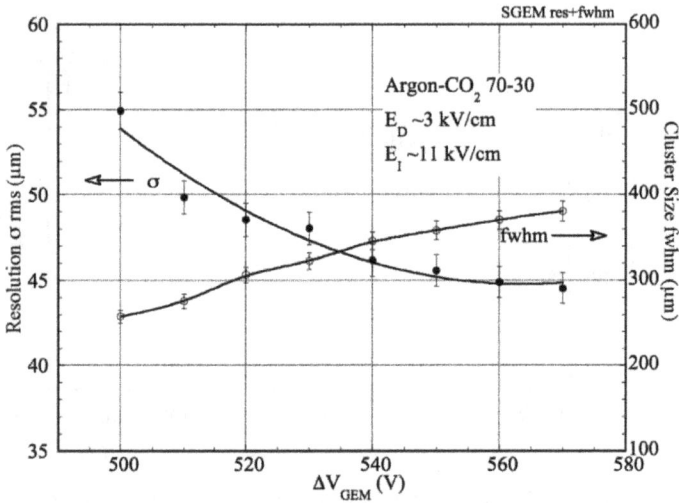

Figure 5.47:   Cluster size and localization accuracy for fast charged particles perpendicular to a single-GEM detector as a function of voltage.

Source: Bressan *et al.* (1999).

of the event with a resolution better than the strips' pitch, as shown in Figure 5.47 for a single-GEM detecting minimum ionizing particles perpendicular to the chamber (Bressan *et al.*, 1999).

For multi-GEM structures, the width of the collected charge increases due to diffusion between the foils, permitting the use of coarser readout strips while ensuring charge sharing.

Figure 5.48 is an example of simulated distribution of electrons propagating through a TGEM structure, showing the lateral spread of the multiplying charge (Bal and Dubey, 2020). The spread over many holes of an originally localized charge explains the higher gains that can be reached with multiple structures, see Section 5.10. Experimental measurements of the collected charge at the anode tend to provide wider distributions, probably due to the spread of the initial ionization charge (Altunbas *et al.*, 2002). Owing to diffusion in the last transfer gap, the distribution of the charge collected at the anode is smoothed, obliterating the hole's structure (Figure 5.48(c)).

Fast, singly charged particles release, in a 3-mm drift gap filled with argon-based mixtures at one bar, around 30 electron–ion pairs.

(a)

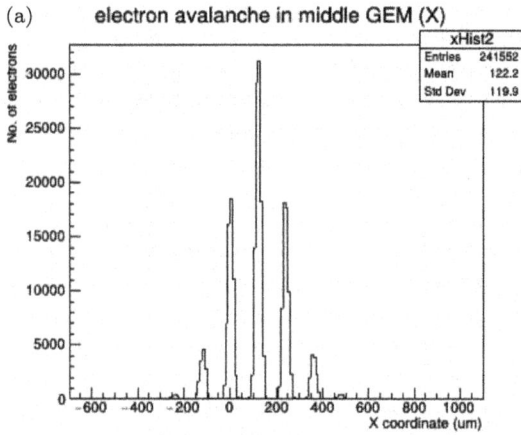

(b)

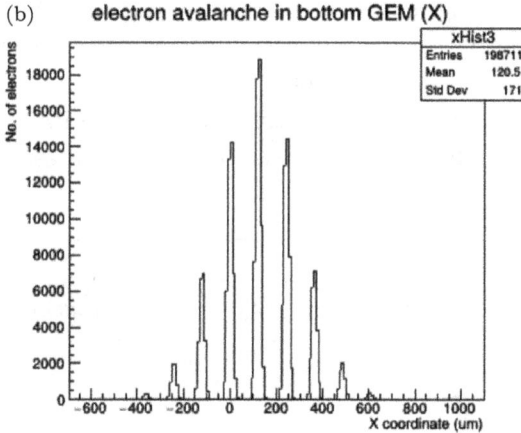

(c)

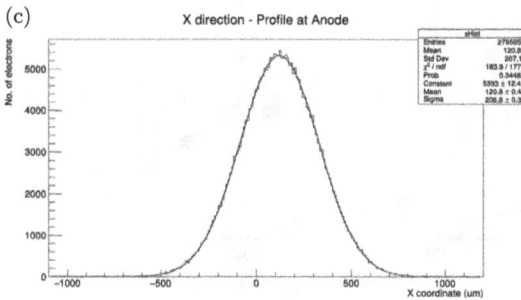

Figure 5.48: Simulated charge propagation through a TGEM structure, for a localized ionization amplified by the first GEM. Distributions along a row of holes at the bottom electrode of the second GEM (a), third GEM (b) and at the anode (c) (Courtesy A. Bal, A. Dubey).

With a TGEM operated at an effective gain of 10^4, this results in a collected charge at the anode of $\sim 3 \times 10^5$ electrons, or ~ 5 pC. For a two-dimensional readout scheme, recording the induced signal distribution on sets of parallel strips, this charge is shared between the two coordinates, and then again between several adjacent strips used for localization. For the 400-μm strips' pitch adopted by COMPASS, the average number of strips sharing the charge (or cluster size) is between two and three. The detection efficiency is then determined by the ratio of the signal to the noise of the front-end electronics used to instrument the detector.

An example of pulse height spectrum for minimum ionizing particles measured using a fast analogue circuit to record the induced signals is given in Figure 5.49; a discrimination threshold set at about 1/20 of the peak value ensures a detection efficiency close to 100%, Figure 5.50 (Altunbas *et al.*, 2002).

Figure 5.51 shows the distribution of residuals, the difference between predicted and measured coordinates in the TGEM

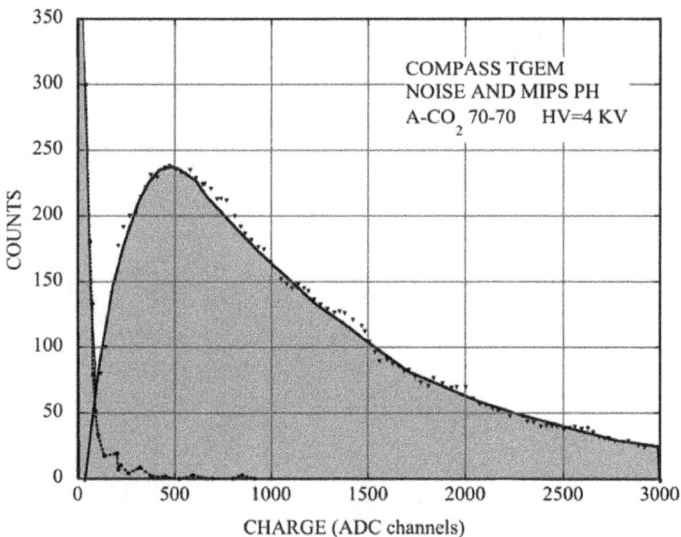

Figure 5.49: Detected charge for fast particles and noise recorded with the COMPASS TGEM tracker.

Source: Altunbas *et al.* (2002).

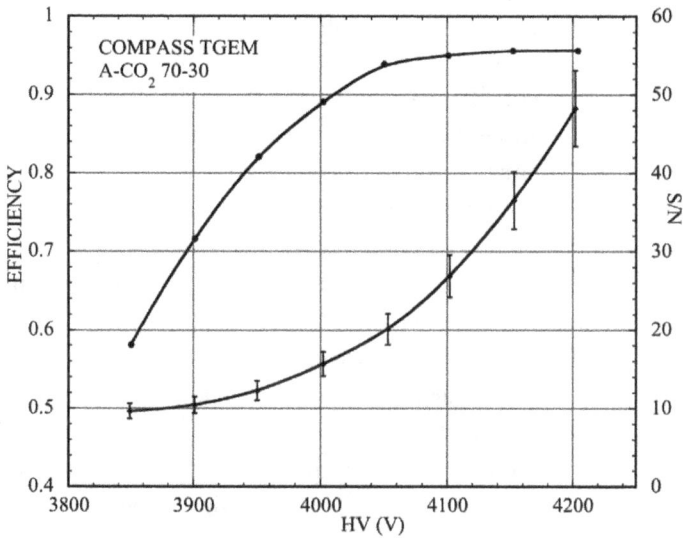

Figure 5.50: Efficiency and signal over noise in the COMPASS chambers as a function of overall applied voltage.

Source: Altunbas *et al.* (2002).

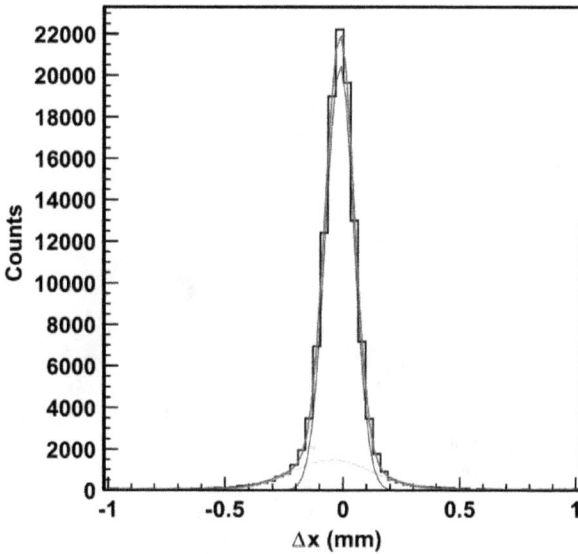

Figure 5.51: Residuals distribution for a single coordinate, with ~70-μm rms.

Source: Ketzer *et al.* (2004).

COMPASS spectrometer; the width of the distribution, 70-μm rms, corresponds to the single coordinate position accuracy measured in realistic beam conditions (Ketzer *et al.*, 2004). This result is determined by the statistics of the detection processes of fast charged particles and the charge spread due to diffusion.

Owing to the large energy loss spread typical for fast particles, the correlation between charges collected on the two coordinates permits in most cases to resolve reconstruction ambiguities for multi-track events, Figure 5.52 (Ketzer *et al.*, 2004).

A simulation taking into account the various dispersive components for a TGEM detector used to track high-energy electrons (energy loss statistics, diffusion, geometry) indicates that the best space resolution is achieved for a strip pitch between 200 and 300 μm; the experimental results, obtained in a high-energy electron beam, confirm this outcome (Kudryavtsev *et al.*, 2017).

Better localization accuracies are obtained with neutral radiation, due to the smaller physical dispersion of the initial ionization. As an example, Figure 5.53 shows the absorption radiography of a

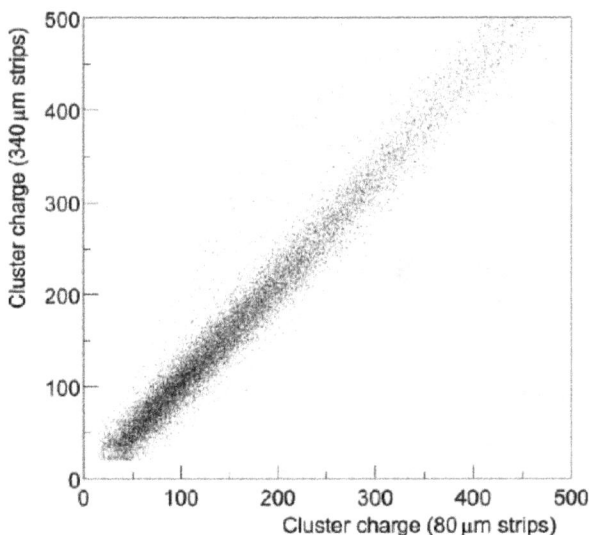

Figure 5.52:   Correlation between the charge recorded on the X and Y coordinates for minimum ionizing particles.

Source: Ketzer *et al.* (2004).

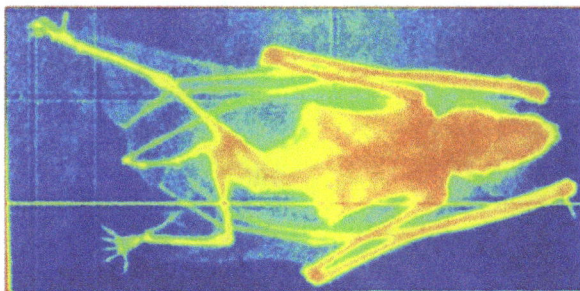

Figure 5.53:   False colour Soft X-ray radiography of a bat. The body size is about 60 × 20 mm.
Source: Bressan *et al.* (1999).

small mammal realized exposing the target to a soft X-ray beam. While of modest medical interest, the image illustrates well the good contrast and resolution properties of the detector.[6]

An intrinsic resolution of 4-μm rms has been achieved with a special GEM-based device making use of micro-pixel solid state readout, Section 8.7 (Bellazzini *et al.*, 2007).

When using digital threshold discriminators for the front-end electronics, as in the TOTEM detector, the position accuracy corresponds to the strips' pitch.

Mixtures of argon and carbon dioxide, which are convenient because of non-flammability and chemical stability, provide moderate time resolutions due to the relatively slow drift velocity of electrons. Faster mixtures, obtained adding to argon various proportions of carbon tetrafluoride (CF_4), have been extensively tested (Alfonsi *et al.*, 2004); as seen in Figure 5.54, the faster gas provides a factor-of-two improvement in the time resolution, an essential advantage for the operation of the detectors in a high-rate environment and to resolve events generated in successive collisions at CERN's LHC, 25 ns apart. Figure 5.55, from the same reference, provides the detector efficiency recorded for fast particles in a 20-ns window and different gas mixtures as a function of gain.

---

[6]The small bat was found (dead and dried) in the author's country barn, and has been used for decades in the GDD laboratory as reference target for imaging studies.

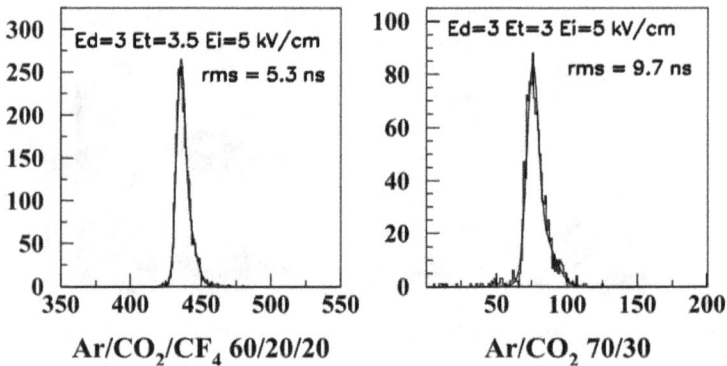

Figure 5.54:   Time resolution for fast particles measured with two gas mixtures.
Source: Alfonsi *et al.* (2004).

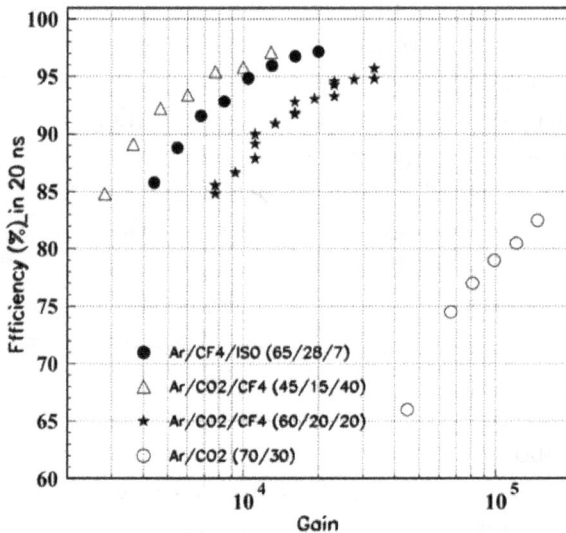

Figure 5.55:   Detection efficiency for fast particles in a 20-ns window in several gases.
Source: Alfonsi *et al.* (2004).

Based on these studies, a system of medium-sized TGEM chambers with a fast gas filling has been used for fast triggering in the LHCb experiment at CERN, described in Section 5.6 (Alfonsi *et al.*, 2007). It should be noted, however, that the use of CF_4 requires special precautions, due to the chemical reactivity of fluorinated compounds liberated in the avalanches in the presence of moisture (see Section 5.9).

Dedicated primarily to a better understanding of the energy resolution of the detector at high rates, phenomenological models have been developed to study the dependence of the X-ray energy discrimination of GEM detectors as a function of geometry, gases and operating conditions (Causa *et al.*, 2015; Azevedo *et al.*, 2015). The outcomes are exploited to improve the results obtained in very-high-rate plasma diagnostics, see Section 8.6 (Song *et al.*, 2016).

5.8 Gain Stability and Uniformity

The detailed shape, regularity and surface conditioning of the holes affect the stability and uniformity of the gain. In most structures, an initial increase of gain is observed immediately after the application of voltage; this increase is minor for near cylindrical holes, while it can be significant for other shapes, Figure 5.56 (Benlloch *et al.*, 1998). The "charging-up" process, a function of the irradiation rate, is due to the deposit and accumulation of positive ions and electrons on the insulating walls of the holes. Figure 5.57 is an example of gain increase and stabilization as a function of time due to charging up for two irradiation rates; after reaching a plateau within a few minutes, the gain remains subsequently stable. This behaviour can be explained by the property of insulators to gather charges until all field equipotentials become parallel to the surface, and no more charges can be collected. Removing the source, the discharge time is much longer, due to the very high resistivity of the insulators (Figure 5.58) (Altunbas *et al.*, 2002). Figure 5.59 is a comparison of the gain stability of a GEM manufactured by laser drilling with cylindrical holes and a standard foil.

Addition of small amounts of water vapour has been found to reduce the charging-up process, but it is generally not recommended as it can increase the discharge probability (see Section 5.10); this effect may however influence experimental comparisons between different structures, if the moisture levels are not well monitored.

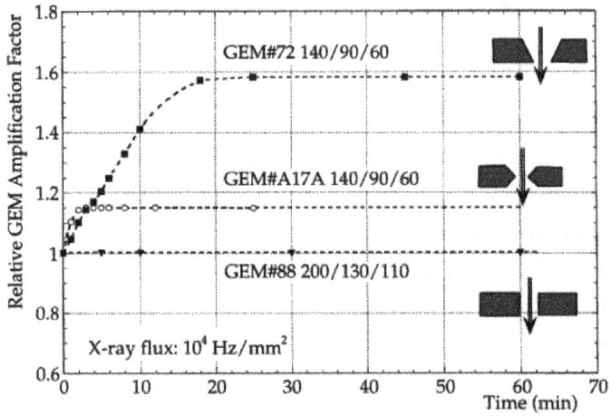

Figure 5.56:   Gain evolution at startup due to the insulator charging up for three shapes of the GEM holes.

Source: Benlloch *et al.* (1998).

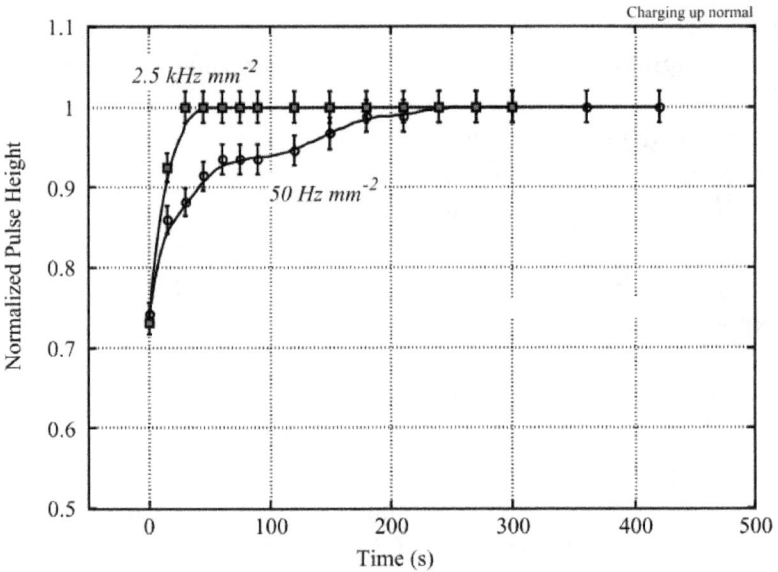

Figure 5.57:   Normalized gain increase and stabilization of a "standard" GEM with double-conical holes at different irradiation rates.

Source: Altunbas *et al.* (2002).

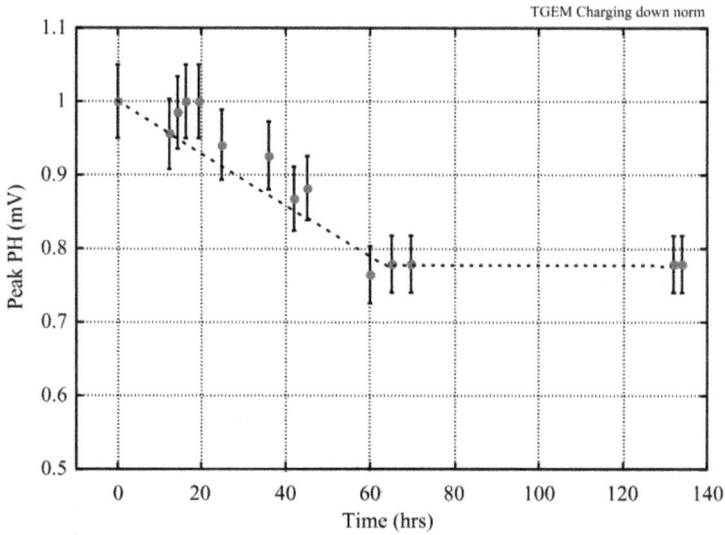

Figure 5.58:   Gain decrease (charging down) of a standard GEM after removal of the irradiation source.

Source: Altunbas *et al.* (2002).

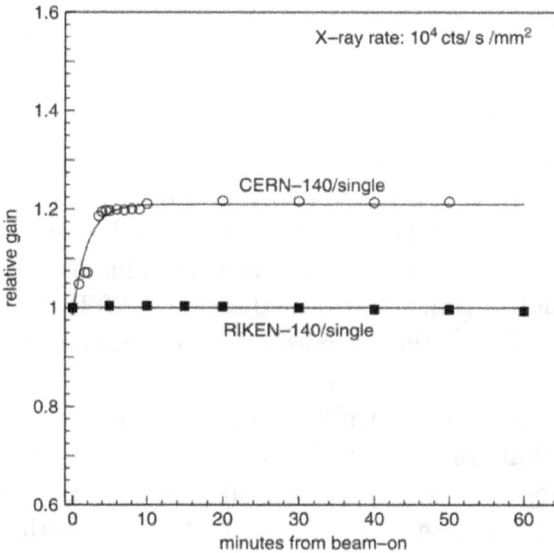

Figure 5.59:   Gain evolution as a function of time for a standard CERN GEM foil and for a laser-drilled RIKEN device.

Source: Tamagawa *et al.* (2006).

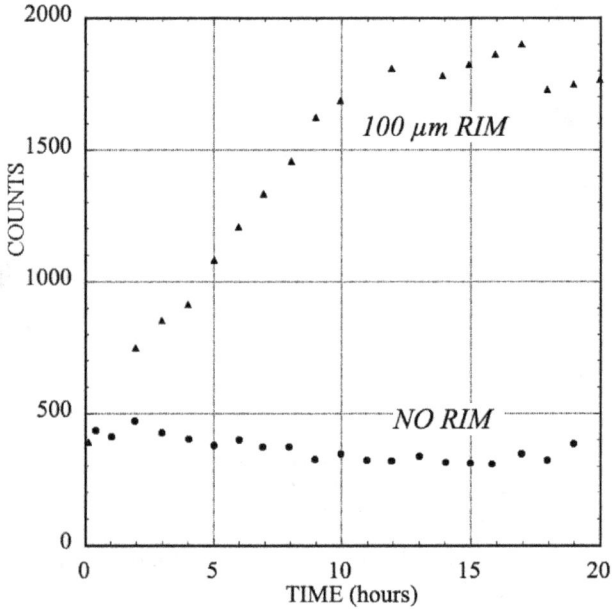

Figure 5.60:   Charging up for TH-GEMs with and without insulating rim around the holes.

Source: Redrawn from data in Alexeev *et al.* (2012).

The increase in gain due to charging up is rather conspicuous for the TH-GEM, having long and narrow insulating holes; an example is given in Figure 5.60 for two choices of the insulating rims around the holes (Alexeev *et al.*, 2012). The general observation is that structures having longer paths between metallic electrodes, such as the double-conical thin GEM and the rimmed TH-GEM, permit to reach higher gains at the expense of more pronounced charging-up processes.

Early attempts to reduce the charge accumulation on the insulators with thin, high-resistivity carbon coatings have not been pursued, probably due to the difficulty of obtaining uniform and defect-free deposits (Beirle *et al.*, 1999). Progress with diamond-like carbon (DLC) thin layer depositions may provide a way to solve the charging-up problem (Abbaneo *et al.*, 2017b), see Chapter 8. A comparison of gain evolution with time of a standard and several

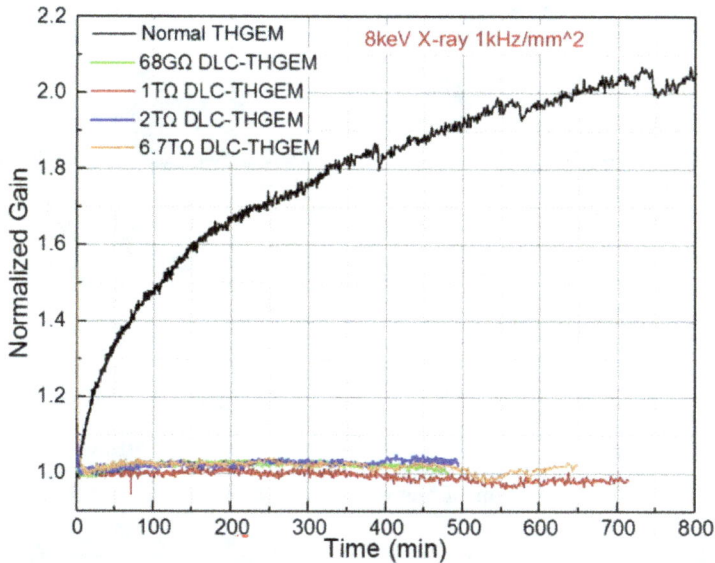

Figure 5.61:   Time evolution of gain for standard (top curve) and several DLC-coated TH-GEMs.

Source: Song *et al.* (2020).

DLC-coated TH-GEMs is given in Figure 5.61 (Song *et al.*, 2020). Although very promising, the process requires the development of technologies for coating large areas.

A detailed simulation of the charging-up processes of TH-GEM devices is given in Pitt *et al.* (2018).

The processes taking place in gaseous detectors can be mimicked using dedicated simulation programs; a review of the methods used to compute charge transports and dynamic gain modification due to the deposition on the dielectrics is given in Tikhonov and Veenhof (2002). An example of computed electron drift lines through a GEM hole before is shown in Figure 5.62(a) and after charging up of the dielectric is shown in Figure 5.62(b); the corresponding time evolution of the collected electron charge on the various electrodes is given in Figure 5.63 (Alfonsi *et al.*, 2012). Simulations show a good agreement between computed and measured gain increase as a function of collected charge (Correia *et al.*, 2014).

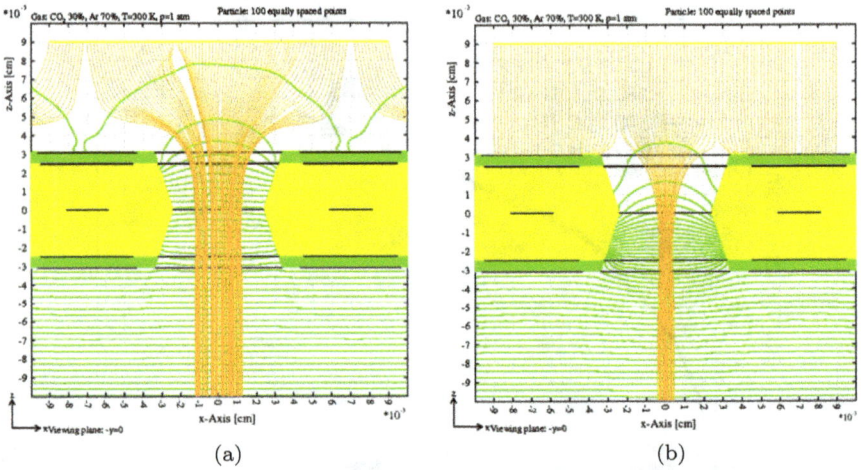

(a)   (b)

Figure 5.62:   Time evolution of the electron drift lines due to the GEM walls'
charging.

Source: Alfonsi *et al.* (2012).

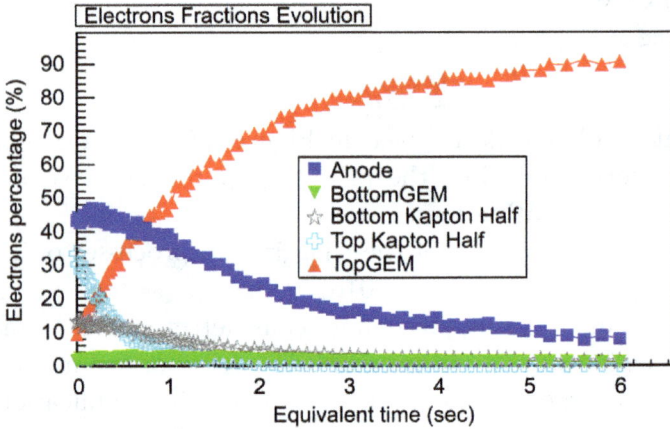

Figure 5.63:   Time evolution of the electron charge on the various electrodes.
Source: Alfonsi *et al.* (2012).

The evolution of gain with time, ultimately reaching saturation, depends on the primary electron density, as shown in Figure 5.64 comparing measurements at different irradiation rates; the behaviour is well reproduced by model calculations (Hauer *et al.*, 2019).

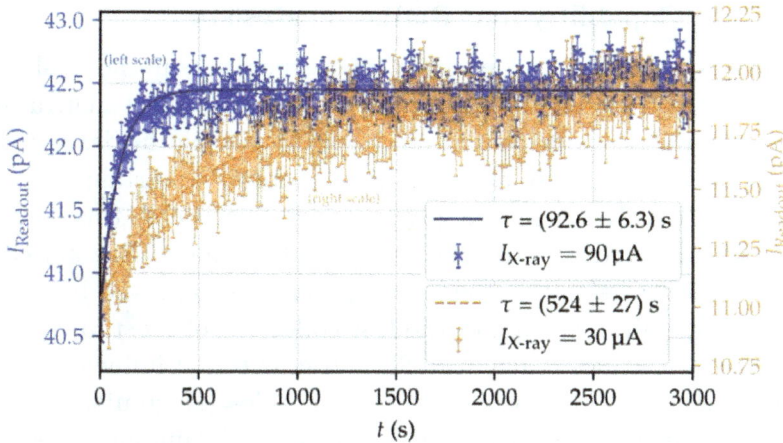

Figure 5.64:   Measured (points with error bars) and computed time evolution of gain at two X-ray generator currents.

Source: Hauer *et al.* (2019). Courtesy P. Hauer (Bonn University).

The charging-up processes and non-uniformities in the artwork (mainly in the diameter of the holes) result in gain variations over the detector area. The local gain of a detector can be measured with exposure to a collimated beam of radiation, soft X-rays from a radioactive source or X-ray generator, after stabilization, and varies typically by 10–15% over the active area of small and medium-sized detectors as those used in the COMPASS tracker (Patra *et al.*, 2017; Altunbas *et al.*, 2002); it can be more conspicuous for larger-sized devices.

A thorough gain calibration and mapping are integral parts of the systematic quality assurance tests for serial productions. Less important for tracking devices, the gain calibrations are more demanding for systems exploiting the measured energy loss for particle identification, such as the ALICE TPC upgrade (Aggarwal *et al.*, 2018).

Systematic studies, both experimental and by simulation, of gain variations in multi-GEM devices as a function of mechanical tolerances, holes' diameters and operating voltages are reported in Abi Akl *et al.* (2016) and Das (2016).

5.9 Rate Capability and Radiation Resistance

In gaseous counters, the positive ions generated in the avalanche process slowly recede to the negative electrodes until neutralized; the accumulated space charge modifies the electric fields, resulting in a reduction of gain and distortions above a certain value of radiation flux. In multiwire proportional chambers, this limit is attained at around 10^4 particles per second per mm^2 (Breskin *et al.*, 1974). In GEM detectors, the rate capability is improved by several orders of magnitude owing to the fast collection of most ions by the electrodes and to the screening effect from external influence due to the confinement of the avalanches in the holes (Buzulutskov *et al.*, 1999). As shown in Figure 5.65, the proportional gain, measured on a single GEM exposed to a soft X-ray generator, remains constant up to a flux above 10^6 s^{-1} mm^{-2} (Benlloch *et al.*, 1998). The measurement is performed in the counting mode for low and intermediate flux, while at the highest rates, it is done in the current mode, matching the gain values at the transition.

A systematic study of rate dependence of gain in multi-GEM devices in a range of operating conditions and geometry has shown

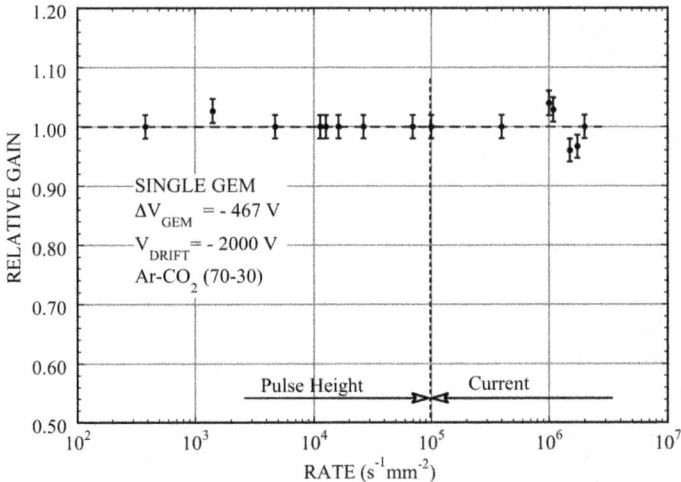

Figure 5.65:   Normalized single-GEM gain as a function of soft-X-ray rate. *Source*: Benlloch *et al.* (1998).

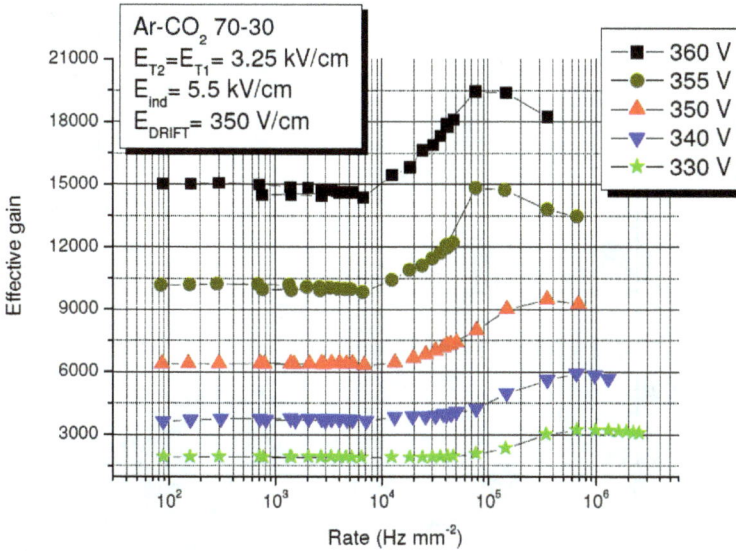

Figure 5.66:   Rate dependence of gain of a TGEM detector on exposure to X-rays in a range of voltages applied to the last GEM.

Source: Everaerts (2006).

a tendency of the gain to increase at very high rates, due to space–charge field distortions induced by the positive ions. Figure 5.66 is an example of measured gain as a function of rate for a TGEM detector at increasing values of total effective gain (Everaerts, 2006). Simulation studies have succeeded in explaining this peculiar behaviour as due to the high density of positive ions accumulating near the bottom of the holes (Alfonsi *et al.*, 2012). As seen in Figure 5.67, at very high rates, while the gain (deduced from the anode electron current) increases, the ion current measured on the cathode and the corresponding ion backflow fraction decrease, a behaviour well reproduced by simulations (Thuiner *et al.*, 2015). Detailed studies on the IBF are given in Section 6.5, in connection with the studies of time projection chambers.

It should be noted that these extreme rates are only met in special applications, as beam and soft X-rays plasma diagnostic, where the gain uniformity is usually not of major concern (Murtas *et al.*, 2010; Pacella *et al.*, 2013), see Section 8.6.

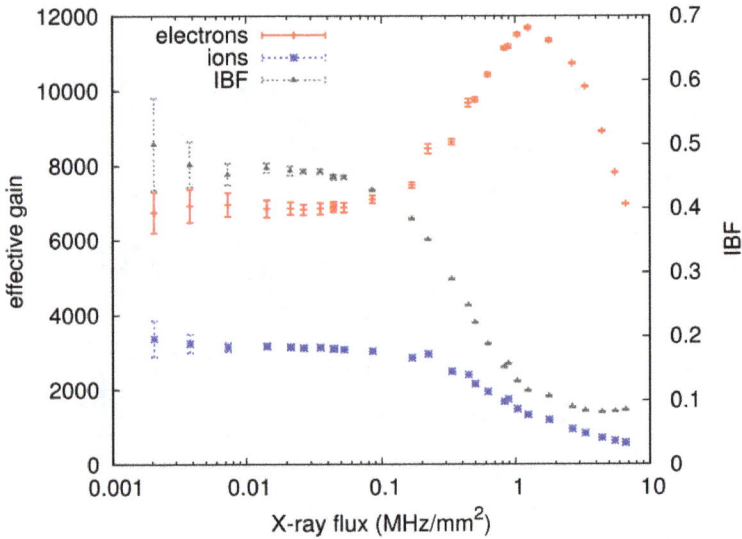

Figure 5.67:   Effective electron and ion gains and ions' backflow fraction as a function of X-ray flux measured with a TGEM.

Source: Thuiner *et al.* (2015).

5.10   Discharges and Breakdown

As most gaseous detectors, GEM devices tend to discharge when the voltage is increased over a threshold value, thus limiting the useful range of operation. Depending on the available energy and the external resistive protections, the discharge can be bland or evolve in a more disruptive spark breakdown. Methods to reduce the stored energy by sectoring the GEM foils were discussed in Section 5.6. While discharges are generally observed in the electrode with the highest amount of charge (the last in a cascade), the discharge can propagate to other electrodes, and in particular to the anode for high values of the induction field.

Disregarding spontaneous breakdowns due to local defects, two major sources of discharges have been identified: high radiation fluxes and large, albeit rare, energy losses by heavily ionizing particles. In either case, a transition from a proportional avalanche regime to a streamer and then to a spark is observed when the released charge exceeds around 10^7 ions and electrons, named the Raether limit after the physicist who extensively studied these processes (Raether, 1964). As described previously, all MPGDs, while safely operating in benign conditions, suffer from a voltage-dependent discharge rate when exposed to high-intensity particle beams, a problem imputed to the occurrence of neutron interactions, electromagnetic showers and other heavily ionizing events.

For a single GEM, the experimental results are well reproduced by a model calculation, expressed both as a function of total charge and charge density, as shown in Figure 5.68; this is probably a consequence of the comparable sizes of the multiplication gap and of the avalanches (Procureur *et al.*, 2012). Figure 5.69 is a comparison between computed and measured discharge probability for different gas mixtures; a fit to the results yields a value for the critical charge in the range 4 to 8 10^6, in agreement with the classic value of the Raether limit (Gasik *et al.*, 2017).

These process have been systematically studied exposing the detectors to alpha particles, either from an external source or from

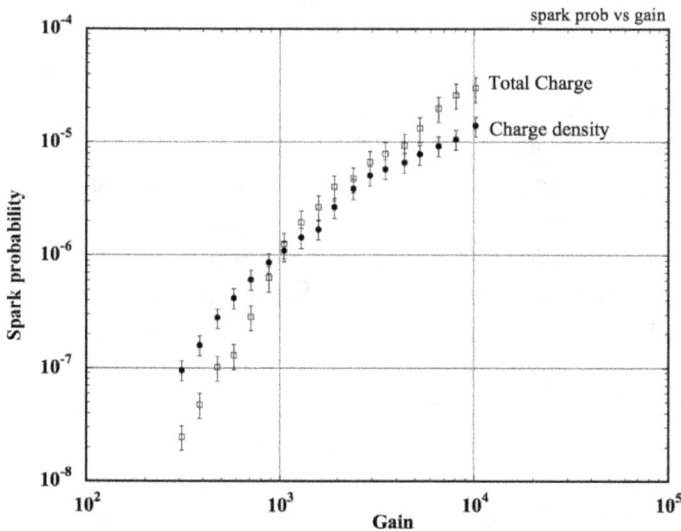

Figure 5.68:   Model prediction of the discharge rate as a function of gain for the total charge (Raether) and the charge density.

Source: Redrawn from data in Procureur *et al.* (2012).

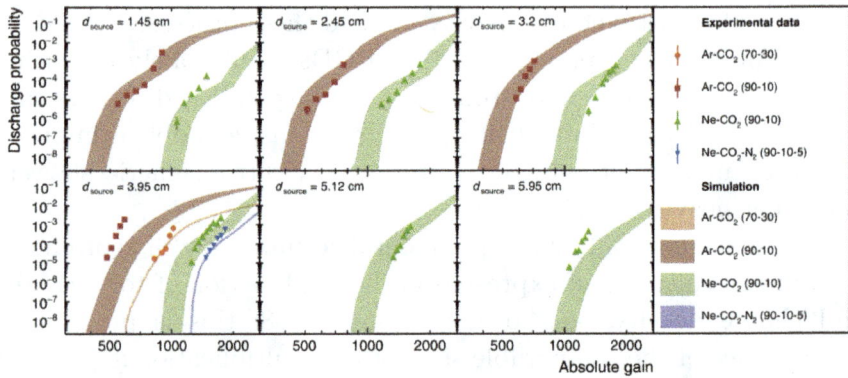

Figure 5.69:   Computed (shaded bands) and measured (data points) discharge probability of a single GEM exposed to an alpha source in different gas mixtures, as a function of gain.

Source: Gasik *et al.* (2017).

the internal decay of isotopes introduced in the gas flow, after a gain calibration realized with exposure to a low-rate soft X-ray source (see Section 2.7). The results from single- and multiple-GEM detectors are shown in Figures 5.23 and 5.25, respectively. Taking into account the average energy loss of the alpha particles, around 500 keV in the narrow drift gap, one can estimate a corresponding increase of the charge threshold, largely exceeding the Raether limit for multiple structures. This can be attributed to the spread of the drifting electron charge due to diffusion between foils, decreasing the density of charge within each hole, as discussed in Section 5.7 and illustrated in Figure 5.48, under the reasonable assumption that the transition to discharge is a local property within the individual holes.

Since for equal overall gains the voltage applied to the individual electrodes is decreased with the number of cascaded foils, the observation is also suggestive of a voltage dependence of the Raether limit (Sauli, 2016). This possibility has been investigated with a detector consisting in a double-GEM structure used to inject a variable amount of charge into a parallel plate counter. Figure 5.70 shows the measured current at discharge as a function of the electric field on the PPC; the result clearly indicates that lower fields permit reaching higher charge densities (Maltsev *et al.*, 2019).

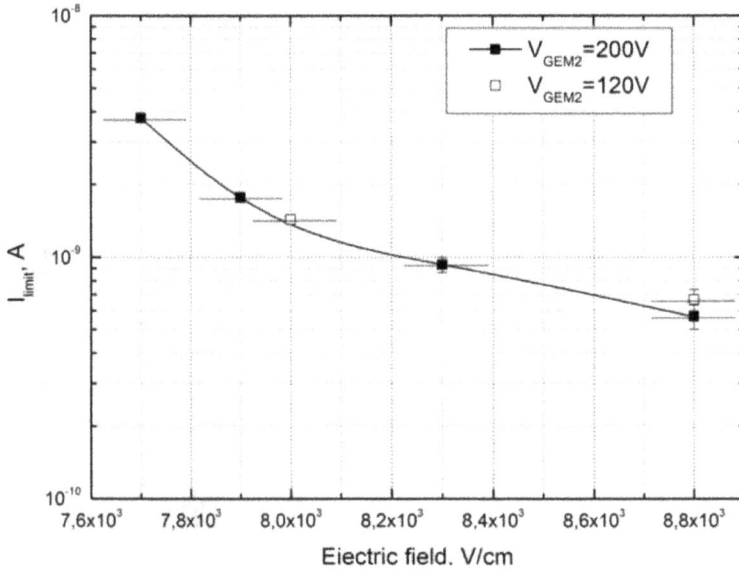

Figure 5.70: Maximum current at discharge as a function of field in a parallel plate counter, measured varying the injected charge with a double-GEM pre-amplification.

Source: Maltsev *et al.* (2019).

The discharge probability for a given overall gain is reduced with a small asymmetry in the applied voltages, favouring a choice of higher gain in the first element of a cascade as seen in Figure 5.71 for a TGEM device. For a given discharge rate, gains almost an order of magnitude larger can be reached with the asymmetric powering scheme adopted for the COMPASS TGEM devices (Ketzer *et al.*, 2002). This choice is, however, conflicting with the requirement of reducing the ions' backflow, a subject covered in Section 6.5.

It should be stressed that the definition of discharge threshold is rather arbitrary, and that the moisture content of the gas filling, in most cases unknown, plays a role in defining the discharge rate, Figure 5.72 (Altunbas *et al.*, 2002). It is therefore often difficult to compare results obtained by different experimenters.

The discharge probability is reduced also in other cascaded devices, for example adding a GEM pre-amplifier to a

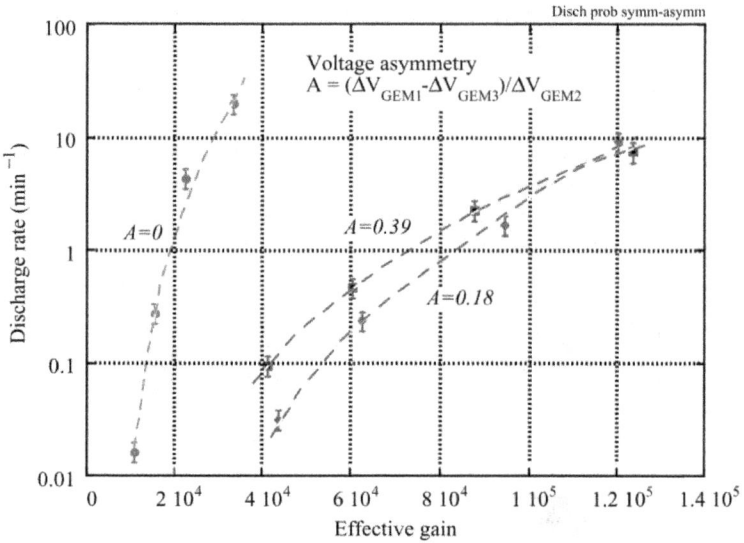

Figure 5.71: Discharge probability on exposure to alpha particles in a TGEM for three values of the asymmetry between the applied voltages, as a function of effective gain.

Source: Altunbas *et al.* (2002).

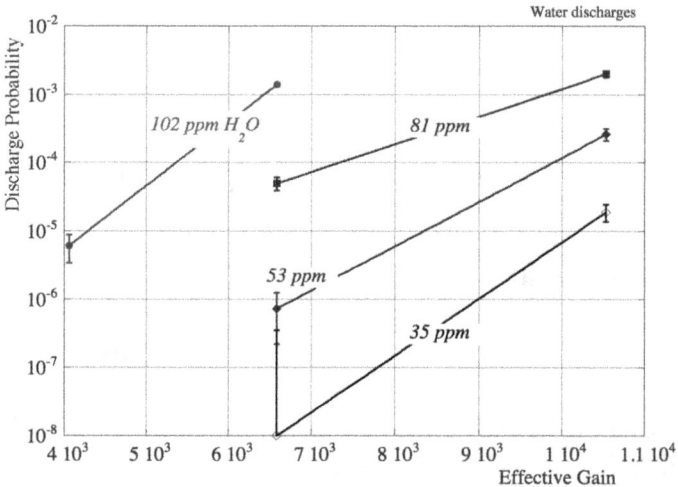

Figure 5.72: Discharge rate as a function of gain on exposure to alpha particles of a TGEM for different moisture contents in Ar–CO$_2$.

Source: Altunbas *et al.* (2002).

MICROMEGAS as discussed in Section 4.8. In this case, the explanation for the improvement based on diffusion, given above, is hardly applicable.

The discharge can be initiated by any GEM foil in a cascade, and propagate through the structure. For high values of the induction fields, it can continue to the anode (Bachmann *et al.*, 2002). The process of this so-called secondary discharge propagation has been studied in detail, motivated by the danger of exposing the sensitive readout electronics to irreversible damages. As seen in Figure 5.73, the discharge propagation probability has a sharp increase above a threshold value of the induction field, and depends on its direction as well as on the way the GEM electrodes are connected to the high voltage (Gasik, 2016; Deisting *et al.*, 2019; Lautner *et al.*, 2019). The formation time of the secondary discharge after the first depends also on the induction field strength, and is surprisingly long for low fields, Figure 5.74; the reasons for this delay, possibly due to ion or photon feedback, are unclear.

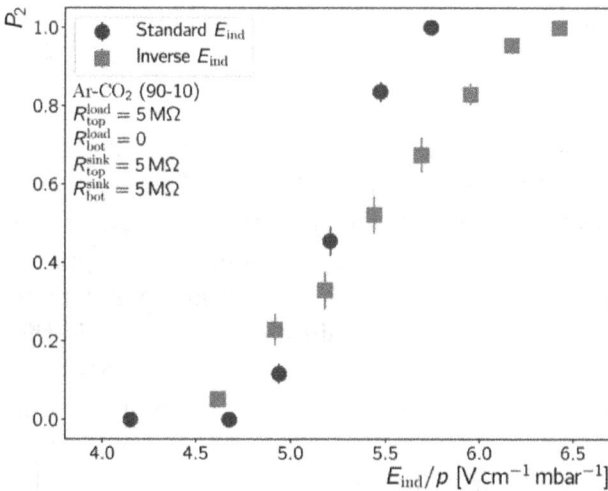

Figure 5.73: Discharge propagation probability between GEM and anode, as a function of induction field strength and polarity.

Source: Deisting *et al.* (2019).

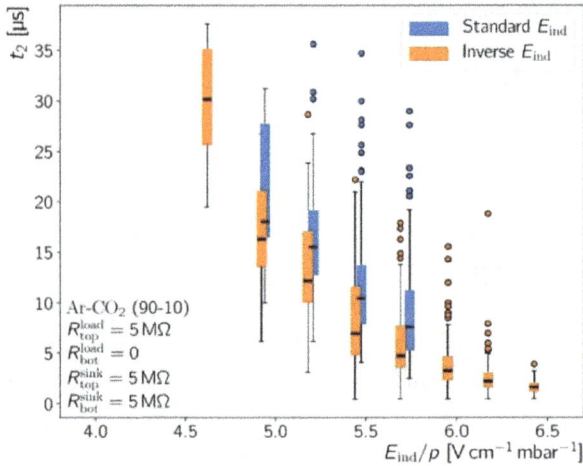

Figure 5.74:   Secondary discharge propagation time as a function of induction field.

Source: Deisting *et al.* (2019).

As will be discussed in Chapter 6, a high value of the induction field is required in TPC-like applications to reduce the drift field distortions induced by the positive ion backflow.

5.11   Ageing

Under sustained irradiation, gaseous detectors can be subjected to the so-called ageing, a permanent damage of the electrodes, resulting in a progressive deterioration of performances. Already observed in the early proportional and Geiger counters, the process has been extensively studied in view of the use of gaseous detectors in high-radiation environments. The main cause of ageing is the growth on the electrodes of thin insulating layers resulting from the polymerization in the avalanche process of organic gas molecules or pollutants released by the manufacturing materials; multiwire and micro-strip chambers, with their thin anodes, are particularly affected. For a review of the ageing processes and materials' outgassing, see for example Capeans (2003).

GEM structures, not depending on the field strength close to thin electrodes for amplification, are expected to have a reduced

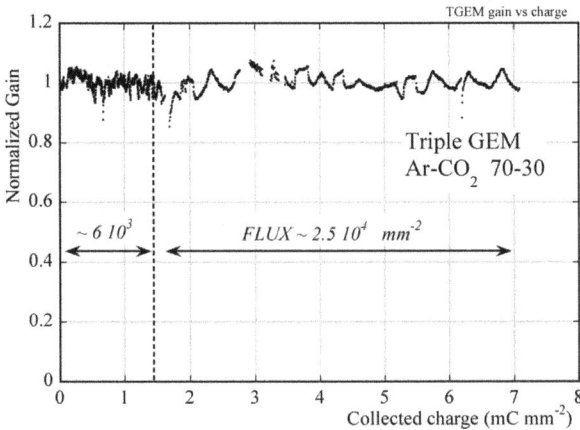

Figure 5.75: Normalized gain as a function of collected charge for a TGEM operated with Ar–CO_2.

Source: Altunbas *et al.* (2003).

sensitivity to the presence of deposits; this has been verified experimentally with long-term irradiations, both in the laboratory and in experiments. Figure 5.75 is an example of measured gain as a function of collected charge in a TGEM operated with an Ar–CO_2 gas filling; no change is seen up to ~7 mC/mm^2, corresponding, at a gain around 10^4, to an integrated flux of ~2×10^{13} minimum ionizing particles per cm^2 (Altunbas *et al.*, 2003).

Motivated by the requirement to improve the time resolution, the addition to the gas mixture of carbon tetrafluoride, owing to its well-known etching properties, extends the detector lifetime up to and above a collected charge of 20 C/cm^2 as shown by the measurement in Figure 5.76 (Alfonsi *et al.*, 2004). The use of CF_4 requires, however, special precautions, due to the reactivity of the fluorine liberated in the avalanches with water to form HF, one of the most aggressive acids for many materials (Alfonsi *et al.*, 2005).

In the development of the forward muon tracker for CMS, exposures to various fields of radiation have confirmed the long-term survivability of large GEM prototypes. Figure 5.77 is an example of gain and energy resolution measured on exposure to high-intensity soft X-rays (Fallavollita, 2019).

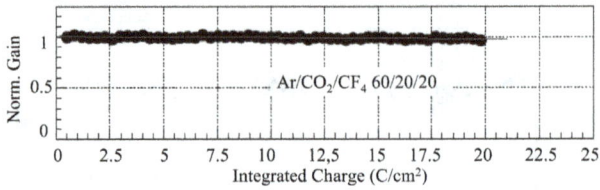

Figure 5.76:   Normalized gain as a function of collected charge for a gas mixture including CF_4.

Source: Alfonsi *et al.* (2004).

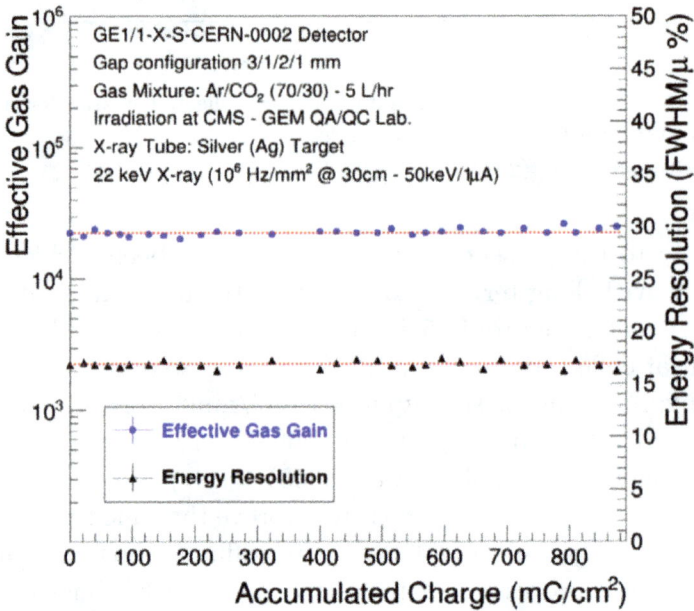

Figure 5.77:   Gain and energy resolution as a function of accumulated charge measured with a CMS GEM prototype.

Source: Fallavollita (2019).

It should be noted that, while GEM devices are more impervious to ageing than other gaseous devices, gain drops have been sporadically observed during long-term operation, imputable to the release in the gas volume of contaminants from silicon sealants improperly cured.

5.12   GEM Applications

The performance, high-rate capability and reliability of GEM devices have fostered numerous efforts to use the technology in a variety of medical and biological applications. A set of TGEM trackers, similar in design to the detectors built for the COMPASS experiment at CERN, has been used to instrument a medical diagnostic tool for hadron therapy, named proton range radiography (PRR). The method involves recording the trajectories of charged particles traversing a body, as well as their residual energy. The correlation between measured position and energy loss permits to obtain a map of the density distribution in the target, a method suggested in the early eighties (Hanson *et al.*, 1981) and developed at the Paul Sherrer Institute (PSI), Villigen, using a scintillating fibre tracker and plastic scintillator stacks for the range determination (Pemler *et al.*, 1999).

The peculiar feature of PRR is to provide a density map as seen by charged particles having the same characteristics of the beams used for therapy, albeit at a much lower intensity; for a review, see (Bucciantonio and Sauli, 2015). Figure 5.78 shows a prototype PRR system, with an active area of 10×10 cm^2, instrumented with two bi-dimensional GEM trackers and a stack of 30 thin plastic scintillators to record the residual range of the protons. Tested at PSI and at the Italian therapy centre CNAO with ~150-MeV protons, the system achieved sub-millimetre space and few percent density resolutions. Figure 5.79 is a two-dimensional projected image of a phantom consisting of a set of holes of various diameters and depth in a plastic slab (Amaldi *et al.*, 2011). The work has continued with the construction of a larger acceptance, 30×30-cm^2 instrument (Bucciantonio *et al.*, 2013a, 2013b).

The advantages of digital recording for X-ray and gamma ray imaging achievable with gaseous detectors have been explored since the early developments of the devices: good resolution, large dynamic range, fast data recording and retrieval, and lower doses than with conventional radiography systems for equivalent contrast. A digital radiography system based on multiwire proportional chambers, developed in the late nineties, has been successfully used in Russia

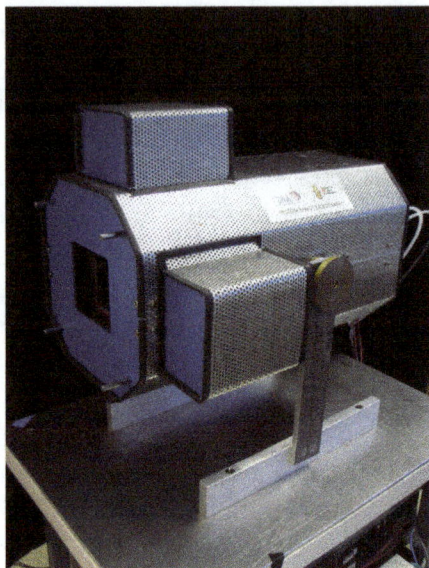

Figure 5.78:   The prototype proton range radiography system.
Source: Picture TERA CERN.

for many years in clinical environments (Babichev *et al.*, 1991, 1998; Martinez-Davalos *et al.*, 1994).

The introduction of the new families of MPGD, easier to manufacture and sturdier than wire chambers, has rejuvenated the development of performing systems capable of competing with standard radiographic equipment; in particular, their large dynamic range and radiation tolerance are suitable for use in portal imaging, a diagnostic method permitting to realize real-time dosimetry during X-ray and gamma ray therapeutic patient exposures. Performance studies of GEM-based detectors suitable for medical imaging are discussed in Anulli *et al.* (2007), Tsyganov *et al.* (2008) and Gutierrez *et al.* (2012).

In the range of energies used for cancer therapy, few hundred keV to multi-MeV, gaseous devices have too small detection efficiencies to be useful. This limitation is countered operating the detectors at high pressures (Despre *et al.*, 2005) or using of internal converters with a design optimized for the energy region of interest.

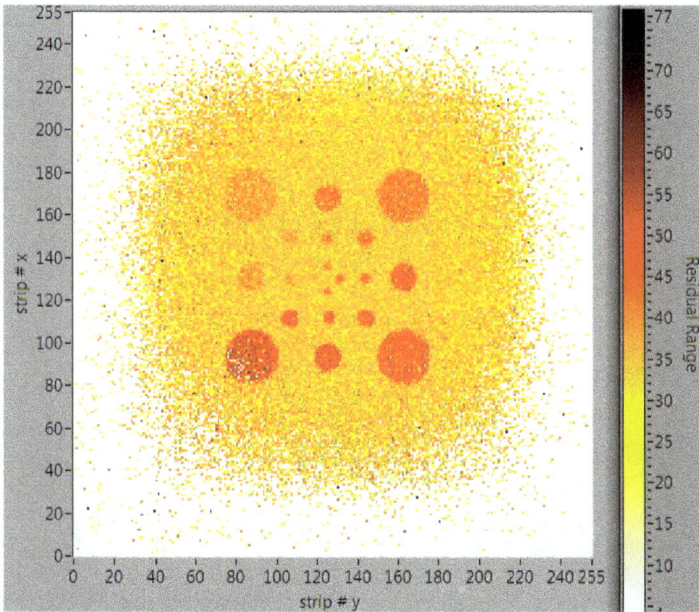

Figure 5.79:   Projected density map of a set of holes in a plastic slab recorded exposing the PRR to 140-MeV protons. The smallest visible hole has a diameter of 1 mm.

Source: Amaldi *et al.* (2011).

In the electronic portal imaging device (EPID), developed at the Karolinska and the Royal Institute of Technology, interleaved metal converters and GEM electrodes ensure conversion and detection of MeV gamma rays during the therapeutic exposure (Figure 5.80); lower-energy X-rays (60 keV), emitted by a generator mounted in parallel with the gamma ray nozzle, convert in the upper gaseous drift layer and give a diagnostic image of the patient prior to the irradiation (Ostling *et al.*, 2003). To provide ambiguity-free position information, the readout is performed with a matrix of individual pads on the anode plane. A custom radiation-tolerant readout electronic system has been developed to withstand the extreme fluxes encountered in a therapeutic session (Ostling *et al.*, 2004). Figure 5.81 is an example of a hard X-ray image of a lamb chop recorder with a prototype EPID.

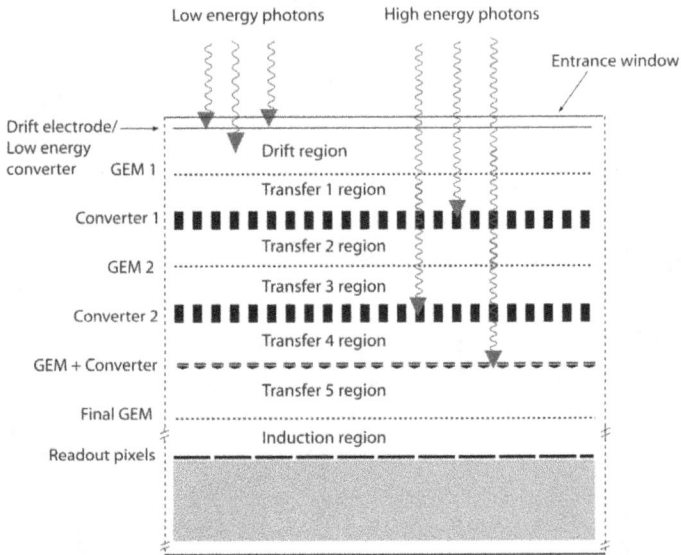

Figure 5.80:   The electronic portal imaging device EPID.
Source: Ostling (2006).

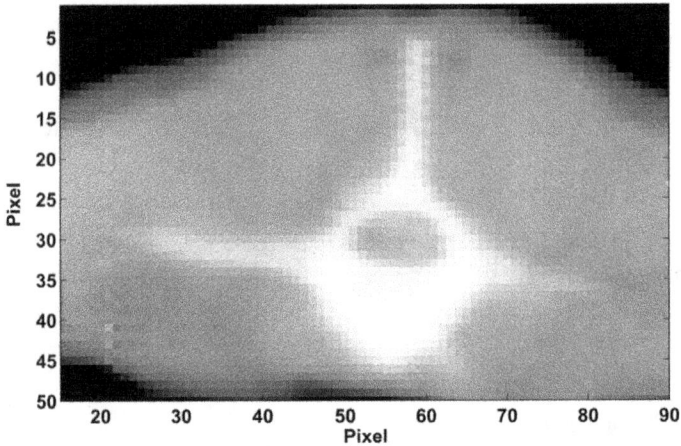

Figure 5.81:   Hard X-ray transmission radiography recorded with the EPID prototype.
Source: Ostling (2006).

Relatively complex and expensive, the MPGD readout system making use of fast electronics has a limited extent of applications. The development of simpler optical recording methods, see Chapter 11, holds the promise of a larger dissemination of these novel diagnostic tools in the clinical environment.

Chapter 6

MPGD Readout of Time Projection Chambers

6.1 Multiwire-based TPCs

The time projection chamber (TPC), introduced in the late seventies by David Nygren and collaborators, is one of the most powerful particle tracking detectors, providing full three-dimensional images of complex ionization patterns left in a gas by radiation; recording the differential energy loss permits also to infer in the particle's velocity. For a description of the early developments and operation of the first TPCs, see Nygren and Marx (1978). In its most common enactment, adapted to the geometry of colliding beam experiments, the TPC consists of a large cylindrical gas vessel, the drift volume, capped at one end with a modular array of multiwire proportional chambers acting as gaseous amplifying devices. Localization is performed recording, in successive time intervals (or slices), the collected charge on wires, strips or pad rows imprinted on the MWPC's outer electrode, hence the name "time projection". The three-dimensional representation of the event is then obtained connecting with a proper reconstruction algorithm the recorded points in the sequence of slices. With increasingly large sizes and improvements, TPCs have been and still are the main components in many experimental setups, see for example Chapter 10 and references therein (Sauli, 2014).

   With an outer diameter of 5 m and a sensitive gas volume close to 100 m^3, the ALICE TPC is the largest system of this kind, Figure 6.1 (Alme *et al.*, 2010). Successfully operated for a decade at CERN's Large Hadron Collider (LHC), the detector consists in

167

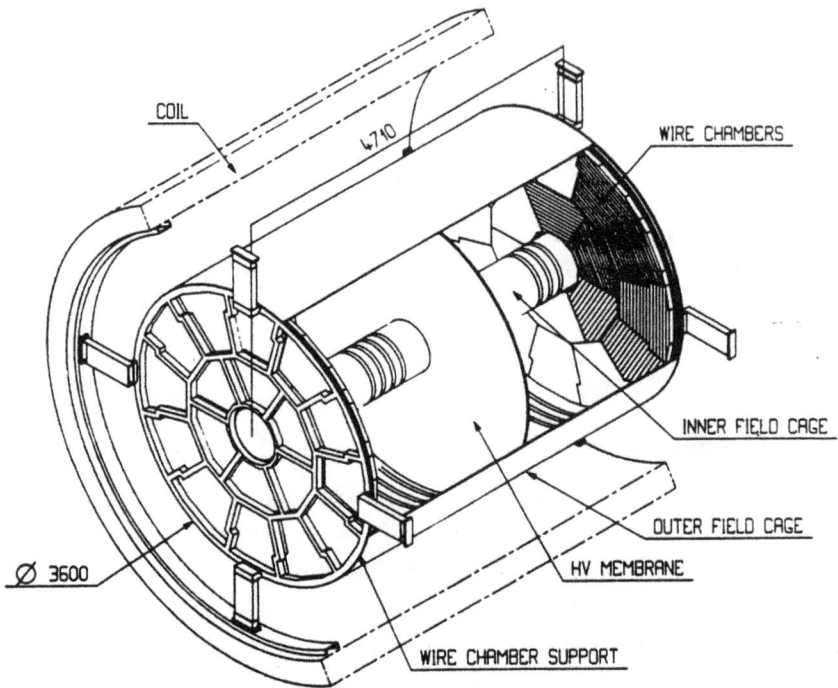

Figure 6.1:   Schematics of the ALICE TPC.
Source: Alme *et al.* (2010).

fact of two identical devices, joined tail to tail and sharing a central common electrode (the cathode) at high negative potential. A set of annular electrodes with graded potentials surrounds the active volume, creating a longitudinal electric drift field; an inner cylinder, also with ring-shaped field electrodes, separates the active volume from the vacuum tubes of the collider.

Multiple measurements of pulse height for each track are utilized to perform identification of the particles, exploiting the dependence of the differential ionization energy loss from the particle's velocity; coupled to the measurement of momentum, deduced from the tracks' bending in the magnetic field, the information provides a detailed characterization of the topology and kinematics of complex events. Illustrating the powerful imaging capabilities of the detector, Figure 6.2 shows an example of a three-dimensional rendering of the

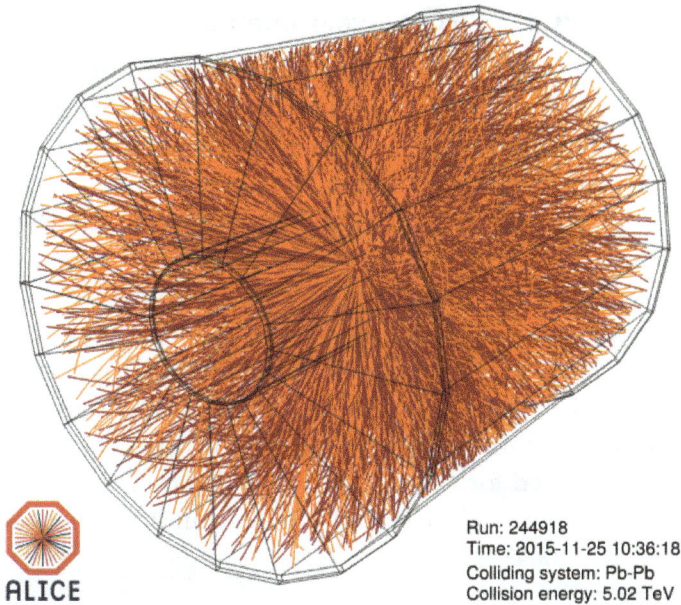

Figure 6.2: A multi-TeV lead–lead collision recorded with the ALICE TPC. *Source*: Picture source CERN.

charged prongs resulting from a collision between two lead ions at multi-TeV energies, with several thousand tracks recorded with the ALICE TPC.

The ever-growing particle rates encountered with the increasing luminosity of the colliders come close to the rate capability of wire-based systems. Based on the recording of the charge induced on strips or pads on the MWPC cathode, the localization suffers from limited two-track resolution; additionally, the positive ions copiously produced by the avalanches around the wires affect the gain of the detector, and their slow backflow in the drift volume induces large distortions of the field and consequently of the recorded tracks. The new generations of micro-pattern devices, with their intrinsically higher tolerance to high particle flux, better resolutions and effective ion backflow suppression, led to the development of MPGD alternatives for the TPC end-cap detectors, described in the following sections.

6.2   Early MPGD–TPC Developments

The attractiveness of replacing the MWPC-based TPC readout with MPGD systems was already discussed in the early developments of the devices. Aside from the superior spatial single- and multi-track resolutions, it was found that the ion backflow is considerably lower than for wire systems, particularly using the MICROMEGAS or multiple cascaded GEMs (see Section 6.5).

Early GEM–TPC prototypes were developed by a CERN–Karlsruhe collaboration, attaching various drift volumes to a double-GEM detector of a design similar to the COMPASS tracker described in Chapter 5, Figure 6.3 (Kaminski *et al.*, 2004). Eight rows of 32 pads, $1.27 \times 12.5\,\text{mm}^2$ each, equipped with fast charge-recording electronics were used for the readout. The detectors were tested in a high-intensity beam at CERN and, inserted in a magnet, at DESY (Kappler *et al.*, 2004).

Figure 6.3:   Early GEM–TPC prototypes.

Source: Kaminski *et al.* (2004).

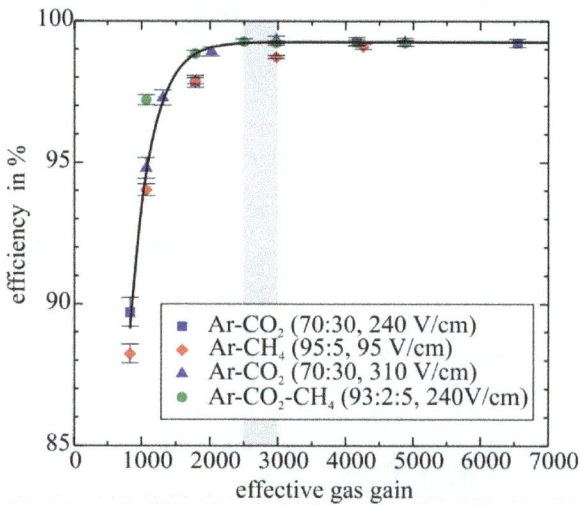

Figure 6.4:   Pad row detection efficiency as a function of gain for the GEM–TPC in several gas mixtures.

Source: Kaminski *et al.* (2004).

Figures 6.4 and 6.5 show the efficiency and the single-point transverse space resolution measured for fast particles as a function of gain in a 4-T magnetic field (Kaminski *et al.*, 2004).

As shown in Figure 6.4, the devices have been tested in a wide range of gases; aside from ensuring good charge multiplication properties, the preferred choice is normally a non-flammable mixture, having low electron diffusion at moderate electric fields. This is the case for the so-called TESLA technical design report (TDR) gas: $Ar–CH_4–CO_2$ in the proportions 93–5–3 (Janssen *et al.*, 2006).

This work has been pursued by many groups, with systematic studies of performances of prototypes (Ableev *et al.*, 2004; Radicioni, 2007; Yu *et al.*, 2005), and in the framework of the development of large-sized devices for the planned International Linear Collider (ILC) detector (Karlen *et al.*, 2005; Carnegie *et al.*, 2005; Roth, 2004; Kobayashi *et al.*, 2007; Ledermann *et al.*, 2007; Attié, 2008; Diener, 2012).

The space coordinates of the tracks in each time slice are deduced recording the collected charge as a function of time on strips or

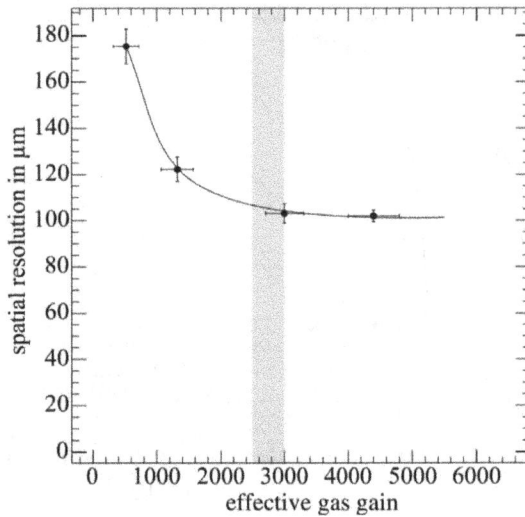

Figure 6.5:   Transverse resolution of the GEM–TPC at 4 T, as a function of effective gain.

Source: Kaminski *et al.* 2004).

pad rows. The signal induction process is described by the pad response function (PDF) whose shape depends on geometry; typically 5–10 mm wide for MWPC-based TPCs, the PRF is much narrower using MPGD devices. Figure 6.6 is an example of the computed correlation between real and reconstructed signal position for several pad widths, using a simple centre-of-gravity (COG) method (Janssen *et al.*, 2006). More sophisticated algorithms have been developed to correct for the distortions induced by the discrete pad size.

Detailed studies have been devoted to the optimization of size and geometry of the readout pads, aimed at reducing their number for large-sized devices (Figure 6.7); a comparison of expected position accuracy as a function of drift distance for several choices of pad row patterns is given in Figure 6.8 (Ledermann *et al.*, 2007). While the outcomes of the studies depend on many factors (filling gas, drift fields and distance, avalanche size), the general conclusion is that the major parameter in determining the localization accuracy is the pad size. Computed for the DESY GEM–TPC prototype, Figure 6.9 shows the dependence of single-point transverse space resolution from

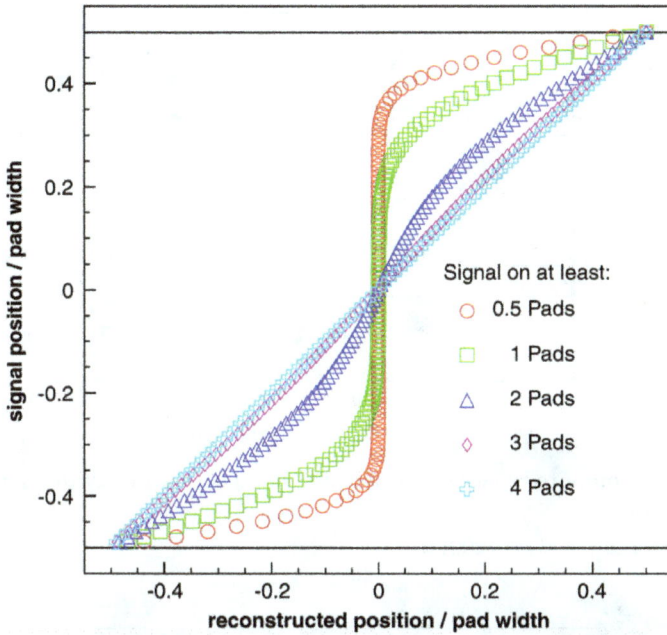

Figure 6.6: Correlation between real and reconstructed positions as a function of the number of pads with a signal over threshold.

Source: Janssen *et al.* (2006).

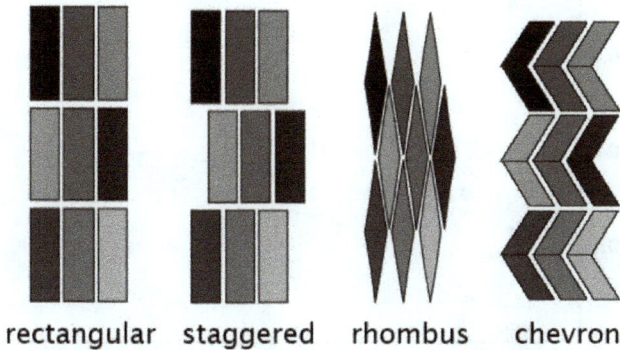

rectangular   staggered   rhombus   chevron

Figure 6.7: Alternative pad rows' geometries.

Source: Ledermann *et al.* (2007).

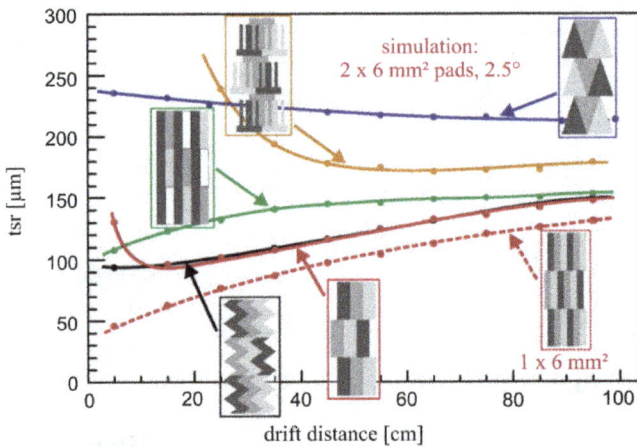

Figure 6.8:   Computed space resolution as a function of drift distance for different pad geometries.

Source: Ledermann *et al.* (2007).

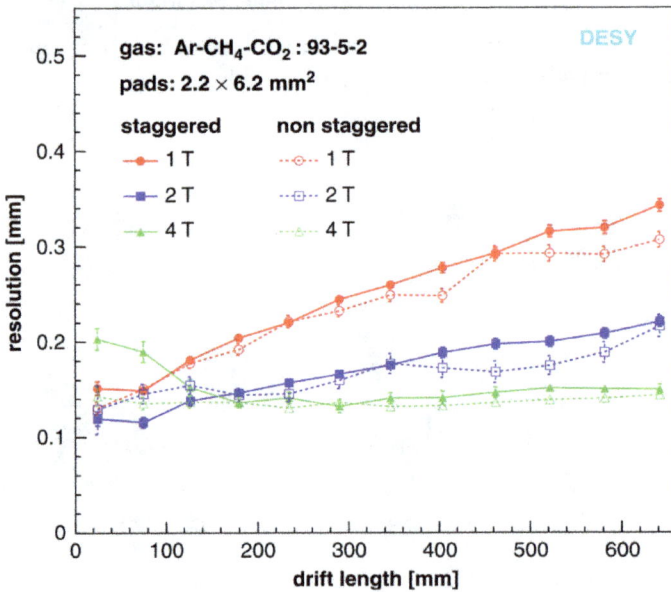

Figure 6.9:   Measured transverse resolution as a function of drift length and magnetic field.

Source: Janssen *et al.* (2006).

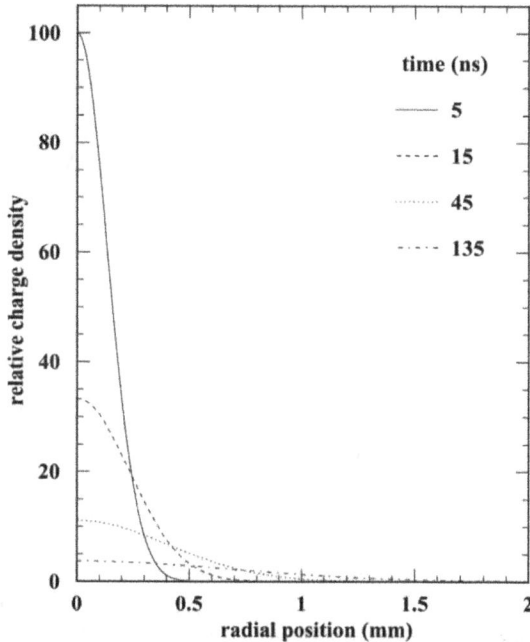

Figure 6.10: Computed time dependence of the detected charge on a resistive anode for different electronics shaping times.
Source: Dixit *et al.* (2004).

drift distance for staggered and non-staggered pad rows and several values of magnetic field (Janssen *et al.*, 2006).

A charge-dispersion localization method based on the use of a resistive anode permits to reduce the number of signal pickup pads in detectors with MPGD readout (Dixit *et al.*, 2004). Figure 6.10 is the computed time evolution of the spatial charge density induced by a point-like charge generated above an anode with a resistivity of 2 MΩ/square. The resulting spread (the pad response function) depends on the input characteristics of the recording electronics, and has typically a width of ~700-μm rms. Permitting to reduce the number of readout channels, the method provides an effective charge interpolation between pads improving the position accuracy, at the expense of a reduced multi-track resolution. Tests with a small-sized GEM–TPC confirm the good spatial resolution that can be achieved

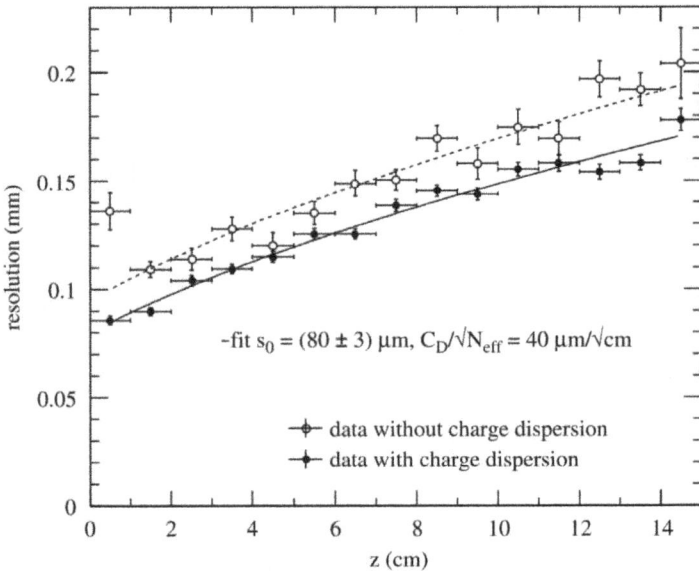

Figure 6.11: Transverse space resolution measured with a GEM–TPC for a standard and charge-dispersion signal recording.

Source: Boudjemline *et al.* (2007).

using the resistive anodes, as shown in Figure 6.11 (Boudjemline *et al.*, 2007).

6.3   MICROMEGAS-based TPCs

A system of modular TPCs with MICROMEGAS readout was built and operated as part of the near detector complex in the Tokay-to-Kamioka (T2K) experiment to study neutrino oscillations. The system includes three medium-sized TPCs, each equipped with a matrix of 12 bulk MICROMEGAS modules on each readout plane, and installed in a large, moderate field magnet (0.2 T) for momentum analysis. In each module, having an area of $36 \times 34\,\mathrm{cm}^2$, the anode is segmented in $\sim$1,700 rectangular pads with individual charge recording, using a dedicated 72-channel ASIC (Baron *et al.*, 2008). Figure 6.12 shows one of the modules (Abgrall *et al.*, 2011). Owing to thorough mechanical construction and calibration procedures, a few percent gain uniformity has been achieved, providing particle

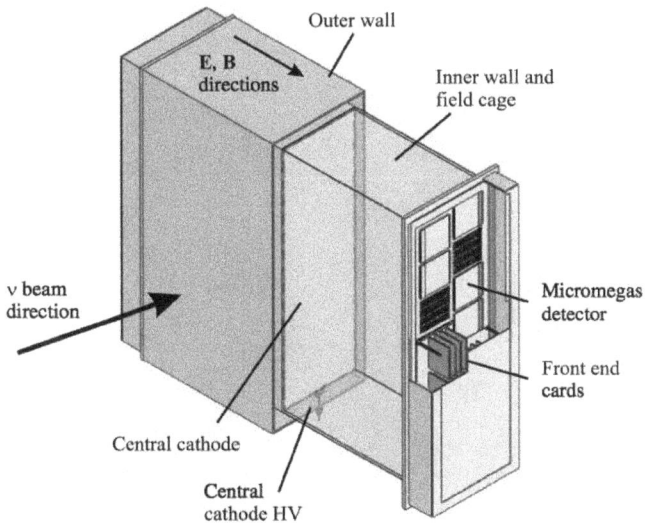

Figure 6.12: Schematics of a MICROMEGAS–TPC module for the T2K experiment.

Source: Abgrall *et al.* (2011).

identification through the measurement of the differential energy loss (Delbart, 2010). Figure 6.13 is an example of a double neutrino interaction recorded in a test run.

The energy spectrum for ^{55}Fe X-rays, recorded on a single pad at a detector gain of 1,500, shown in Figure 6.14(a), has a resolution of 8.2% rms, with a variance across the whole area of less than 3%, adequate to satisfy the requirements of the experiment; a typical plot of detector gain as a function of voltage is shown on the right (Anvar *et al.*, 2009). The single-track space resolution along the drift direction is given in Figure 6.15 as a function of drift distance without and with a 0.5-T magnetic field; the degradation at short distances is caused by events for which all charge is detected on a single pad, preventing interpolation (Arogancia *et al.*, 2009). Particle identification is performed correlating the local measurements of differential energy loss and the reconstructed bending radius of the track in the magnetic field. The result for one of the TPC modules is shown in Figure 6.16 (Abe *et al.*, 2011).

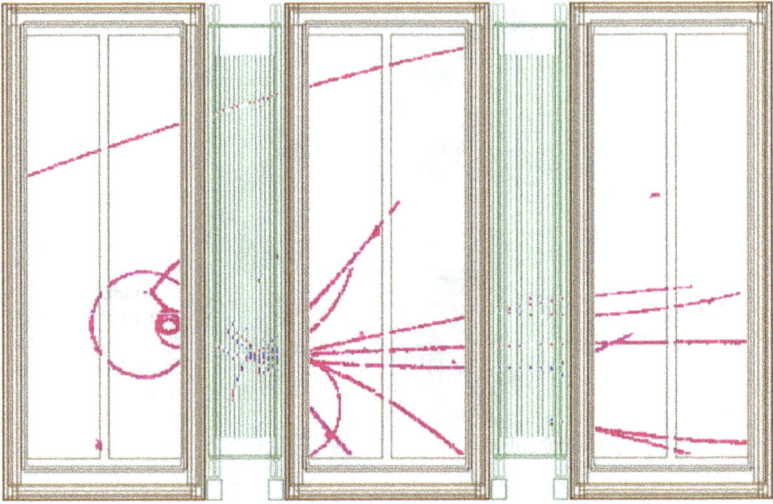

Figure 6.13:   An example of multi-track event recorded with the T2K TPC system.

Source: Abgrall *et al.* (2011).

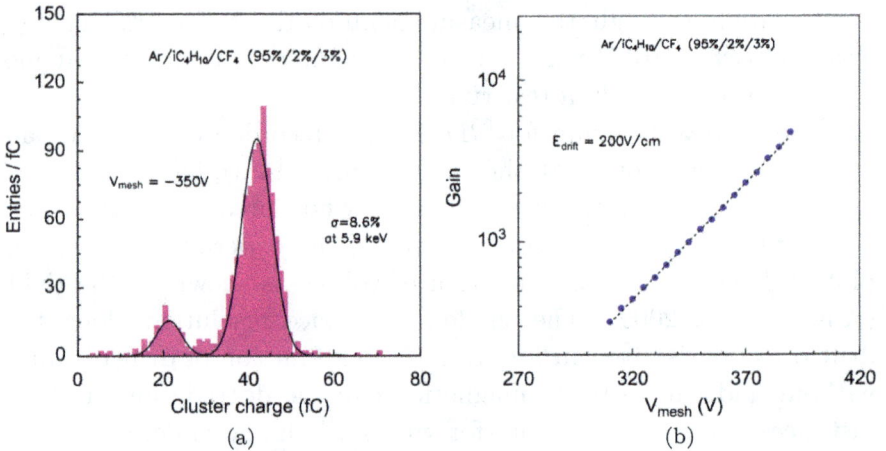

Figure 6.14:   Energy resolution for ^{55}Fe X-rays (a) and gain as a function of voltage (b) of the MICROMEGAS–TPC module.

Source: Anvar *et al.* (2009).

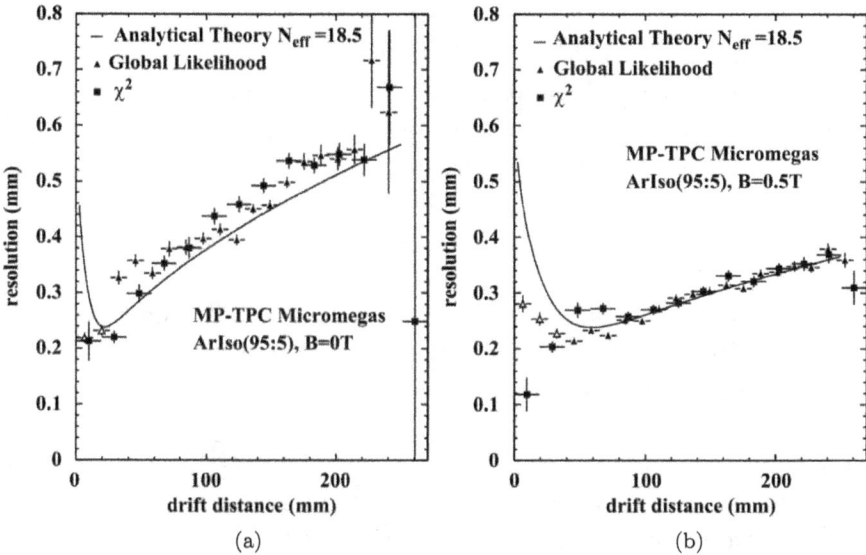

Figure 6.15: Space resolution as a function of drift distance without (a) and with a 0.5-T magnetic field (b).

Source: Arogancia *et al.* (2009).

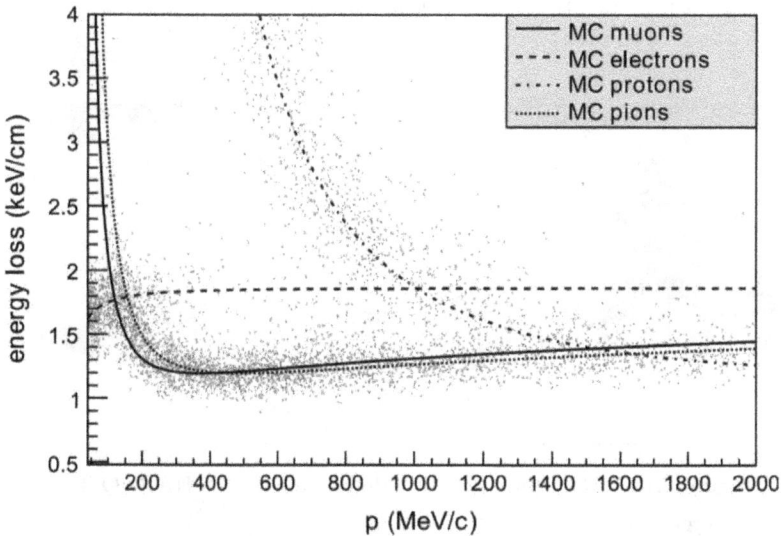

Figure 6.16: Differential energy loss as a function of momentum for tracks generated by neutrino interactions.

Source: Abe *et al.* (2011).

In the framework of the upgrade of the T2K detector complex, a new design of the TPC end-cap readout has been proposed, based on the use of resistive MICROMEGAS, described in Section 4.4. The larger charge spread on the readout pads required the development of an improved reconstruction algorithm, and results in improved space resolution while preserving the energy resolution needed to ensure the particle identification properties of the detector (Attié *et al.*, 2020).

Based on the experience acquired with the T2K chambers, the HARPO MICROMEGAS TPC (Bernard, 2013) is a smaller-volume device dedicated to the detection of gamma rays in the MeV region for astrophysics applications. Capable of operating at pressures up to 5 bars, the cubic device 30 cm on the side has a readout plane patterned with alternating parallel strips and interconnected pads to perform two-dimensional localization (Figure 6.17). Figure 6.18 is an example of tracks produced by the conversion of a 11-MeV photon (Geerebaert *et al.*, 2017). The photon polarization can be deduced from the reconstruction of the plane of conversion.

A small-sized MICROMEGAS–TPC optimized for the detection of neutrons is described in Section 9.4.

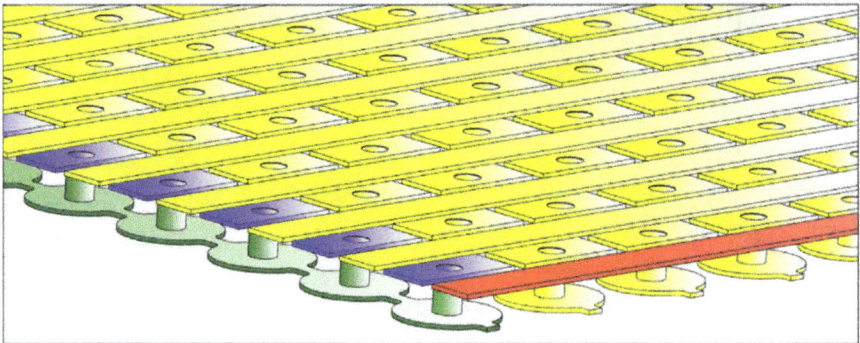

Figure 6.17:   *X-Y* patterned anode readout of the HARPO TPC.
Source: Geerebaert *et al.* (2017).

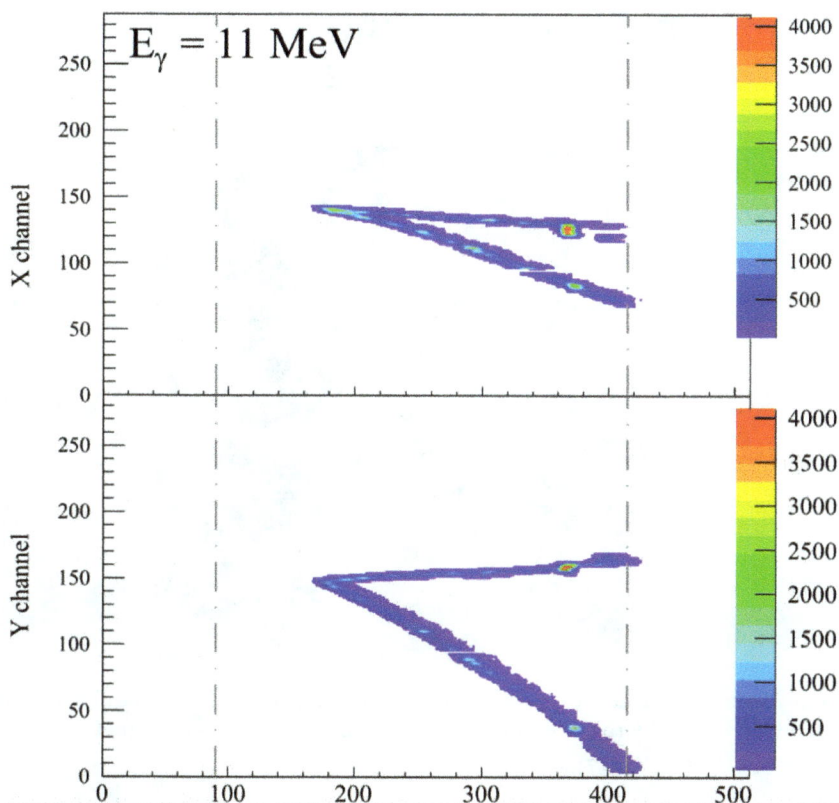

Figure 6.18: Pair conversion recorded with the HARPO TPC.
Source: Geerebaert *et al.* (2017).

6.4 GEM–TPC Developments

In the framework of the International Linear Collider project, a general-purpose assembly end-plate and test facility has been installed in a test beam at DESY permitting the comparison of various detectors' design and operating conditions (Schade and Kaminski, 2011). Figure 6.19 is a view of the setup with several triple-GEM (TGEM) modules mounted on the endplate (Attié *et al.*, 2017); the drift field cage, 57 cm in length, is inserted in a solenoidal

Figure 6.19:　A general-purpose TPC assembly. MPGD test modules can be attached individually to the main drift volume, inserted in a 1-T magnet at DESY. *Source*: Attié *et al.* (2017).

magnet capable of reaching 1 T. The readout system, using a modified ALTRO electronics (Bosch *et al.*, 2003), records the input charge as a function of time on radial pad rows with the geometry shown in Figure 6.20; tracks are sampled about 100 times. Figure 6.21 is the experimental resolution as a function of drift distance, measured in the projected $(r\phi)$ and drift (z) directions, with a gas mixture of Ar–CF_4–iC_4H_{10} (95–3–2).

Originally developed for the PANDA experiment (Arora *et al.*, 2012), the GEM–TPC depicted in Figure 6.22 was extensively tested in the framework of the FOPI spectrometer at GSI Darmstadt (Berger *et al.*, 2017). With ∼30-cm active outer diameter and 73-cm length, the detector fits into the central drift chamber of the experiment. Optimized for operation in high magnetic fields, it is operated with a Ne–CO_2 (90–10) gas mixture; signals are read out on

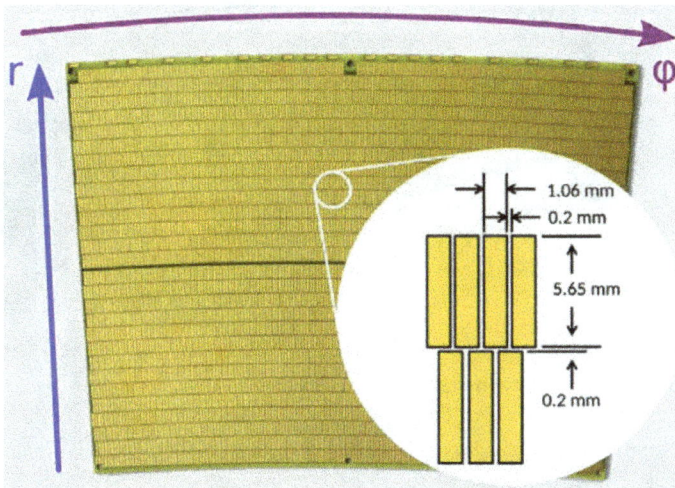

Figure 6.20:   Radial pad rows for the readout of a GEM module.
Source: Attié *et al.* (2017).

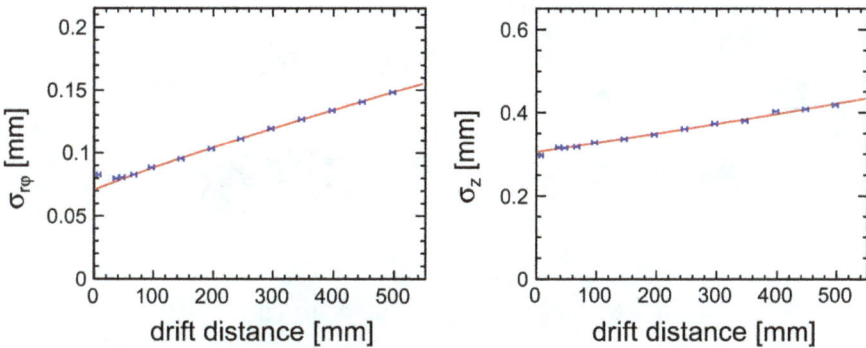

Figure 6.21:   Resolution as a function of drift distance of the ILC GEM–TPC.
Source: Attié *et al.* (2017).

a matrix of ∼10,000 pads, with outer diameter of 1.5 mm, Figure 6.23 (Berger *et al.*, 2017).

A thorough analysis of the drift field structure and distortions and corrections for gain non-uniformities demonstrates the excellent particle identification properties of the device obtained from the

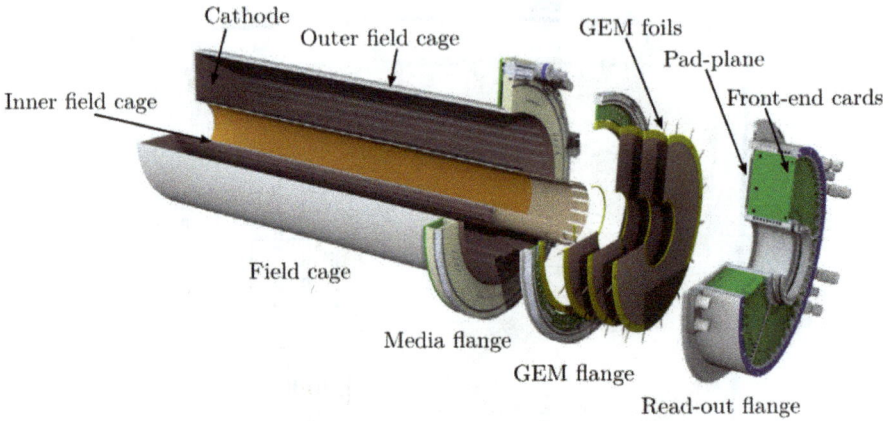

Figure 6.22:   The GEM–TPC for FOPI.

Source: Berger *et al.* (2017).

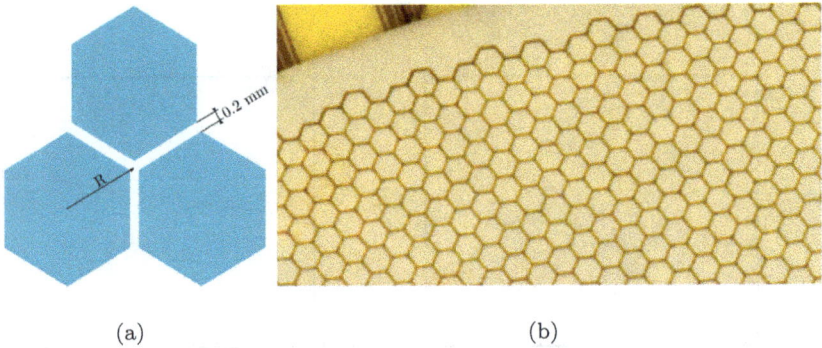

(a)   (b)

Figure 6.23:   Pad readout of the FOPI GEM–TPC.

Source: Berger *et al.* (2017).

correlation between measured momentum and differential energy loss, Figure 6.24 (Böhmer *et al.*, 2014).

Confronted with the challenging requirements of the increased luminosity of the LHC, the ALICE TPC group has started the construction of a GEM–TPC upgrade, after several years of intensive research and development (ALICE Collaboration, 2014). The improved detector makes use of the existing field cage, replacing

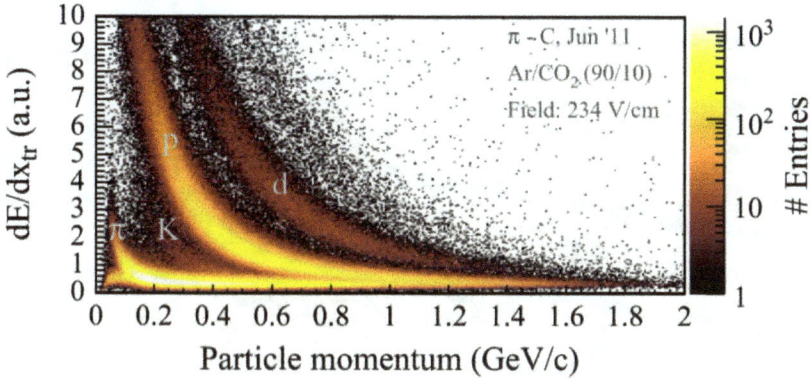

Figure 6.24: Particle identification power of the GEM–TPC.
Source: Böhmer *et al.* (2014).

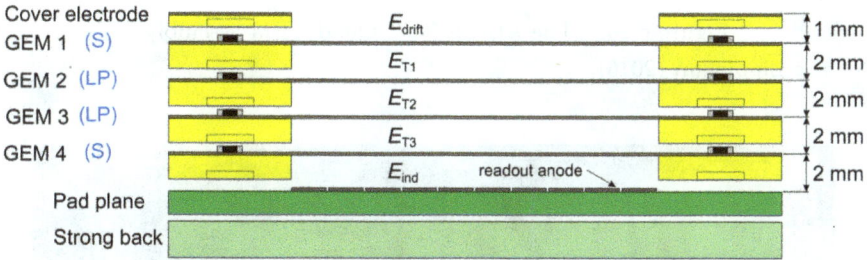

Figure 6.25: Schematics of the ALICE TPC upgrade Quad-GEM chamber.
Source: Aggarwal *et al.* (2018).

the MWPC end-cap with a GEM-based readout. Mainly motivated by the requirement to decrease the ion backflow, the development resulted in the adoption of a Quad-GEM cascade, with alternating electrodes having the hole pattern with a different pitch; see Figures 6.25 and 6.26. Figure 6.27 shows a prototype of large detector panels, consisting of four GEM stacks (Lippmann, 2016). Figure 6.28 is a view of the ALICE TPC end-cap in the course of replacement of the MWPCs by the GEM modules.

The reduction of the ion backflow fraction (IBF) achieved with this geometry is discussed in Section 6.5.

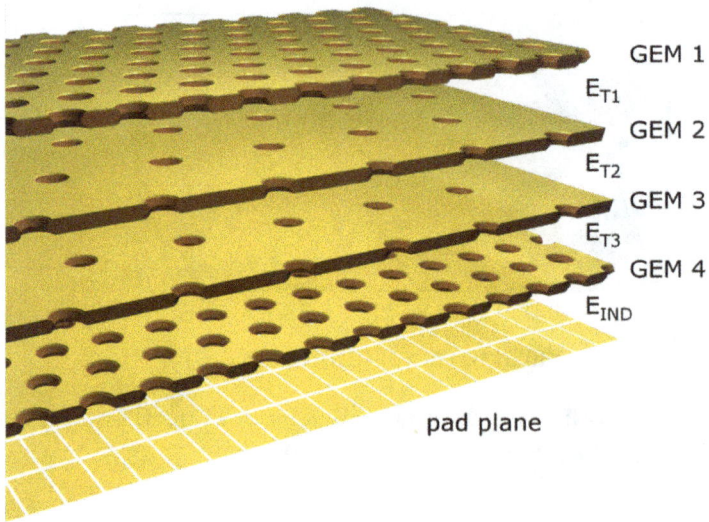

Figure 6.26:   The ALICE TPC Quad-GEM assembly.
Source: Lippmann (2016).

Figure 6.27:   A prototype readout sector for the ALICE GEM–TPC upgrade.
Source: Lippmann (2016).

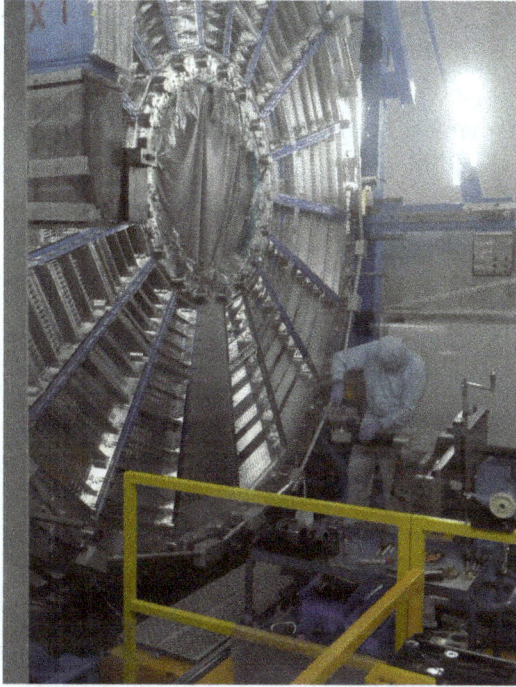

Figure 6.28: The ALICE TPC during assembly of the GEM end-cap modules.

6.5 Positive Ion Back Flow (IBF)

Positive ions produced by electron-molecule collisions during the avalanche amplification process can upset the detector performance in several ways. In photosensitive detectors, the ion bombardment can swiftly deteriorate the quantum efficiency, a subject discussed in Chapter 7. In drift chambers, ions recede along the field lines in the direction opposite to the negative charge; due to their low velocity (about three order of magnitude slower than electrons for equal field values), ions tend to accumulate both in the amplification gap and in the drift volume. An overview of electron drift velocities and ion mobility as a function of field and gases is provided in Chapter 4 in the reference (Sauli, 2014).

The accumulation in the amplification gap generally results in a reduction of field, hence of gain, above a certain value of the detected ionizing particles' flux, as discussed in previous chapters for different types of devices. As indicated there, the narrow-high field multiplication gaps of MPGDs result in faster clearing of the positive charge, thus ensuring a high rate capability. Once injected, however, in the low field region above the amplification, ions recede with a much slower motion and induce modifications both in strength and direction of the field, with ensuing distortions of the electron drift trajectories. The effect is particularly severe in large-volume drift detectors, as TPCs, and has therefore been studied in detail in view of the use of the devices at the ever-increasing particle fluxes.

The transmission of charges through electrodes (or transparency) obeys simple electrostatic laws, convoluted with the losses due to the physical aspect ratio of the electrode, or optical transparency, and to diffusion. For a constant and uniform radiation flux, the IBF is defined as the ratio of the (positive) ion current collected on the drift electrode to the negative (electron) current on the anode.[1] In the simplest case of a metallic mesh separating drift and amplification, as for MICROMEGAS, the IBF is equal or smaller than the ratio of fields, owing to the diffusion losses due to the capture of electrons by the mesh, and is therefore gas and geometry dependent. For multi-GEM and hybrid devices, the IBF is a more complex function of the various multiplication and inter-electrode fields.

Figure 6.29 shows the measured IBF as a function of ratio of fields between amplification and drift regions for a MICROMEGAS manufactured with an electroformed Ni mesh having 500 lines per inch (Colas *et al.*, 2004); the continuous curve is the theoretical prediction without diffusion losses (the ratio of fields). The result is not affected by a magnetic field up to 2 T. For a smaller pitch micro-mesh with 1,500 lines per inch, Figure 6.30, the IBF approaches the theoretical value given by the ratio of fields.

Emulating the multi-GEM structures, described in the following, a dual-mesh MICROMEGAS permits to considerably reduce the IBF. Figure 6.31 shows the computed drift lines for electrons and

---

[1]In some works, the IBF is improperly called Ion Feedback

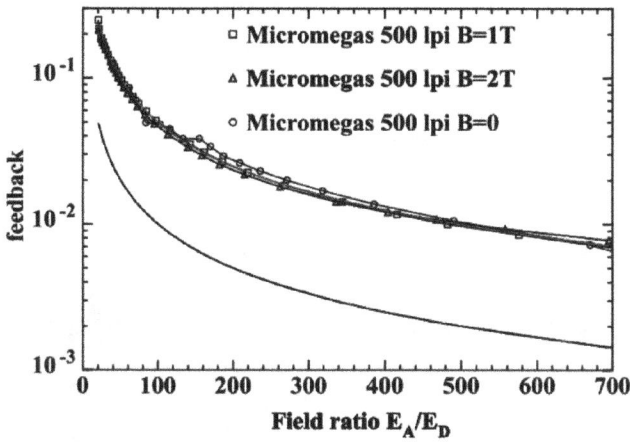

Figure 6.29: IBF as a function of the ratio of amplification and drift fields using a coarse pitch mesh. The points are experimental values, and the lower line represents the simple ratio of fields.

Source: Colas *et al.*, 2004).

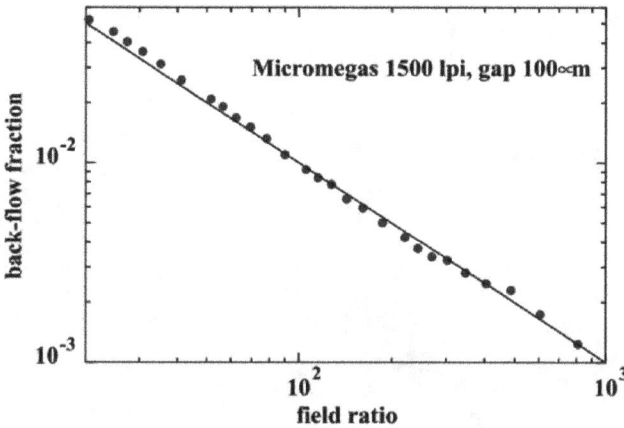

Figure 6.30: IBF for a smaller pitch mesh; the line is the field ratio prediction.
Source: Colas *et al.* (2004).

ions in the double structure. The measured values are presented in Figure 6.32 as a function of gain for a single and double device; the IBF is decreased from around 1% to below 10^{-4} (Jeanneau *et al.*, 2010).

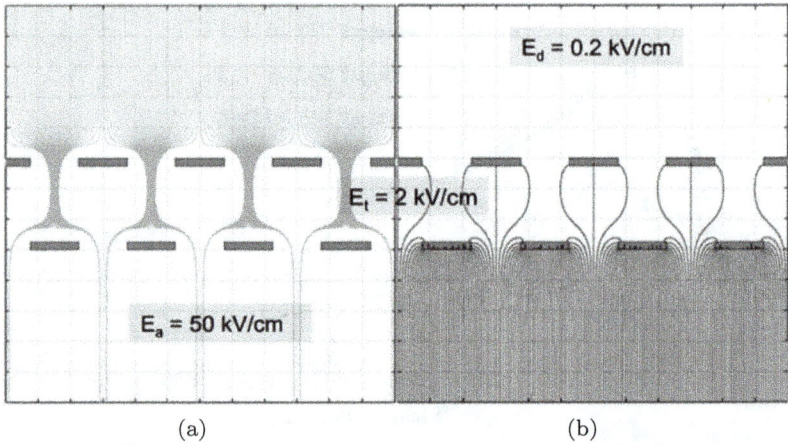

Figure 6.31:   Computed electrons (a) and ions' drift lines (b) for the double-mesh MICROMEGAS.

Source: Jeanneau *et al.* (2010).

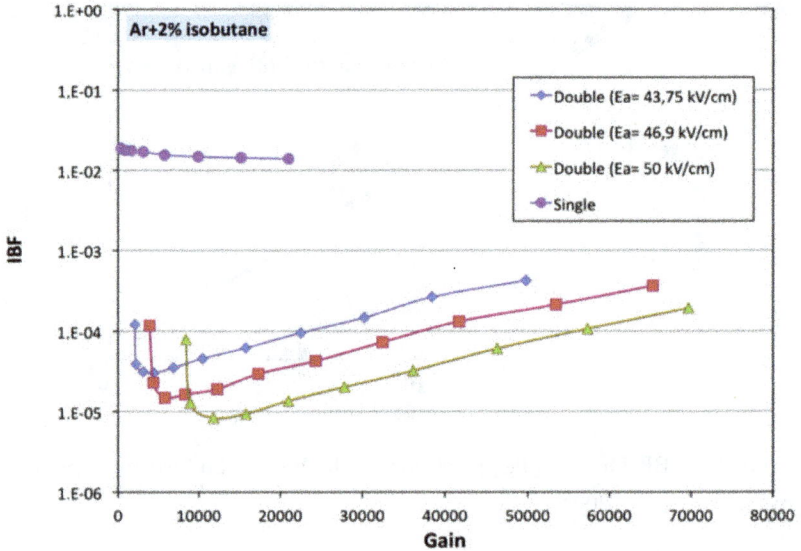

Figure 6.32:   IBF as a function of gain in the single and double MICROMEGAS.

Source: Jeanneau *et al.* (2010).

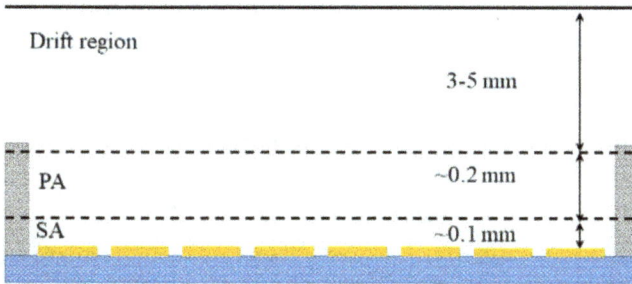

Figure 6.33:   Schematics of the dual-mesh detector.
Source: Zhang *et al.* (2018).

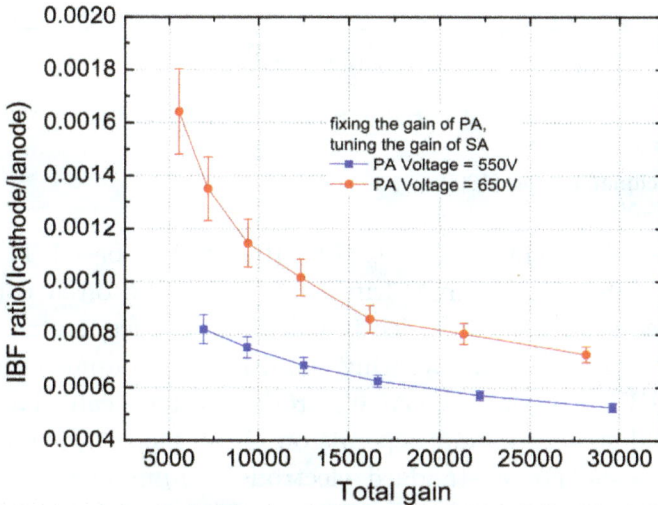

Figure 6.34:   Ion backflow as a function of gain in the dual-mesh detector for two values of the pre-amplification voltage.
Source: Zhang *et al.* (2018).

With similar double-micro-mesh structures operated as independent amplifiers, Figure 6.33, a value of the IBF below 0.1% has been reached, with good energy resolution, Figure 6.34 (Zhang *et al.*, 2018). The technology used for the fabrication of the double structure, thermal bonding to insulating frames, is limited however to small detector sizes.

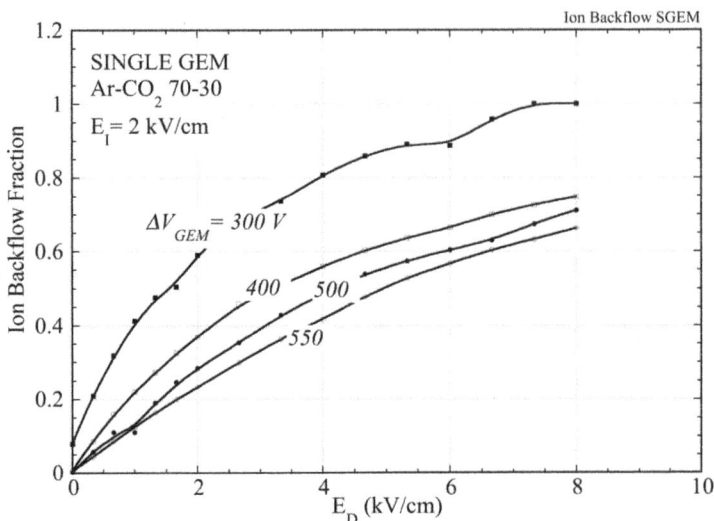

Figure 6.35:   IBF for a single GEM as a function drift field and several values of GEM voltage.

Source: Bachmann *et al.* (1999).

It should be noted that for MICROMEGAS-based TPCs, the value of multiplication and drift fields obeys the often conflicting requirements on gain and electron drift properties, restricting the choice of filling gas and operating conditions to minimize the IBF.

The IBF of a single-GEM electrode is intrinsically larger, due to the field structure, and depends on GEM geometry, voltage and external fields. For a standard electrode (70-μm holes at 140-μm pitch), Figure 6.35 shows the IBF as a function of drift field and GEM voltage, for a fixed induction field (2 kV/cm) (Bachmann *et al.*, 1999).

In multiple structures, fine tuning of the fields in the transfer gaps between electrodes permits to strongly reduce the IBF. Figure 6.36 is an example of IBF and electron transparency measured as a function of drift field with a double-GEM structure, for typical values of the GEM and transfer fields; at a drift field of 500 V/cm, ensuring full electron collection, the IBF is around 5%.

The IBF depends only little on the gas filling, Figure 6.37, but can be reduced further with proper choice of the GEM foil geometry

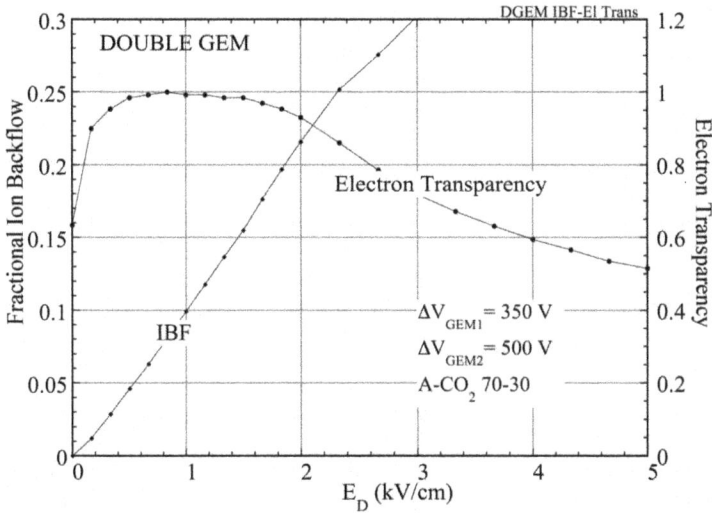

Figure 6.36: IBF and electron transparency for a double GEM.
Source: Bachmann *et al.* (1999).

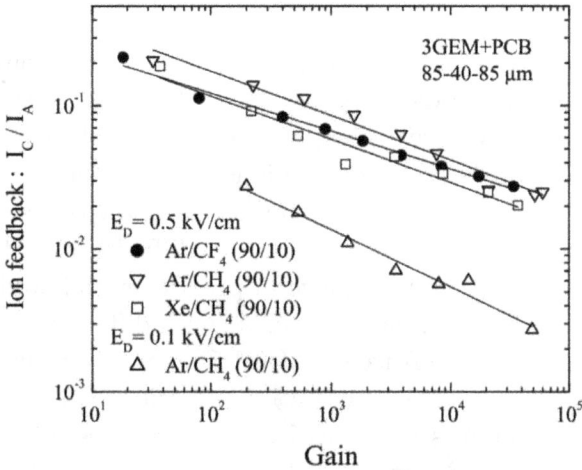

Figure 6.37: IBF for a TGEM in different gas mixtures and two values of drift field.
Source: Bondar *et al.* (2003).

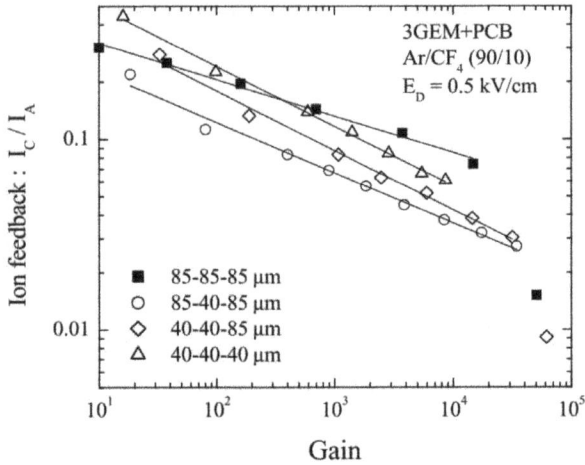

Figure 6.38:   IBF as a function of gain in TGEM structures; the inset indicates the holes' diameter in the three cascaded foils.

Source: Bondar *et al.* (2003).

and fields, as shown for several TGEM structures in Figures 6.38 and 6.39 (Bondar *et al.*, 2003).

While electrons and ions drift in opposite directions along the same field lines, methods to reduce the IBF exploiting their different diffusion properties have been proposed (Sauli *et al.*, 2006). With two cascaded GEMs having misaligned holes, the backward flow of ions is reduced, while a substantial fraction of electrons finds its way through the structure owing to diffusion. While a regular misalignment is difficult to achieve in practice with standard foils, it can be easily realized alternating GEMs with different hole pitches, a solution adopted by the Alice GEM–TPC upgrade group (Ball *et al.*, 2014). As a result of a systematic investigation, the baseline of the detector design mounts four GEM foils, with holes' pitch of $140\,\mu m$ (LP) and $180\,\mu m$ (HP), assembled in the sequence LP–HP–HP–LP, as described in Section 6.4 (Figure 6.26).

Resulting in a reduction of the electron transparency, a small IBF implies a decrease in energy resolution, Figure 6.40 (Gasik *et al.*, 2017); as a compromise, at an IBF of 1%, the resolution of around 11% satisfies the particle identification requirements of the detector.

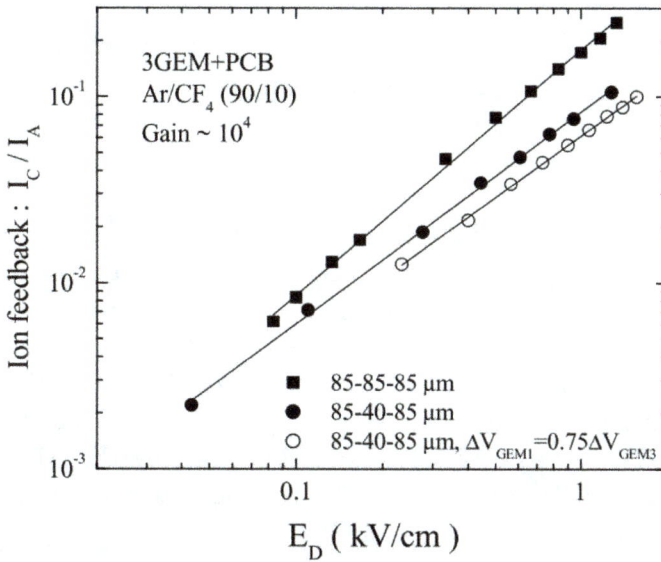

Figure 6.39: IBF as a function of drift field in TGEM structures.
Source: Bondar *et al.* (2003).

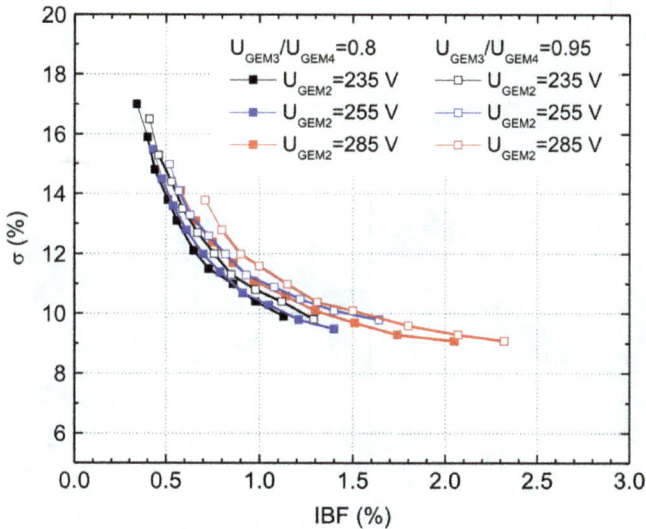

Figure 6.40: IBF and energy resolution for the ALICE–TPC Quad-GEM detector in a range of GEM voltages.
Source: Gasik (2017).

The process of charge transmission through multi-GEM devices has been studied experimentally and by simulations (Bhattacharya *et al.*, 2017). Systematic studies of IBF and resolution for TGEM device, built with a cascade of large, standard and small pitch holes, confirm the outcomes of previous work, in particular the reduction of the backflow to below 1% with an energy resolution for soft X-rays of 12% or better (Natal Da Luz *et al.*, 2018).

As an alternative, a combination of two GEM electrodes with a MICROMEGAS as the final multiplier has been studied (Aiola *et al.*, 2016). With a thorough optimization of the gain sharing and of the transfer fields between the three elements, the authors could achieve in Ne–CO_2 an IBF below 0.4% with an energy resolution for 5.9-keV X-rays of 12%, a result confirmed by the ALICE TPC works, Figure 6.41 (Ratza *et al.*, 2018). The outcome compares with the values achieved by the Quad-GEM configuration adopted for the detector.

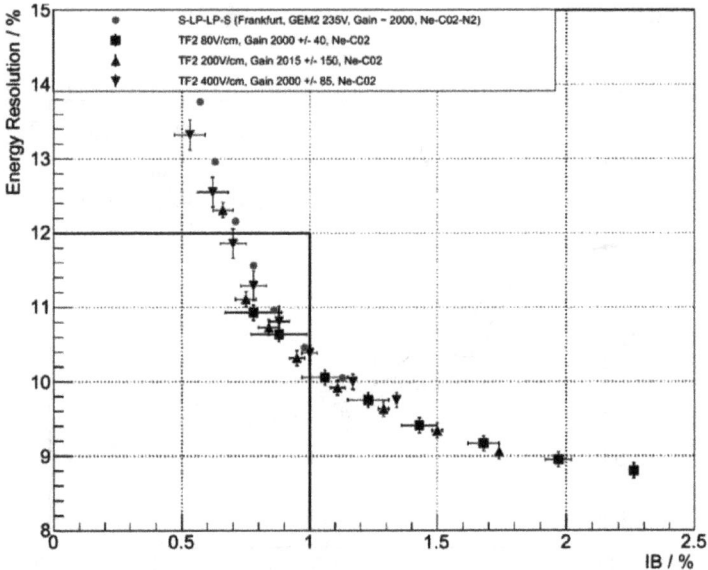

Figure 6.41: IBF and energy resolution in Ne–CO_2 (90–10) of a double-GEM+MICROMEGAS hybrid detector, compared to the four-GEM ALICE TPC baseline.

Source: Ratza *et al.* (2018).

A dedicated experimental and simulation setup to study the charge flow in MICROMEGAS-based TPCs is described in Bhattacharya *et al.* (2017).

When allowed by the timing operation of the TPC, the flow of ions into the drift volume can be completely blocked with a controlled voltage electrode, or gate, as used in the first-generation MWPC-based TPCs (see, for example Chapter 10 and Sauli (2014)). GEMs constitute a particularly elegant way to implement a gating electrode, as they require rather small voltages to be activated or blocked, particularly when using large hole sizes; see Figure 6.42 (Sauli *et al.*, 2006). As a follow-up of this work, an optimization of parameters and geometry initiated in the framework of the high-precision TPC for the Linear Collider project (Gros *et al.*, 2013) led to the realization of a "gating" GEM electrode with large optical transparency, Figure 6.43 (Kobayashi *et al.*, 2019). As seen in the computed results of Figure 6.44, the transmission for ions and

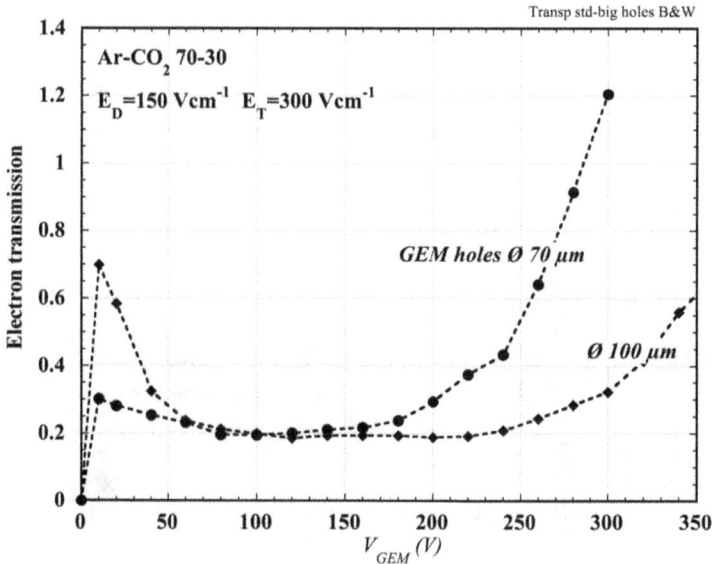

Figure 6.42: Electron transparency as a function of GEM voltage for two GEM holes' diameter.

Source: Sauli *et al.* (2006).

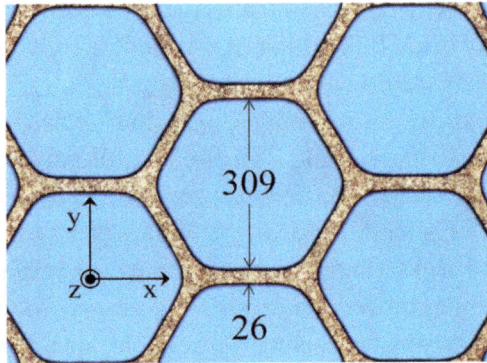

Figure 6.43:   GEM gate, with maximized optical transparency.
Source: Kobayashi *et al.* (2019).

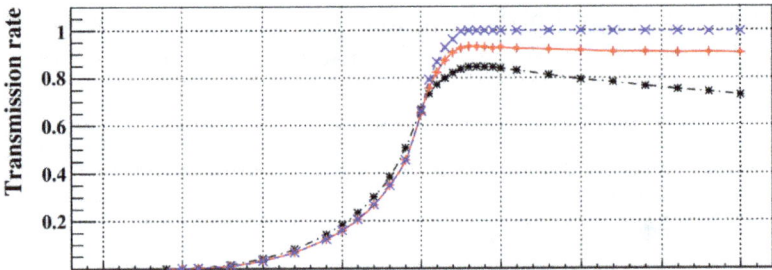

(a)

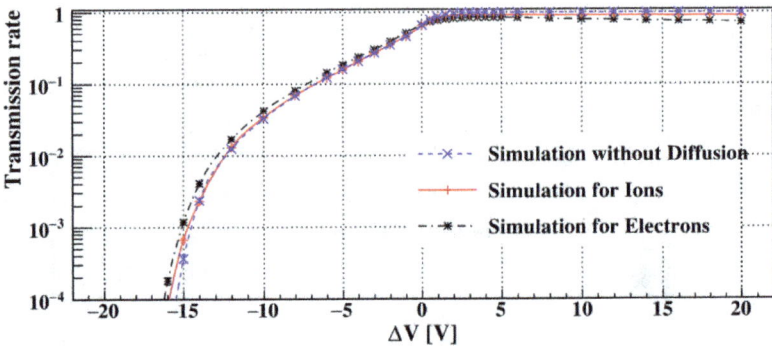

(b)

Figure 6.44:   Computed charge transmission through the GEM gate.
Source: Kobayashi *et al.* (2019).

electrons can be controlled with a few volts of difference between the two sides of the GEM; for electrons, the transparency is slightly reduced by diffusion losses. With properly applied voltages, the foil transmits electrons without amplification and can be actively closed to ions. Confirmed by the experimental results, the study has been extended to include the presence of an external magnetic field.

The presence of an external magnetic field, parallel to the electrons' drift direction in TPC devices, alters the drift and diffusion properties of charges and can affect the IBF fraction. This is seen in the measurement shown in Figure 6.45, obtained with a small TPC with TGEM end-plate readout, and operated with the TDR gas mixture; a simulation study suggests that the IBF reduction is actually due to an increase of the electrons' transparency, the ions being essentially unaffected by the magnetic field (Killenberg *et al.*, 2004). This result, obtained in the framework of the ILC detectors' development, has been challenged by later work by the PANDA TPC development group, and attributed to an effect of field distortions due to the excessive irradiation rate used for the measurement.

Innovative MPGD structures based on the micro-hole and strip plates (MHSP) permit to reduce the IBF further, at the cost of

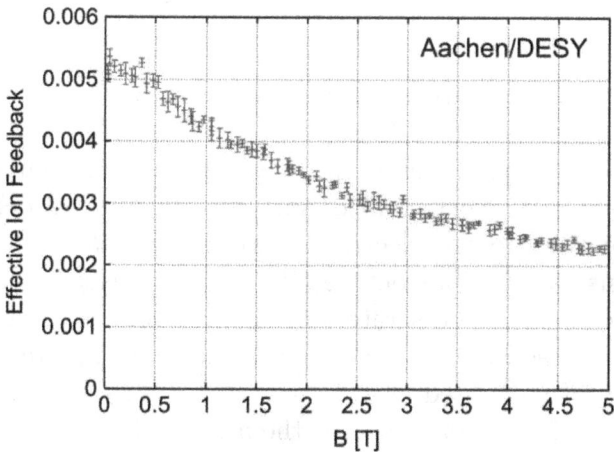

Figure 6.45: IBFs at increasing values of magnetic field.
Source: Killenberg *et al.* (2004).

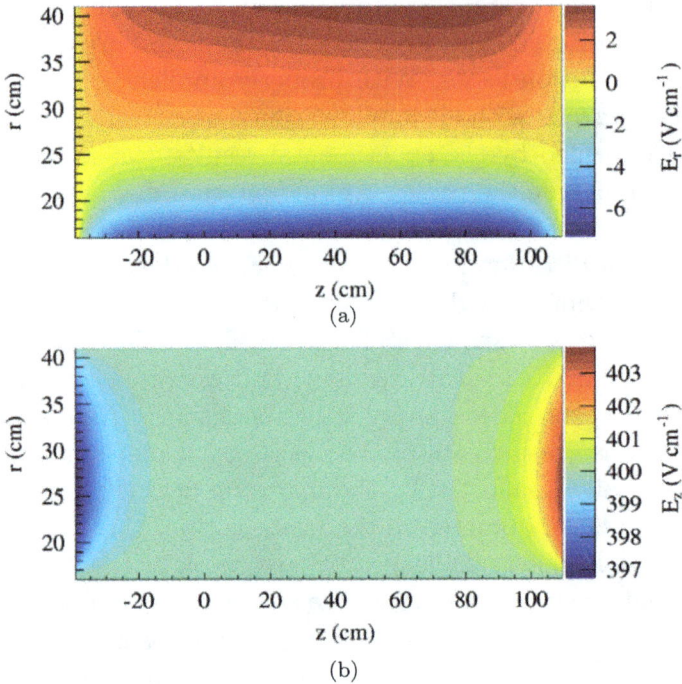

Figure 6.46:   Drift field distortions in typical operating conditions computed for the PANDA TPC.

Source: Böhmer *et al.* (2013).

an increased complexity of manufacturing; they are described in Section 8.2.

The positive space charge build-up in the drift volume induces distortions in the electric field and consequently alters the trajectories of the drifting electrons. This effect has been estimated with various assumptions on the IBF, ions and electrons' mobility, detector geometry, events rate and distribution. For the PANDA-FOPI TPC described in the previous section, Figure 6.46 gives an example of computed electric field distortions for $\sim 10^7$ proton-antiproton collisions per second at the planned International Facility for Antiproton and Ion Research (FAIR) in Darmstadt (Böhmer *et al.*, 2013); in the worst case, the distortion results in an error close to 10 mm in the position of ionization trails. Software corrections based on the model calculation are possible, but are time consuming and bound to introduce large errors.

Similar effects have been estimated for the ALICE TPC, thus motivating an extensive research effort to reduce the ions' backflow (Ketzer, 2013; Berger *et al.*, 2017; Divani Veis *et al.*, 2018).

Short of finding a better solution to reduce the IBF at high particle fluxes, the relevance of these distortions sets a limit to the deployment of TPC-based detectors in high-rate particle physics.

Chapter 7

UV Photon Detection and Localization

7.1 Statistics of Single-Electron Avalanches

While the most probable value of the multiplication is described by Townsend's exponential expression, statistical fluctuations in the individual ionizing collisions with the molecules induce large dispersions around the average. The process has been extensively studied in the early developments of gaseous proportional counters, as it governs the energy resolution of the devices, see (Alkhazov, 1969) and references therein.

For an avalanche initiated by a single electron and developing in a uniform field along a length x, the most probable avalanche size is

$$\bar{n} = e^{\alpha x},$$

where α is the first Townsend coefficient.

Statistical considerations lead to an approximate expression for the probability of an avalanche to reach a size n in a gap of thickness s, the so-called Furry law (Sauli, 2014, Section 5.4)

$$P(n, s) = \frac{1}{\bar{n}} \left(1 - \frac{1}{\bar{n}}\right)^{n-1} \approx \frac{e^{-\frac{n}{\bar{n}}}}{\bar{n}}.$$

The expression can be extended by suitable averaging to the general case of non-uniform fields as those of proportional counters (Byrne, 1969; Alkhazov, 1970). It has the particularity to be maximum for $n = 1$, meaning somehow counterintuitively that the highest probability for an electron is not to multiply at all over the full gap, Figure 7.1.

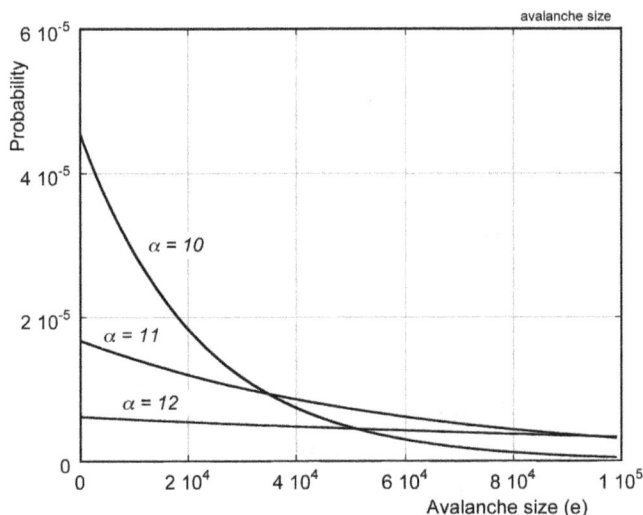

Figure 7.1:   Single-electron avalanche-size probability for several values of the Townsend coefficient.

It was observed experimentally that the single-electron avalanche distribution can evolve from a pure exponential to a peaked shape at large amplification factors (Curran and Craggs, 1949; Schlumbohm, 1958; Fonte *et al.*, 1999). Figure 7.2, from the latter reference, is an example of evolution of the single-electron avalanche size for two values of gain, measured with a parallel plate counter.

Taking into account that at large gains, and therefore small values of the mean ionization path, a non-negligible fraction of the electron path is needed to acquire enough energy to experience an ionizing collision, one can deduce the following general expression for the avalanche probability, named Polya distribution (Byrne, 1969):

$$P(n, \theta) = \left[\frac{n}{\bar{n}}(\theta + 1)\right]^{\theta} e^{-\frac{n}{\bar{n}}(\theta+1)},$$

function of a parameter θ, related to the variance of the distribution:

$$f = \left(\frac{\sigma_n}{\bar{n}}\right)^2 = \frac{1}{(1+\theta)}.$$

For $\theta = 0$, the expression reduces to the simple exponential obtained above, Figure 7.3.

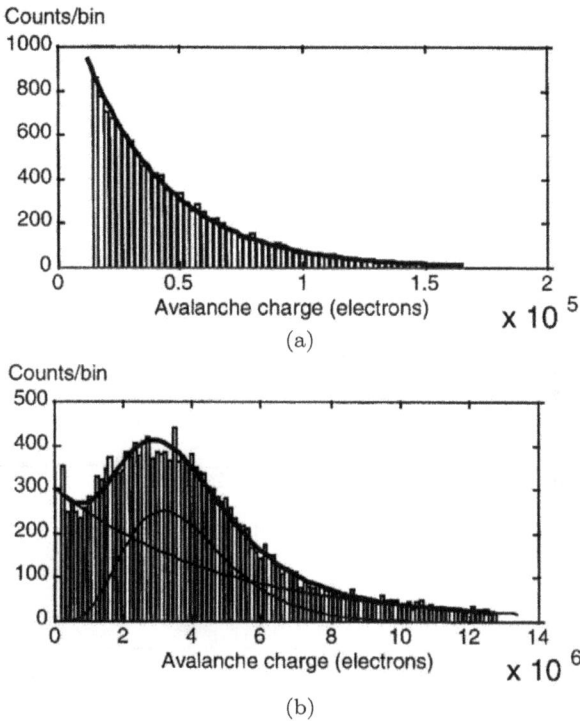

Figure 7.2: Experimental single-electron avalanche-size distribution at moderate and high gains.

Source: Fonte *et al.* (1999).

The field-dependent transition from a Furry to a Polya distribution favours structures like the MICROMEGAS to obtain peaked pulse height distributions (Derré *et al.*, 2000; Zerguerras *et al.*, 2009). Figure 7.4 is an example of single-electron avalanche-size distribution measured with a MICROMEGAS in helium–isobutane at a gain of 5.7×10^3 (Zerguerras *et al.*, 2015). Peaked distributions have been obtained also with the double-mesh MICROMEGAS, which permits to reach gains well above 10^5, Figure 7.5 (Zhang *et al.*, 2018).

In multi-GEM devices, the multiplying field is relatively low in each element of the cascade, permitting to reach overall values of gain well above 10^5 in a large variety of gases, including pure argon, Figure 7.6 (Buzulutskov *et al.*, 2000). The statistics of single-electron

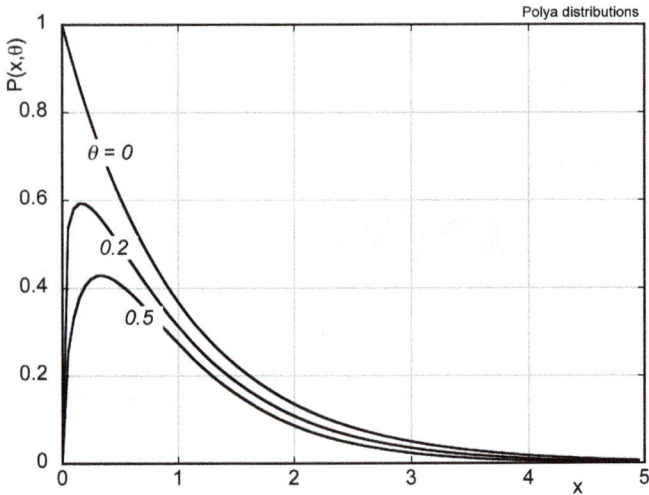

Figure 7.3:   Computed single-electron avalanche distribution for several values of the Polya parameter θ.

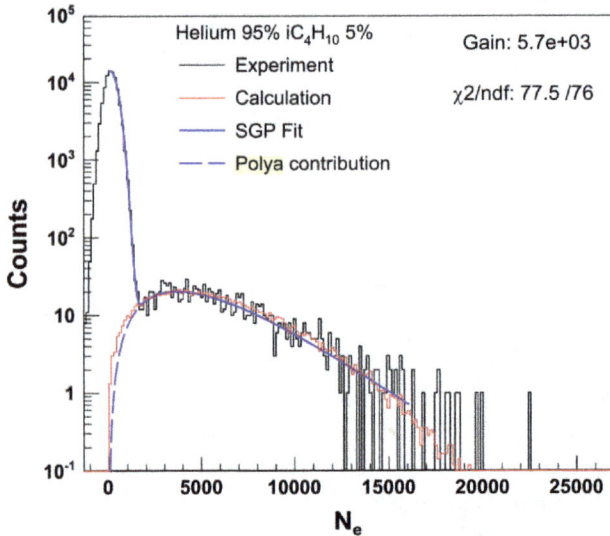

Figure 7.4:   Single-electron avalanche size measured with a MICROMEGAS. A fit separates the true Polya distribution from the Gaussian-shaped noise (SGP).

Source: Zerguerras *et al.* (2015).

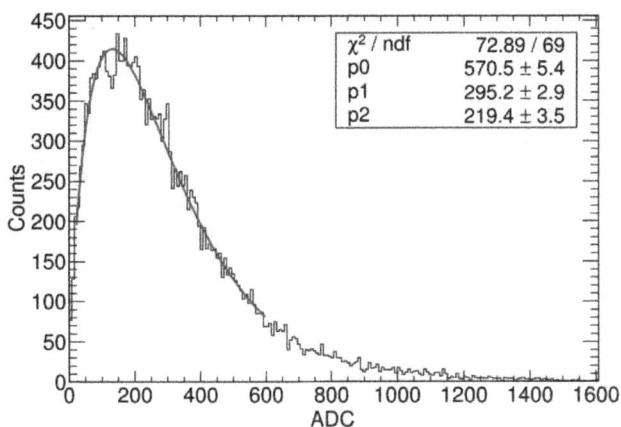

Figure 7.5: Single-electron response of a double MICROMEGAS.
Source: Zhang *et al.* (2018).

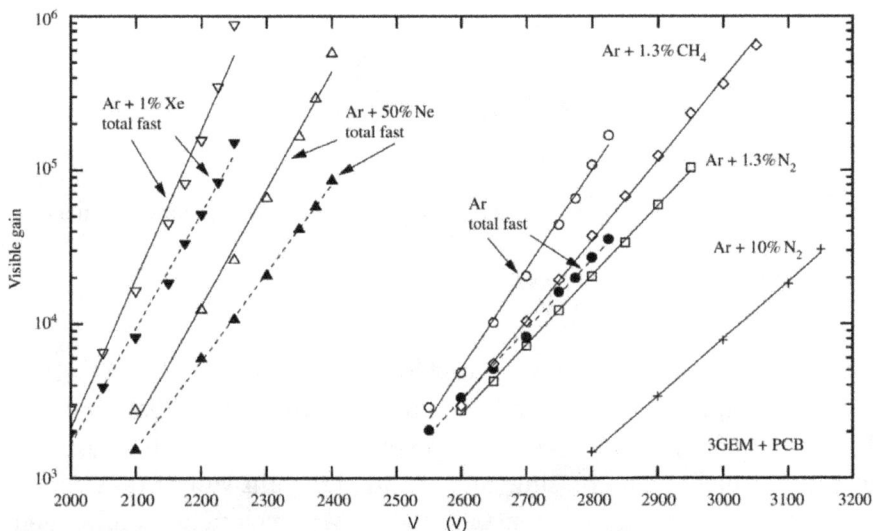

Figure 7.6: Single-electron gain for a TGEM with CsI photocathode in various gases, as a function of total voltage.
Source: Buzulutskov *et al.* (2000).

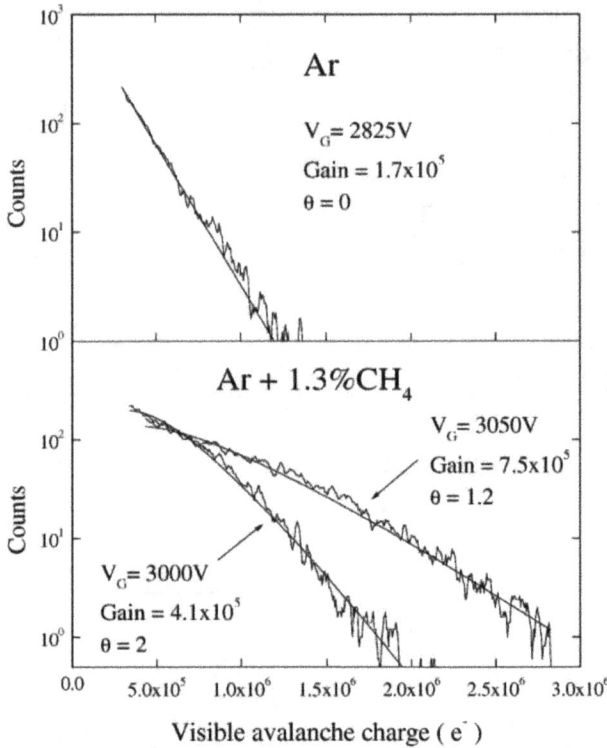

Figure 7.7:   Single-electron response of a TGEM in pure and poorly quenched argon.
Source: Buzulutskov *et al.* (2000).

avalanches is, however, dominated by the low value of the field in the first element of the cascade, leading to exponential distributions. Figure 7.7 shows examples of single-electron response in argon and argon–methane at increasing gains. The high gain values that can be reached in pure argon are attributed to an avalanche confinement in the holes, preventing the spread of photon-induced secondary processes (Buzulutskov *et al.*, 1999). The observation is relevant in view of the realization of sealed counters that could be easily damaged with the use of organic quenchers, see Section 7.2.

High gains, suitable for single-electron detection, have been obtained operating a triple-GEM (TGEM) with a reflective CsI

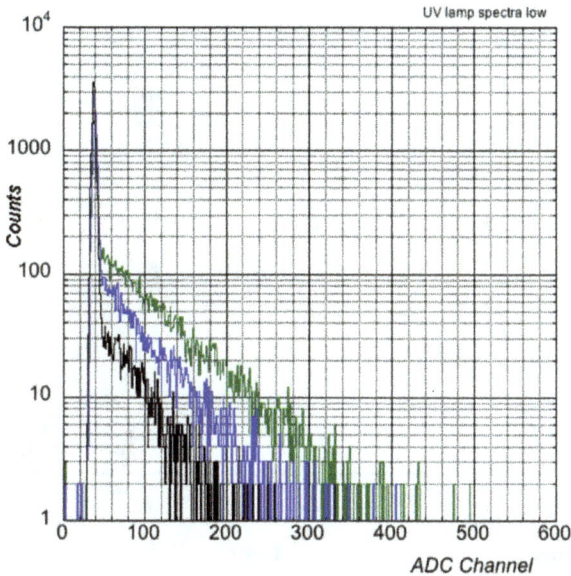

Figure 7.8: Single-electron response of a TGEM with CsI reflective photocathode at increasing values of gain.

Source: Meinschad *et al.* (2004).

photocathode in CH_4 and CF_4, pure or in mixtures with argon, permitting to attain large quantum efficiencies, Figure 7.8.

Initial work with a small-sized thick-GEM (TH-GEM) seemed to hold the promise that efficient single-electron detection could be reached with a single amplifying stage. Further work indicated, however, that high stable gains could be obtained only with cascaded multiple electrodes.

Exponential distributions are obtained also with the RICH detectors combining moderate gain TH-GEM pre-amplifiers with a MICROMEGAS as the last amplification stage (see Section 7.4.), confirming that the key ingredient for obtaining a peaked distribution is the value of field in the first multiplication stage; see Figure 7.9 (Agarwala *et al.*, 2018).

Desirable to confirm the single-electron response, a peaked distribution is not necessarily needed to achieve a good detection efficiency, as this depends more on the ratio between signal and noise,

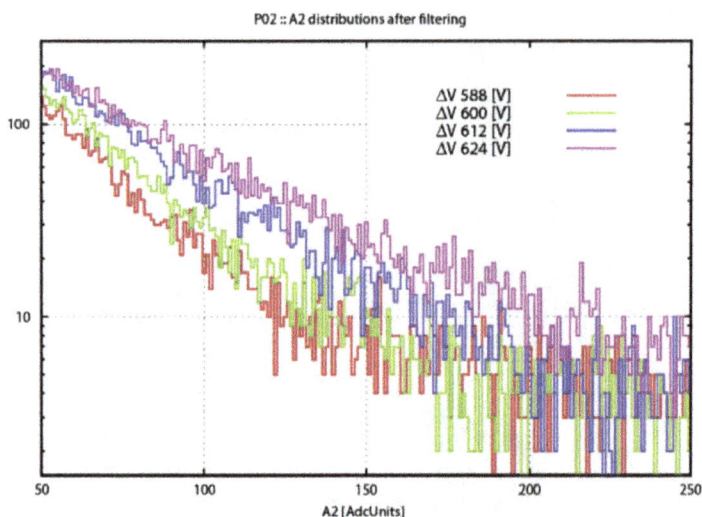

Figure 7.9:   Single-electron response in the COMPASS RICH-1, cascading two TH-GEMs and a MICROMEGAS. The distributions are measured at increasing MICROMEGAS gains.

Source: Agarwala *et al.* (2018).

and can be ensured for exponential distributions with a sufficiently low electronic threshold.

7.2   Gaseous and Solid Photocathodes

Gaseous proportional counters have only occasionally been used for the detection of low-energy photons, due to the limited extension of the spectral response between the ionization threshold of photosensitive materials and the cutoff of the window separating the detector from the source. Photons in the energy range from ultraviolet to infrared can be converted to free electrons in the gas with the use of suitable gaseous or solid photocathodes; under the effect of the applied electric fields in MPGDs, they can be multiplied in an avalanche process.

The development of single-photon, position-sensitive gaseous detectors was largely driven, in the late seventies of the previous century, by a proposed method to identify charged particles in the GeV range based on the recording of photons emitted by the Cherenkov effect, the so-called Ring Imaging Cherenkov (RICH) Counters (Séguinot and Ypsilantis, 1977), but has received numerous applications in other fields. For large systems, such as those used in particle physics, the amplification should be 10^4 or above to permit the use of cheap, large-scale integrated electronics. As charge multiplication is accompanied by the emission of photons, the detectors are designed to minimize the efficiency of reconversion of secondary photons emitted by the avalanche process that would lead to signal spread and gain divergencies. For devices using internal solid photocathodes, another requirement concerns the likelihood that positive ions released in the avalanches can damage the photosensitive layer; particular detector geometries have been devised to reduce the ions' feedback.[1]

All gases absorb photons above an energy threshold value, whose value increases with the complexity of the molecules. When the photon energy exceeds the ionization threshold, an electron can be released in the gas and exploited for detection. An early survey of the spectral response and possible combinations of window-photosensitive gas is given in Carver and Mitchell (1964). Further works identified triethyl amine $(C_3H_5)_3N$ (TEA) as a suitable vapour for UV photon detection (Holroyd *et al.*, 1987; Séguinot *et al.*, 1980). A liquid with a vapour pressure of 55 torr at room temperature and a photoionization threshold of 7.45 eV, TEA can be used in conjunction with fluoride windows to cover a sufficiently wide spectral region, Figure 7.10. Used as an additive to a gas mixture transparent in the region of spectral sensitivity, TEA has an absorption length below 1 mm, permitting the realization of efficient and fast UV photon detectors.

The development of the multi-step detector scheme, effectively suppressing the photon feedback processes between two cascaded amplification stages, permits to reach gains in excess of 10^6

---

[1]The term "ion-feedback" is used for processes that can damage the photocathode, as against the term "backflow" used in the context of drift chambers.

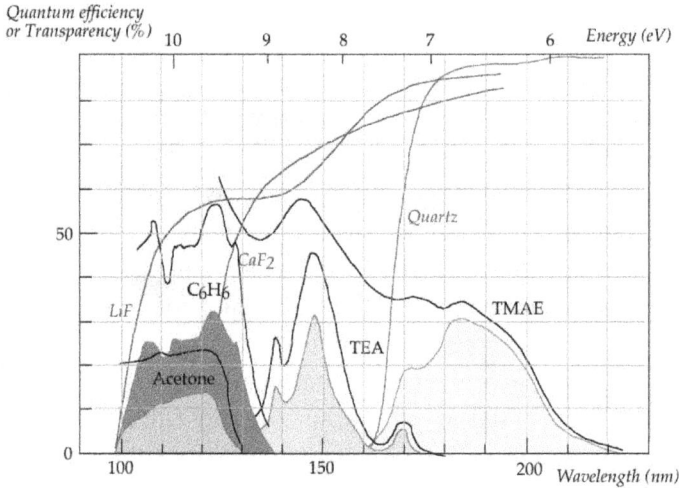

Figure 7.10:   TEA and TMAE quantum efficiency and windows transparency as a function of photon energy and wavelength (compilation from different sources).

and perform single-photon localization with sub-millimetre accuracy (Charpak and Sauli, 1978). Based on this conception, the first-generation RICH counters were successfully operated with a TEA–CaF_2 combination (Adams *et al.*, 1983; McCarty *et al.*, 1986).

Less convenient to use due to its chemical reactivity and low vapour pressure, TMAE (tetrakis dimethyl amino ethylene, $C_2[(CH_3)_2N]_4$), with the exceptionally low photoionization threshold of 5.5 eV, permits the use of cheap quartz windows, a significant advantage for the implementation of large-sized detectors (Anderson, 1980). TMAE was used in the second-generation RICH detectors: OMEGA and DELPHI at CERN, SLD at SLAC. For a detailed historical summary of MWPC-based photon detectors, see for example (Séguinot and Ypsilantis, 1994).

With the development of gas-compatible solid photocathode materials, vapour-based photosensitive devices have been largely abandoned. Investigated as a photocathode material in gaseous counters in the early nineties, caesium iodide (CsI) has a photoionization threshold of 5.7 eV and can withstand short-term exposures to air, an essential property to permit the construction of large detector

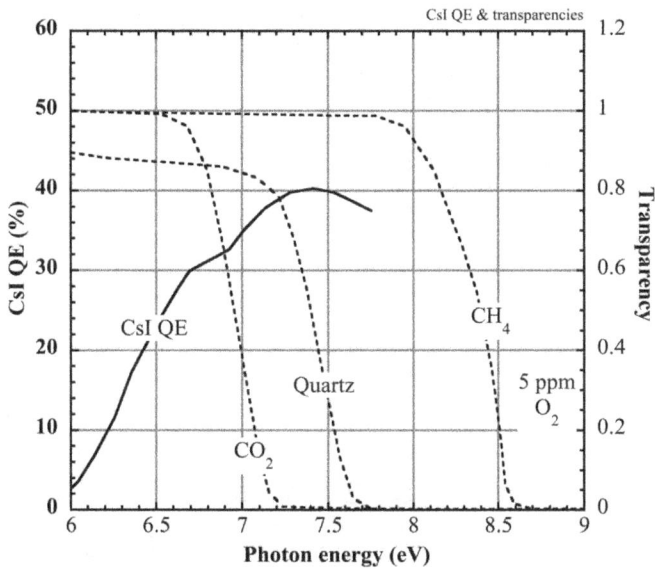

Figure 7.11: CsI quantum efficiency as a function of photon energy and transparencies of quartz windows and some gases (from different sources).
Source: Séguinot *et al.* (1990).

systems (Dangendorf *et al.*, 1990; Séguinot *et al.*, 1990; Charpak *et al.*, 1992).

The quantum efficiency and long-term stability of operation of CsI have been extensively studied (Lu and McDonald, 1994; Berger *et al.*, 1995; Breskin, 1996; Va'vra *et al.*, 1997; Hoedlmoser *et al.*, 2006). The result depends on many factors: deposition method, nature of the substrate and thickness of the layer (Friese *et al.*, 1999; Nitti *et al.*, 2009). Figure 7.11 shows an example of measured dependence of CsI quantum efficiency from photon energy (Séguinot *et al.*, 1990), as well the transparencies of quartz window and of two gases that can be used in the counters.

Most measurements of the CsI photocathodes' quantum efficiency have been made in vacuum. In the presence of a gas, the QE depends on the applied surface extraction field (Breskin *et al.*, 1994); with close to 90% of its value in vacuum for methane at atmospheric pressure above few hundred volt cm^{-1}, it requires higher

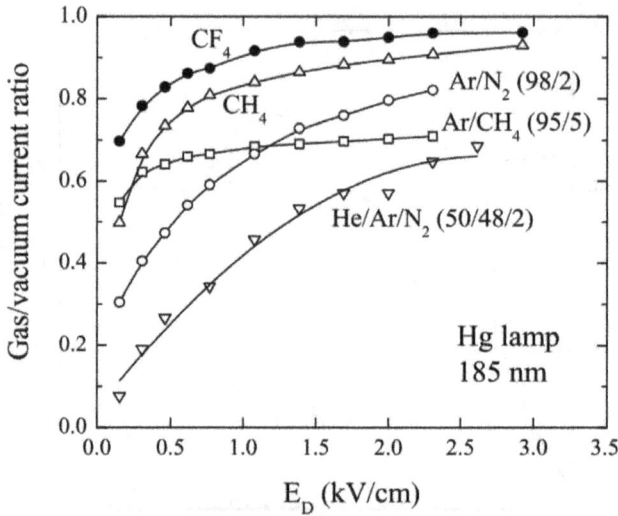

Figure 7.12:   CsI quantum efficiency at 185 nm relative to vacuum as a function of the extraction field in several gases at atmospheric pressure.
Source: Breskin *et al.* (2002).

fields and hardly reaches 60–70% in mixtures containing argon or neon, Figure 7.12, owing to the large elastic cross section of noble gas molecules, backscattering the photoelectrons (Azevedo *et al.*, 2010). This property restricts the design and the choice of the gas mixture used in single-photoelectron detectors.

Caesium iodide has been used as a photocathode material in the MWPC-based RICH detectors, as in the COMPASS experiment (Albrecht *et al.*, 2005), HADES (Fabbietti *et al.*, 2003), ALICE (Gallas, 2005) and others.

Alternative compounds suitable for extending the CsI spectral sensitivity have been studied. Figure 7.13 gives the QE of several photocathodes in the VUV region, suitable for solar-blind detector applications, and Figure 7.14 provides the QE in the visible range of bialkali photocathodes, bare or coated with thin protection layers of caesium bromide and caesium iodide in an attempt to increase their survival in a gaseous environment (Breskin *et al.*, 2000).

In conventional photodetectors, the sensitive layer is deposited on the inner side of the window separating the detector from the

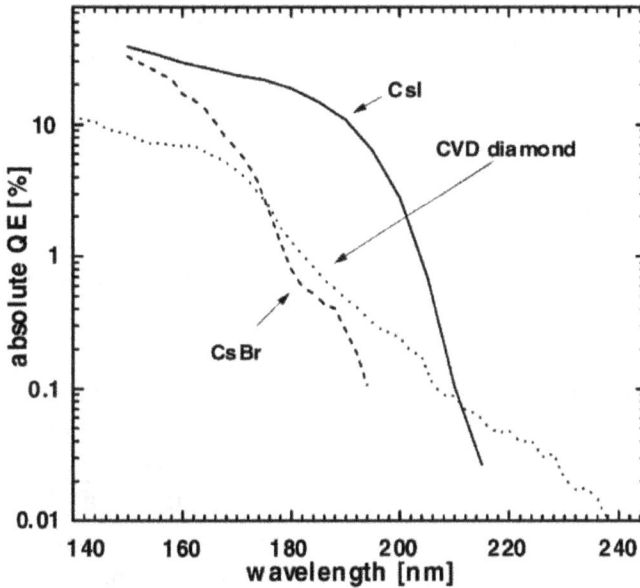

Figure 7.13: Quantum efficiency in vacuum as a function of wavelength of caesium iodide, caesium bromide and CVD diamond.

Source: Breskin *et al.* (2000).

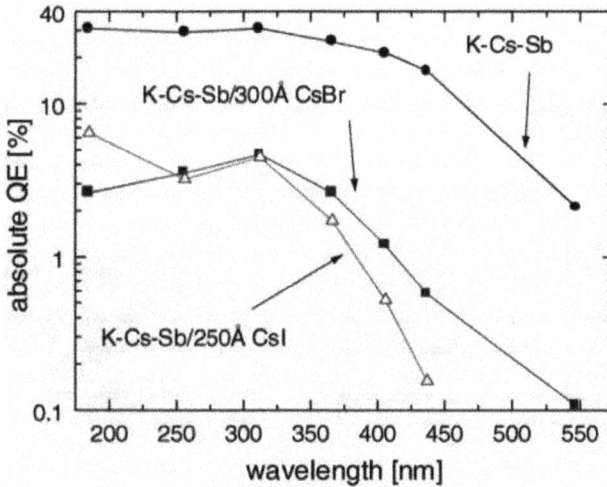

Figure 7.14: Absolute quantum efficiency in vacuum of K–Cs–Sb, bare and coated with protection layers.

Source: Breskin *et al.* (2000).

source, the so-called transmissive or semi-transparent geometry; the thickness of the layer has to be thoroughly controlled to optimize the ratio between quantum efficiency and self-absorption. To prevent charging up, the window itself has to be coated with a thin transparent conducting layer.

In most gaseous detectors' designs, photons released by the charge multiplication process can hit the photocathode, resulting in feedback processes prone to induce divergencies at high gains. Positive ions produced in the avalanches, receding to the cathode, can also damage the delicate photosensitive layer.

A multi-GEM structure, Figure 7.15(a), providing an effective screening between the last amplification stage and the cathode, mitigates the photon and ion feedback and permits to reach higher gains. A better detector conception, making use of a photosensitive layer deposited on the upper electrode of the first GEM in a cascade facing the window (reflective photocathode) eliminates completely the photon feedback, and helps in reducing the ion feedback, Figure 7.15(b). The smaller area of the layer, due to the optical transparency of the electrode (typically 50–60%) is compensated by the larger quantum efficiency of the reflective photocathode;

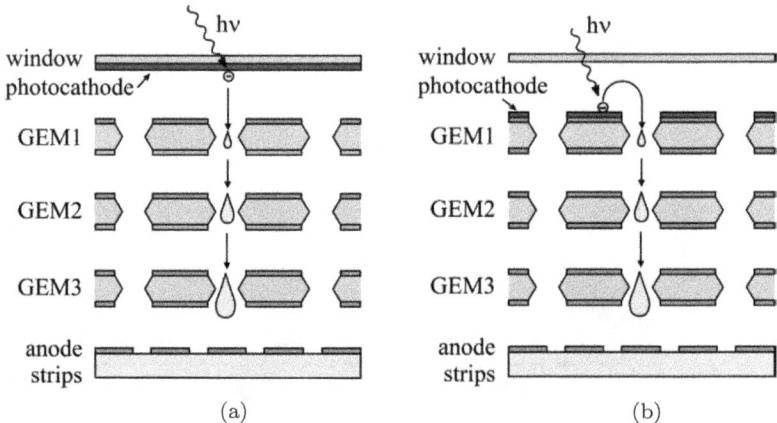

Figure 7.15:   Schematics of a multi-GEM photon detector with semi-transparent (a) and reflective photocathodes (b).

Source: Mörmann *et al.* (2003).

Figure 7.16: A small-sized sealed GEM gaseous photon detector.
Source: Breskin *et al.* (2002).

the relative independence of the QE from thickness makes also the deposition process less critical.

In view of commercial applications, the efficiency and long-term endurance of sealed gaseous photomultipliers have been studied with small-sized prototypes. Figure 7.16 is an example of detector mounting multiple cascaded GEMs as amplifying devices (Breskin *et al.*, 2002). Figure 7.17 shows the gain as a function of voltages applied to each of a TGEM assembly with a CsI reflective photocathode, measured at one atmosphere in two argon–methane mixtures and pure carbon tetrafluoride (Mörmann *et al.*, 2003).

The fractional ion feedback, expressed as ratio between cathode and anode currents, decreases with gain and the number of cascaded GEMs, Figure 7.18 (Breskin *et al.*, 2002).

Much effort has been devoted to the development of photocathode materials capable of extending the spectral response of gaseous counters into the visible region. Aside from possible commercial applications, their use in RICH devices would permit to reduce the resolution-limiting chromatic dispersions, and the use of cheaper glass windows. The progress has been hindered by the extreme sensitivity of these materials to pollutants, particularly oxygen and

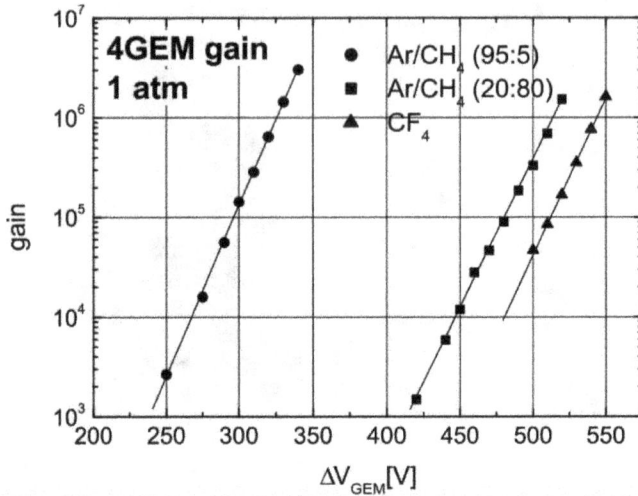

Figure 7.17:   Gain as a function of voltage in a sealed TGEM counter with reflective CsI photocathode in several gases.

Source: Mörmann *et al.* (2003).

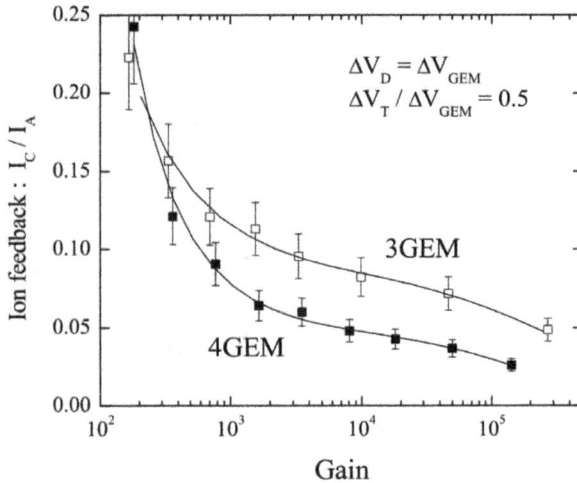

Figure 7.18:   Fractional ion feedback as a function of gain for triple and quadruple GEMs with semi-transparent CsI photocathodes.

Source: Breskin *et al.* (2002).

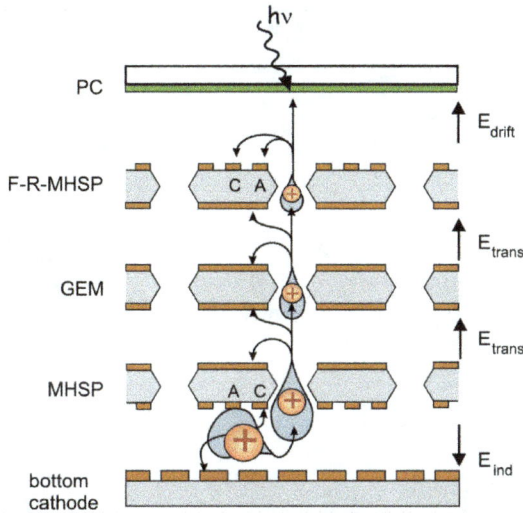

Figure 7.19:   Cascaded GEMS and MHSPs structures for ion backflow reduction. *Source*: Lyashenko *et al.* (2009).

water, and by the presence of secondary photon- and ion-induced processes, only partly solved with the use of MPGDs.

Replacing the first and last GEM in a cascade with special multiplying structures having on one side independently powered strips, interleaved with the rows of holes, the so-called micro-hole and strip plate (MHSP) as in Figure 7.19, permits to reduce the ion backflow below $\sim 10^{-4}$ at a gain of 10^5, fulfilling the requirements for long-term stable operation (Veloso *et al.*, 2000; Maia *et al.*, 2004; Lyashenko *et al.*, 2009; Breskin *et al.*, 2010). With the described detector configuration, the quantum efficiency of a semi-transparent bialkali K–Cs–Sb photocathode operated in argon-methane at atmospheric pressure is 50% of its value in vacuum, Figure 7.20.

An evolution of the MHSP concept, the COBRA structure, has been developed with the aim of reducing the ion backflow in TPC-like detectors; it is described in Section 8.2.

With a small-sized sealed double-mesh MICROMEGAS and a bialkali photocathode, Tokanai *et al.* (2009) succeeded in obtaining with a gas filling of Ar–CH$_4$ (90–10) at one bar a QE in the visible range only a factor of two lower than in vacuum, Figure 7.21.

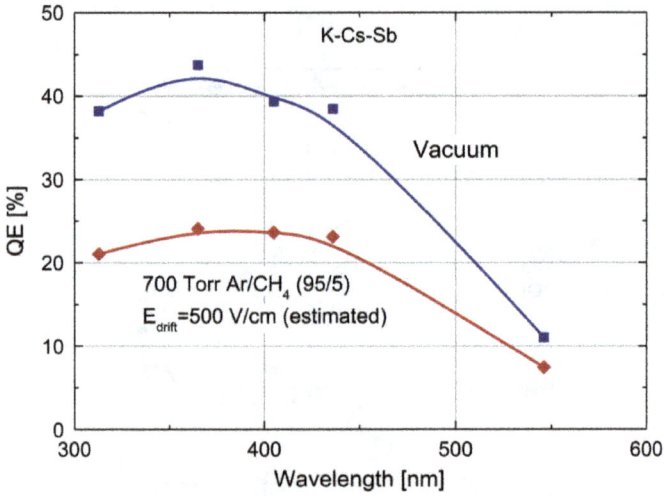

Figure 7.20:   Quantum efficiency of an MHSP–GEM sealed detector with bialkali photocathode in vacuum and in Ar–CH$_4$.

Source: Lyashenko *et al.* (2009).

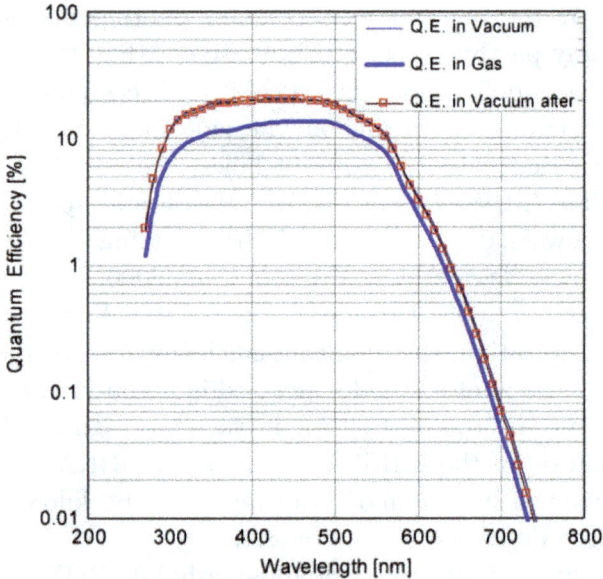

Figure 7.21:   Quantum efficiency as a function of wavelength measured with a sealed double-mesh MICROMEGAS with a bialkali photocathode.

Source: Tokanai *et al.* (2009).

The QE is fully recovered evacuating the detector, demonstrating that the photocathode is not damaged by exposure to the gas.

For comprehensive reviews of gaseous photon detectors, see for example Di Mauro (2014) and Nappi (2020).

7.3 Cherenkov Ring Imaging

The first large-scale applications of single-photon sensitive detectors are for particle identification in high-energy Physics exploiting the Cherenkov light emitted by fast charged particles in suitable radiators, a technique named RICH Counters or Cherenkov Ring Imaging Detectors (CRID). Proposed by Jacques Séguinot and Thomas Ypsilantis in the late seventies (Séguinot and Ypsilantis, 1977), detection and localization of UV photons with large-area devices permits to cover a region of energy where the separation of tracks through the measurement of the differential energy loss becomes inoperable.

The basic scheme of a large solid angle RICH counter is shown in Figure 7.22: a spherical mirror of radius R reflects photons emitted by charged particles in the radiator by the Cherenkov effect in a ring pattern on the focal plane, of radius $R/2$. Detecting and recording the positions of the photons permits to reconstruct the radius of the ring r and deduce the particle velocity, through the expression

$$r = \frac{R}{2} \tan \theta,$$

where θ is the Cherenkov angle, determined by the particle velocity β and refraction index n of the radiator

$$\cos^{-1} \theta = \frac{\beta}{n}.$$

The region of spectral sensitivity is composed between the transparency cutoff of the window and the photoionization threshold of the detector. The choice of the radiator is dictated by the range of particle momenta to be covered, with the requirement to be transparent in the region of sensitivity. For a detailed theory of ring imaging Cherenkov counters and an exhaustive summary of possible

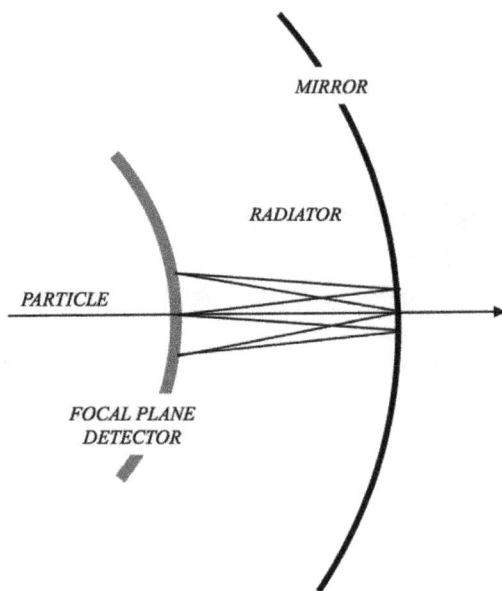

Figure 7.22:   Mirror focusing RICH.

radiators, windows and photoionization media, see Ypsilantis and Séguinot (1994) and Nappi and Séguinot (2005).

From the very beginning of their development, multiwire proportional chambers, filled with a photosensitive vapour, appeared to be a suitable device to amplify and detect single electrons and apt to instrument the large sensitive areas required by experiments.

The development of solid state, gas-compatible photocathodes has substantially changed the conception of RICH detectors, ensuring better performance and easier operation. Figure 7.23 shows schematically a RICH module, with the mirror reflecting the photons emitted by a charged particle onto an MWPC with the lower cathode coated with the photosensitive layer.

The CsI–MWPC RICH-1 detector of the COMPASS experiment at CERN consisted of a radiator vessel filled with C_4H_{10}, with an array of mirrors reflecting the Cherenkov photons on a matrix of eight CsI-coated MWPCs, Figure 7.24 (Albrecht *et al.*, 2003). The detector operated at CERN from 2003 to 2015, before undergoing a

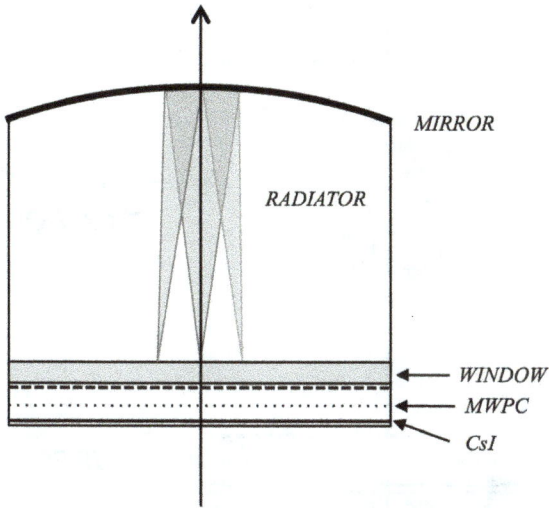

Figure 7.23: Schematic of a reflective CsI RICH module.

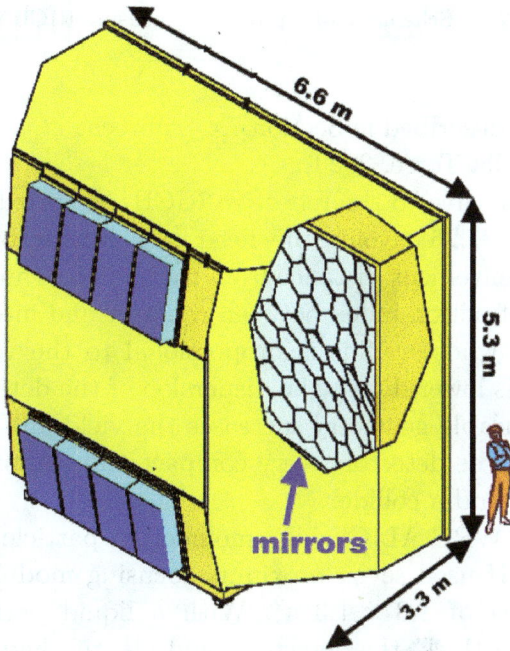

Figure 7.24: The COMPASS RICH-1 system.
Source: Albrecht *et al.* (2003).

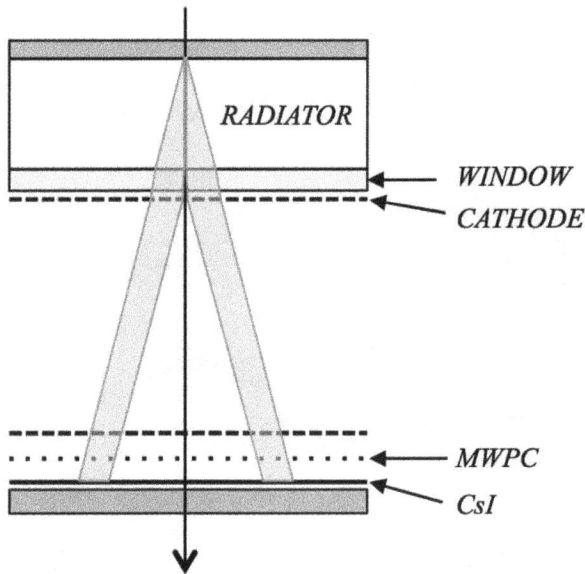

Figure 7.25:   Schematic of a proximity focusing RICH module.

major upgrade described in Section 7.4 (Albrecht *et al.*, 2003; Abbon *et al.*, 2011; Dalla Torre, 2020).

As an alternative to the reflective RICH, the proximity focusing counter, Figure 7.25, avoids the need of a mirror by placing the detector at a convenient distance from the radiator, and exploits the difference in refraction index between radiator and interface to form a ring pattern whose radius is proportional to the velocity. While the resolution is lower due to the dispersion of the detected photons, the more favourable geometry increases the yield and simplifies the construction of the detectors, very compact and compatible with an installation around a collider.

The CsI–MWPC ALICE high-momentum particle identification (HMPID) RICH used seven proximity focusing modules, each with a sensitive area of $1.3 \times 1.3\,\mathrm{m}^2$. With a liquid perfluoro hexane radiator, 15 mm thick, the detector could identify hadrons from few hundred MeV/c up to several GeV/c (Cozza *et al.*, 2003). Operated successfully at CERN's LHC between 2009 and 2018, the system is being upgraded to cope with the increased luminosity of the collider (Acconcia *et al.*, 2015; Volpe, 2020; Gauci *et al.*, 2020).

For a review of MWPC-based RICH counters using photosensitive vapours and solid state photocathodes, see Dalla Torre (2011) and Sauli (2014, Chapter 14).

7.4 MPGD-based Cherenkov Detectors

While meeting the particle identification requirements of the experiments, MWPC-based detectors appeared in the long term to suffer from high particle background, limited resolution and damages to the photosensitive layer due to discharges and positive ion feedback (Alexeev *et al.*, 2017; Hoedlmoser *et al.*, 2007). The development of micro-pattern gaseous counters, having better resolutions and being less prone to ageing has led to the redesign of the COMPASS RICH detector, replacing the MWPCs with MPGD sensors capable of withstanding the higher luminosity upgrade of the LHC.

Both MICROMEGAS and GEM have been demonstrated to efficiently detect and localize single photoelectrons ejected from a sensitive layer deposited on the cathode or on the top layer of the multiplying structure, as discussed in Section 7.2. Multi-GEM structures, Figure 7.15, are particularly suitable as they permit to attain large gains while reducing both the ion and photon feedbacks. Localization is performed recording the collected charge on patterned anode strips or pads (Sauli *et al.*, 2004).

With a CsI-coated TGEM, a single-photoelectron position accuracy around 160-μm FWHM (including the 100 μm source width) has been measured exposing the detector to a collimated UV photon beam in two positions, 200 μm apart, and computing event per event the centre of gravity of the charge recorded on a set of parallel strips at 200-μm pitch, Figure 7.26 (Meinschad *et al.*, 2004). Two-dimensional localization is performed using sets of perpendicular strips.

To resolve the ambiguities arising in the detection of several simultaneous photons, as in the case of RICH, the anode can be patterned with a matrix of pads, individually read out. An alternative

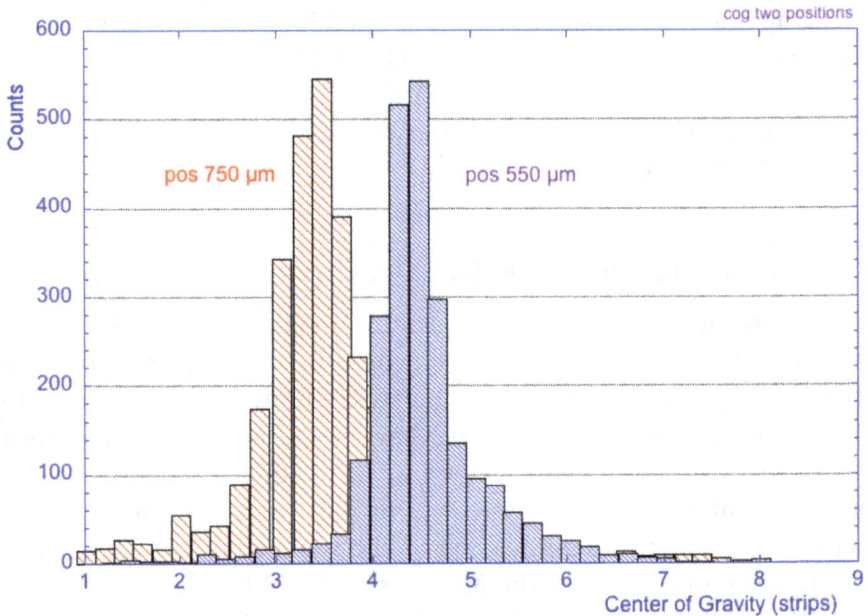

Figure 7.26:   Localization of two UV photon beams, 200 μm apart. *Source*: Meinschad *et al.* (2004).

scheme is to interconnect rows of hexagonal pads at 60° to each other, with the so-called hexaboard geometry, providing three projections for each event. Figure 7.27 shows an event with two detected photons recorded with the hexaboard pattern; the two-point resolution is around a millimetre (Sauli *et al.*, 2004).

A systematic research comparing various geometries led for the COMPASS RICH upgrade to the choice of a hybrid design, with two cascaded TH-GEM multipliers followed by a MICROMEGAS as a final stage of amplification, Figure 7.28 (Agarwala *et al.*, 2018). This geometry ensures stable operation at the high gains needed for single-photoelectron detection, while reducing to about 3% the ion feedback to the CsI photosensitive layer deposited on the top electrode of the first multiplier (Di Mauro, 2014; Tessarotto, 2017). The TH-GEMs have 400-μm holes at 800-μm pitch, and are manufactured by mechanical drilling on a metal-coated 400-μm-thick

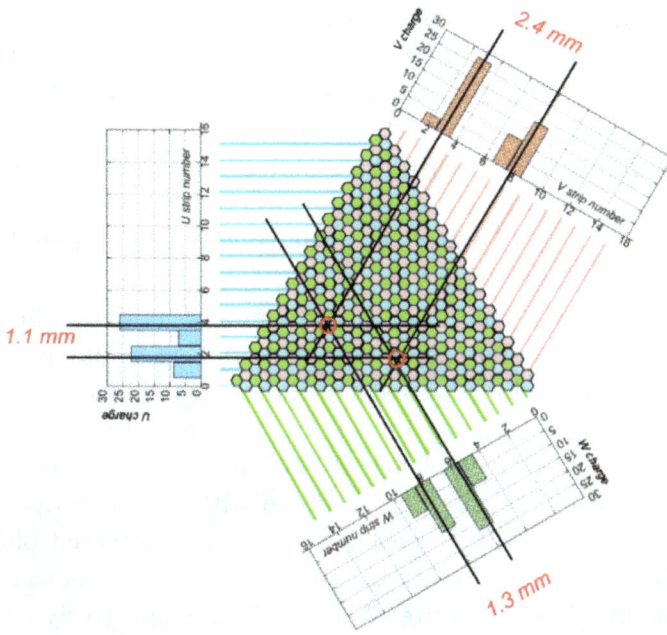

Figure 7.27: A two-photon event recorded with the hexaboard readout pattern.
Source: Sauli (2005).

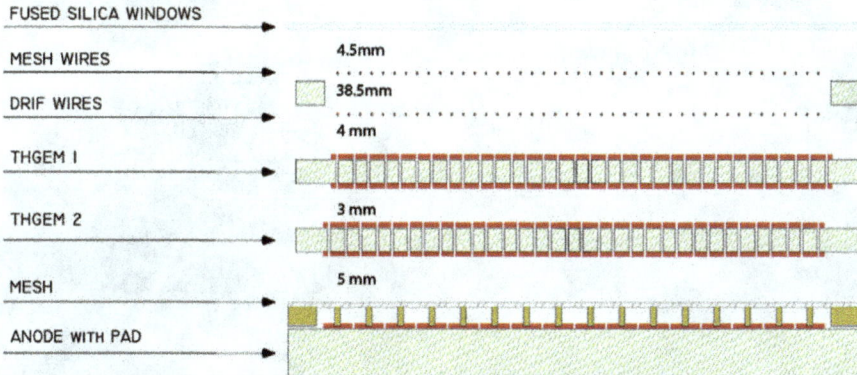

Figure 7.28: Schematic of the COMPASS RICH-1 hybrid MPGD.
Source: Tessarotto (2017).

printed circuit board. A rimless execution ensures minimal gain shifts due to charging up (Alexeev *et al.*, 2012) (see Sections 5.4 and 5.8 of Chapter 5). An offset between the holes in the two GEMs helps in reducing the ion feedback. The MICROMEGAS final amplifier, manufactured at CERN with the bulk technology, has a 128-μm gap and 18-μm woven stainless steel wire mesh cathode with 63-μm pitch. An array of 300-μm-diameter insulating pillars at 2-mm pitch ensures the gap uniformity.

Each detector is composed of two identical modules, $60 \times 30 \text{cm}^2$, Figure 7.29; the charge signals are collected on a matrix of square pads, at 8-mm pitch. Four hybrid photodetectors of the described design have been installed on the existing radiator, replacing the MWPC, and operated in the 2016 runs. Figure 7.30 shows an example of the Cherenkov ring recorded with the upgraded detector (Agarwala *et al.*, 2018); the average number of detected photons as a function of the Cherenkov angle and the single-photon angular resolution are given in Figures 7.31 and 7.32, respectively (Agarwala *et al.*, 2019).

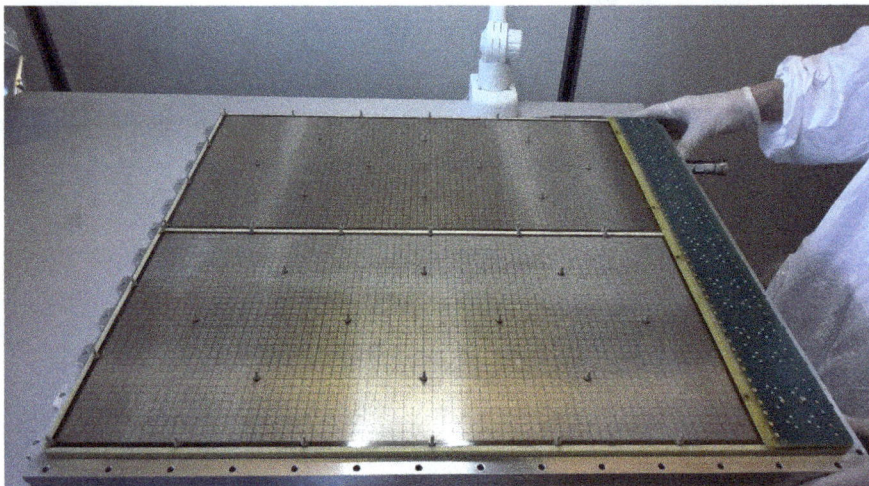

Figure 7.29:   A double hybrid module during assembly; the pillars ensuring the gap thickness between MICROMEGAS and the second TH-GEM.

Source: Agarwala *et al.* (2019).

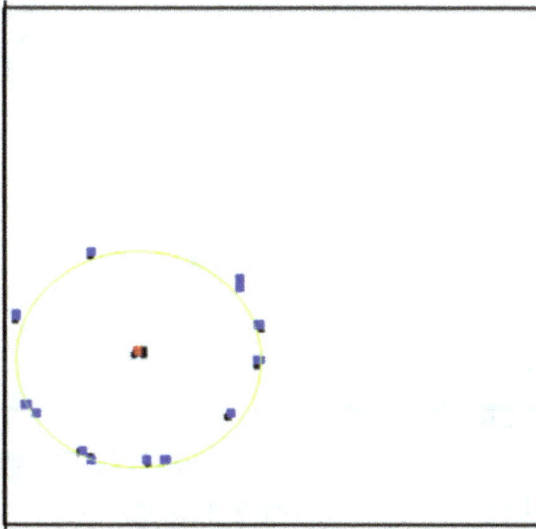

Figure 7.30: Example of Cherenkov ring recorded with the COMPASS RICH-1 hybrid detector.

Source: Agarwala *et al.* (2018).

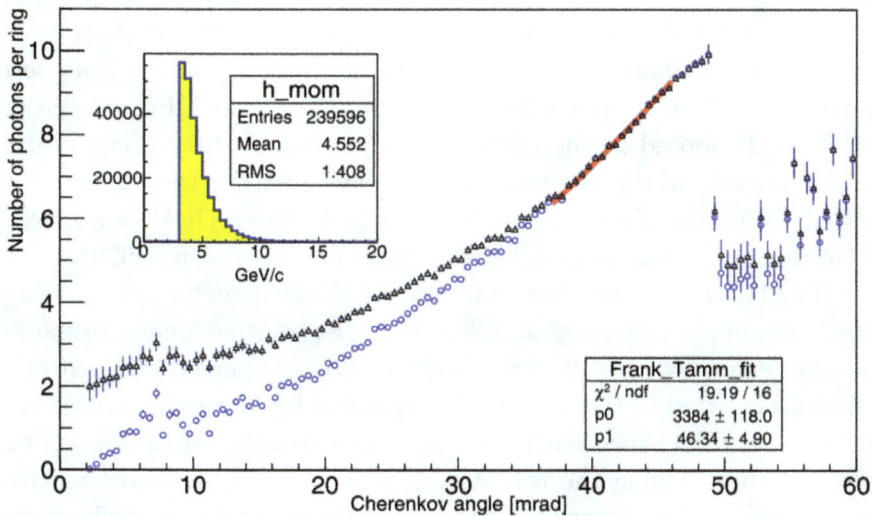

Figure 7.31: Number of detected photons as a function of Cherenkov angle. The open dots result from a statistical correction to the data.

Source: Agarwala *et al.* (2019).

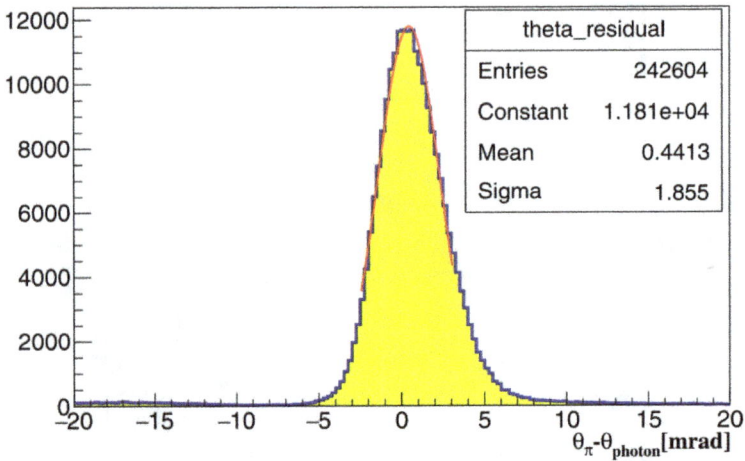

Figure 7.32:   Single-photon angular resolution.
Source: Agarwala *et al.* (2019).

7.5   Hadron Blind Detectors

The development of gaseous position-sensitive detectors with high-efficiency solid photocathodes opened the possibility to build windowless threshold Cherenkov counters sensitive to electrons with a good rejection of heavier particles, the so-called hadron blind detectors (HBDs). Proposed in the early nineties of last century (Giomataris and Charpak, 1991), the concept was developed for the experiment PHENIX at the Relativistic Heavy Ions Collider (RHIC) at BNL (Aidala *et al.*, 2003; Fraenkel *et al.*, 2005; Tserruya *et al.*, 2020).

The HBD consists of a cylindrical TPC-like detector surrounding the beam intersection region, filled with CF_4 at atmospheric pressure as Cherenkov radiator, Figures 7.33 and 7.34 (Anderson *et al.*, 2011). Photons emitted in the 50-cm-thick radiator by charged particles are detected and localized with a set of three cascaded GEM amplifiers, with the first, facing the radiator, coated with a CsI photosensitive layer, Figure 7.35 (Tserruya, 2006); the photons are contained within a wide spot, or blob, of a radius depending on the Cherenkov angle. A reverse electric field in the TPC collects the direct ionization charge; the photon-generated amplified signal is recorded on a matrix

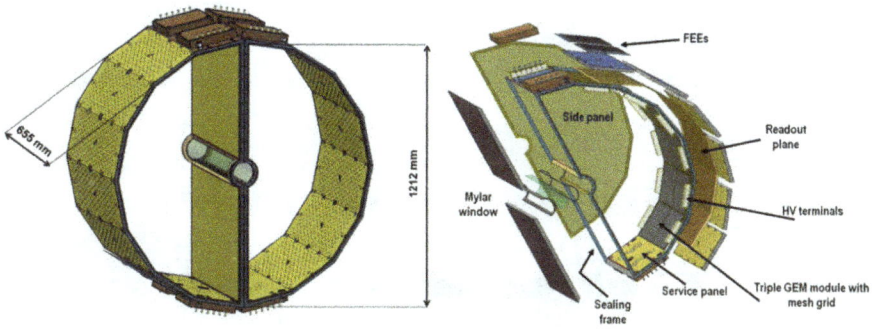

Figure 7.33:   Schematic of the HBD.
Source: Anderson *et al.* (2011).

Figure 7.34:   A sector of the HBD in the glove box before installation on the radiator.
Source: Anderson *et al.* (2011).

of hexagonal pads with an area of $\sim$6 cm^2 each, comparable with the area of the Cherenkov blob.

A photon background due to the particle-induced scintillation of CF$_4$ overlaps with the true signal (Hesser and Dressler, 1967;

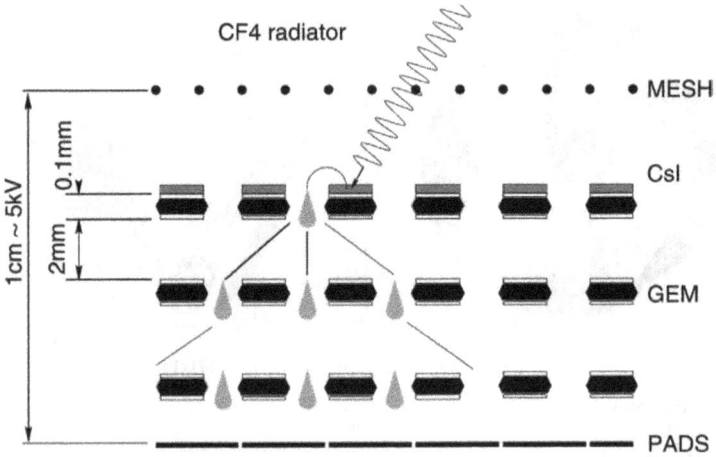

Figure 7.35:   Schematic of the CsI-coated TGEM photon detector. *Source*: Tserruya (2006).

Pansky *et al.*, 1995); methods to reduce this background with shadings or with the addition of small quantities of absorbers were investigated (Aidala *et al.*, 2003)(Blake *et al.*, 2015) but not adopted in the final design. For a detailed analysis of the primary and field-induced scintillation spectra of CF_4, see Chapter 11.

In the range of energies covered by the experiment, electrons and positrons with momenta above the Cherenkov threshold of the radiator emit UV photons that are detected by the photosensitive GEM chambers, while hadrons are below threshold. Due to the reverse electric field configuration, only a small fraction of the direct ionization, released close to the GEMs, is detected. Figure 7.36 shows the recorded charge spectra for hadrons, with direct (TPC-like) and reverse drift field, while Figure 7.37 shows the charge spectrum recorded for $e^+ - e^-$ pairs. The hadron-electron rejection factor depends on the threshold on the signal amplitude, Figure 7.38.

Similar in conception, the HBD developed for the J-PARK E16 experiment to discriminate electrons from pions makes use of a CsI-coated TGEM detector with finely segmented pad readout, capable of resolving the circular shape of the Cherenkov blob and

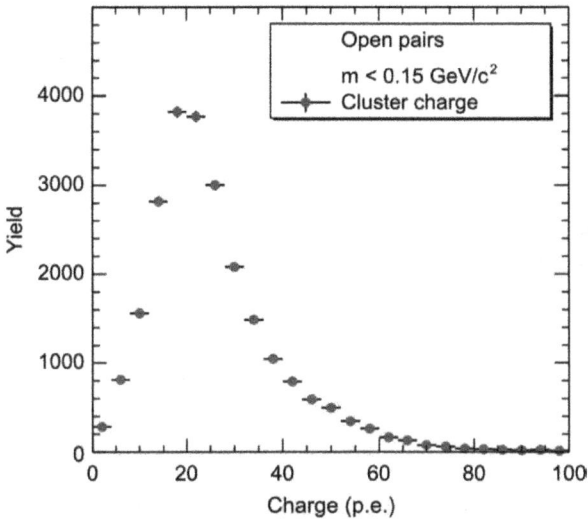

Figure 7.36: Recorded charge, expressed in number of photoelectrons, for electrons and positrons.

Source: Anderson *et al.* (2011).

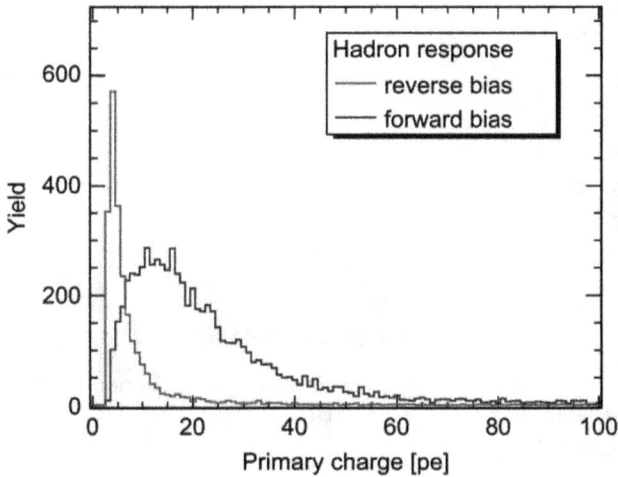

Figure 7.37: Charge signal for hadrons, in the reverse (smaller peak) and forward drift voltage configurations.

Source: Anderson *et al.* (2011).

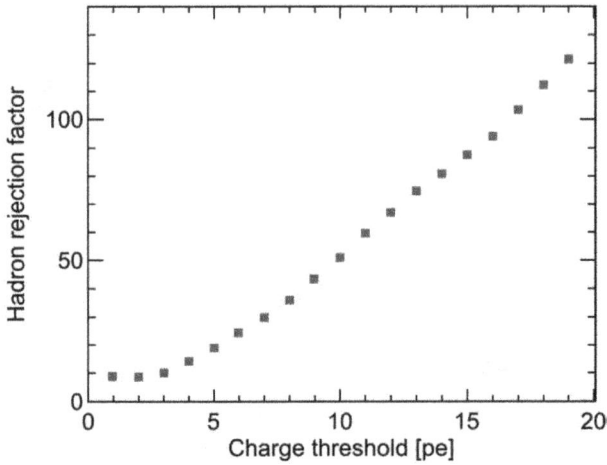

Figure 7.38:   Hadron rejection factor as a function of charge threshold.
Source: Anderson *et al.* (2011).

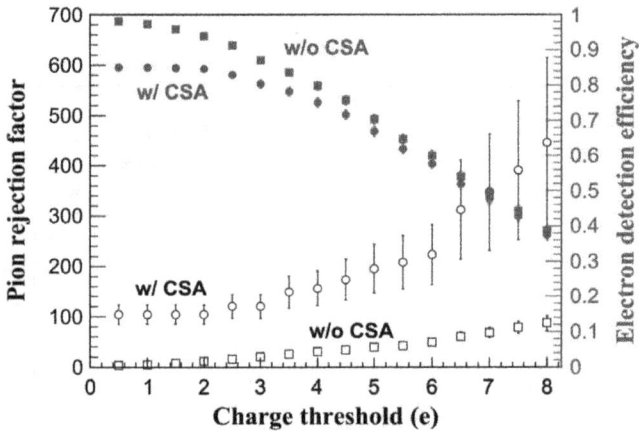

Figure 7.39:   Electron detection efficiency and pion rejection factor as a function of charge threshold for the J-PARK HBD.
Source: Kanno *et al.* (2016).

improving the particle identification power (Kanno *et al.*, 2016). Figure 7.39 shows the electron detection efficiency and pion rejection factor as a function of the charge threshold, deduced with a cluster analysis exploiting the charge sharing information from the pads.

Chapter 8

Advanced Micro-Pattern Devices

8.1 Introduction

Sprouted by the increasingly demanding experimental requirements and the interest of numerous worldwide detector research groups, many novel MPGD designs or improvements of previous devices have been pursued. They are described here loosely in order of their introduction. As the field is in continuing evolution, the reader is encouraged to peruse the literature for the most recent updates on the various devices.

8.2 Micro-Hole and Strip Plate

As reviewed in Chapter 2, the first successful application of the gas electron multiplier (GEM) was as charge pre-amplifier for cascaded GEM-MSGC detectors, providing large safe gains while preserving the good rate and localization properties of the MSGC. A structure combining the two functionalities in a single device, named micro-hole and strip plate (MHSP) was introduced soon after, aiming at the reduction of the positive ion backflow in photosensitive detectors, as discussed in Chapter 7 (Veloso *et al.*, 2000). As shown in Figure 8.1, the MHSP detector has a GEM-like electrode with one side patterned with thin anodes surrounded by cathode strips, much as in the MSGCs. Ionization electrons released in the upper region drift into the holes and multiply in the high field; the charge is further amplified and collected on anode strips on the bottom side of the electrode. In analogy with the MSGC operation, a large part of the positive

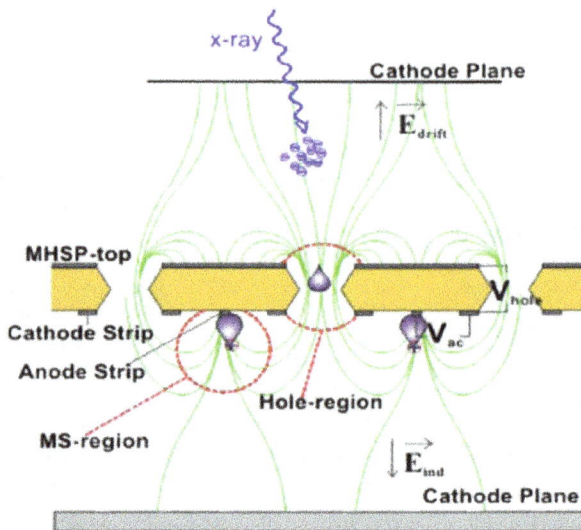

Figure 8.1:   Schematic diagram of the MHSP.
Source: Veloso *et al.* (2007).

ions is collected by the neighbouring cathode strips, thus effectively reducing the fractional ion backflow into the drift gap. Figure 8.2 shows the measured fraction of ions propagating through the holes as a function of gain for several values of the anode strips voltage (Maia *et al.*, 2004). The fast ion clearing also ensures a high rate capability, Figure 8.3 (Veloso *et al.*, 2007). Two-dimensional localization is achieved recording the induced signals on the anodes and on the strips of a facing cathode, or engraving readout strips perpendicular to the anodes on the MHSP top.

Combined with one or more GEM pre-amplification electrodes, the MHSP permits to reach the high gains needed for detection of single electrons in gaseous photomultiplier structures, while ensuring a substantial reduction in the number of ions reaching the photocathode (Breskin *et al.*, 2010). The use of the MHSP to reduce the ions' backflow in the time projection chambers (TPCs) was described in Section 6.5.

More complex, and reminiscent of previous developments by the CERN group (Charpak *et al.*, 1980), the photon-assisted cascaded

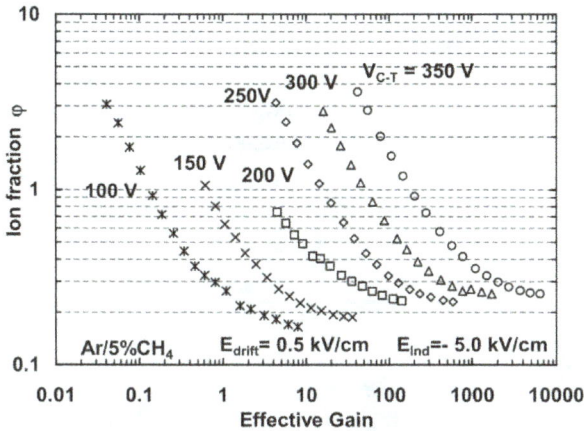

Figure 8.2:   Fractional ion backflow through the MSHP as a function of gain for several values of the anode voltage.

Source: Maia *et al.* (2004).

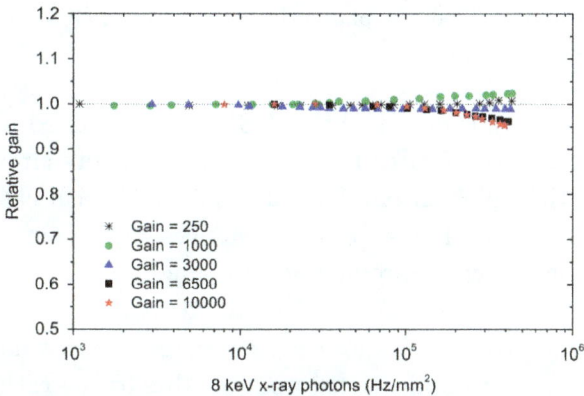

Figure 8.3:   Normalized gain as a function of rate for several values of the MHSP gain.

Source: Veloso *et al.* (2007).

electron multiplier (PACEM) relies on the transfer of the signals from the charge amplification section to a second device, mediated by ultraviolet scintillation photons emitted by xenon, Figure 8.4 (Veloso *et al.*, 2007). With this design, the ion backflow hitting the photocathode is completely eliminated.

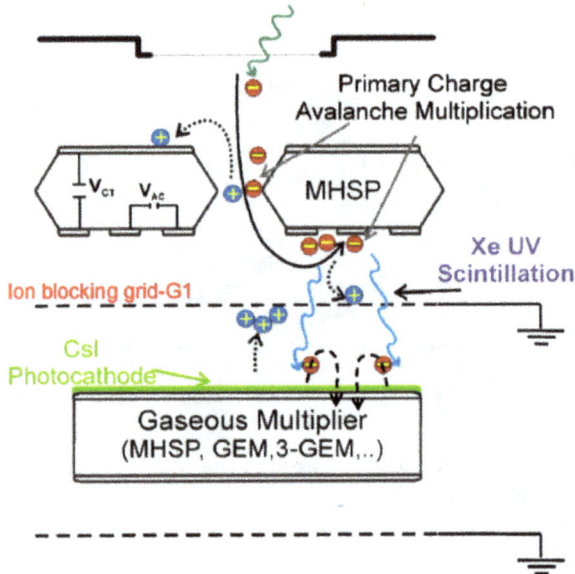

Figure 8.4:   Schematic of the photon-assisted cascaded electron multiplier. *Source*: Veloso *et al.* (2007).

A coarser evolution of the MHSP, the so-called COBRA or thick-COBRA (TH-COBRA) structure, implements circular anodes surrounding the holes on one face of a THGEM with cathode strips winding in between, Figure 8.5; it achieves high gains and strong positive ion backflow suppression with a single device (Amaro *et al.*, 2010; Veloso *et al.*, 2011; Terasaki *et al.*, 2013).

While very promising, these sophisticated devices have received a limited amount of applications, possibly due to the rather intricate manufacturing process. A noticeable exception is the detectors used for energy dispersive X-ray fluorescence (EDXRF) analysis used for material sample analysis and cultural heritage investigations. Figure 8.6 shows the imaging system making use of a two-dimensional MSHP (Veloso and Silva, 2018). X-rays in the 5-10-keV range from an X-ray generator illuminate the sample under analysis; lower-energy fluorescence photons emitted by the sample convert to charge in the xenon-filled drift tube and are recorded by the detector in position and energy. A selection on the two-dimensional image based on the

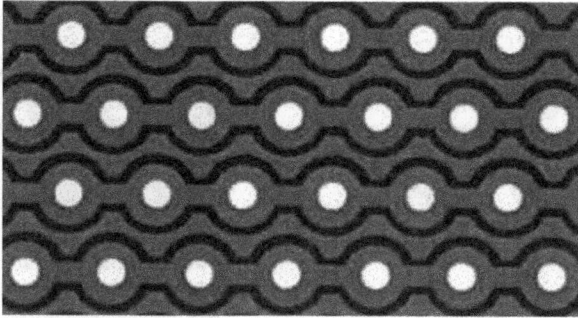

Figure 8.5:   Structure of the TH-COBRA electrode.
Source: Veloso *et al.* (2011).

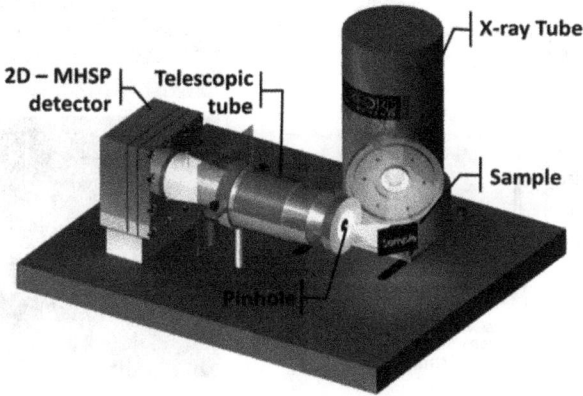

Figure 8.6:   Setup for the EDXRF imaging..
Source: Silva *et al.* (2013).

region of interest in the recorded X-ray spectra permits to map the elemental distribution in the sample (in this case a tooth), Figure 8.7 (Silva *et al.*, 2013).

With a similar setup, making use of a triple GEM with electronic two-dimensional readout, Zielińska *et al.* (2013) demonstrated the performances of imaging gaseous detectors for the analysis of pigments in paintings for cultural heritage studies.

Intrinsically simpler to implement and operate, systems based on the optical GEM readout have been developed in view of applications in the same fields, see Chapter 11.

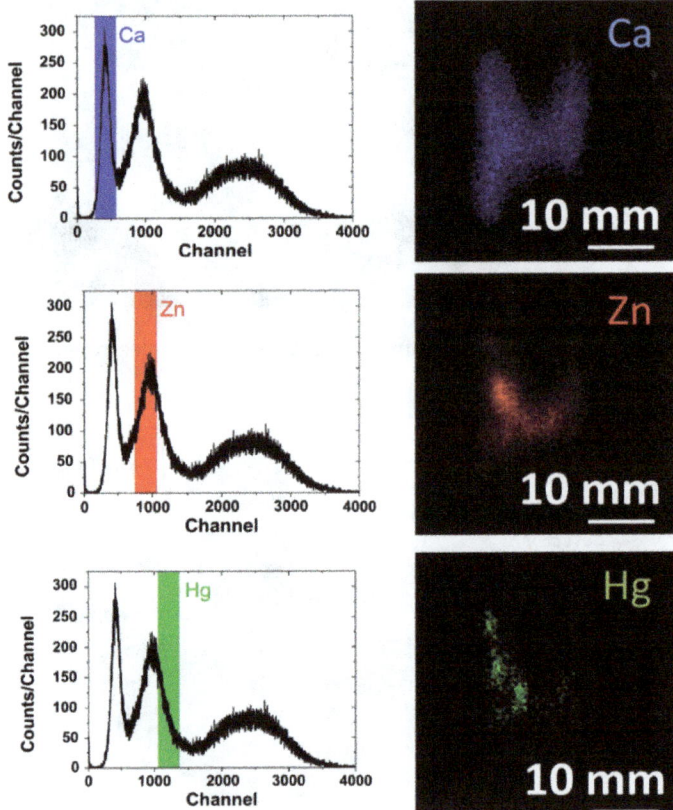

Figure 8.7:   EDXRF image of a tooth; a selection of different regions in the X-ray recorded spectra permits the positional identification of the constituents.
Source: Silva *et al.* (2013).

8.3   Micro-Pixel Detector

A scaled-up version of the MICRODOT, the micro-pixel (μ-PIC) detector is manufactured on a thin printed circuit board, with a typical cell size of $400\,\mu$m, Figures 8.8 and 8.9 (Ochi *et al.*, 2001). Parallel cathode strips on the top side of the support are etched to obtain rows of circular anode pixels, connected through the board to a set of parallel readout strips on the back side of the circuit. On application of a voltage between anode and cathode, each pixel acts as an individual counter. In detecting soft X-rays, the μ-PIC has

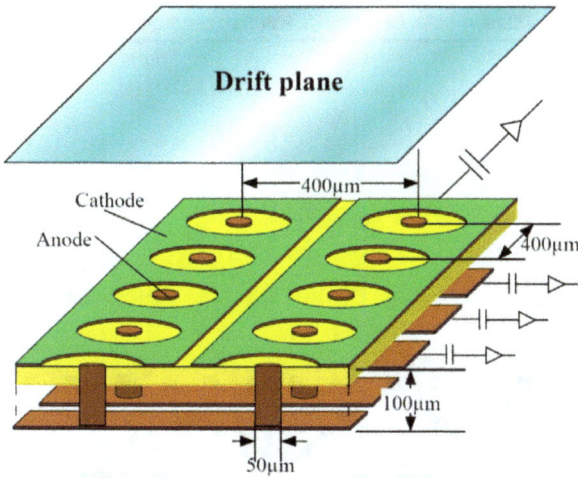

Figure 8.8:   The micro-pixel detector (μ-PIC).
Source: Ochi *et al.* (2001).

Figure 8.9:   Close-up view of the μ-PIC detector.
Source: Ochi *et al.* (2001).

good energy resolution and can reach gains close to 10^4. As in other detectors having insulating surfaces between anodes and cathodes, the gain increases reaching a constant value after the initial power-up (Ochi *et al.*, 2002). Owing to the technology employed, the device

Gas: Ar90%, CH$_4$ 10%, T = 300 K, p = 1 atm

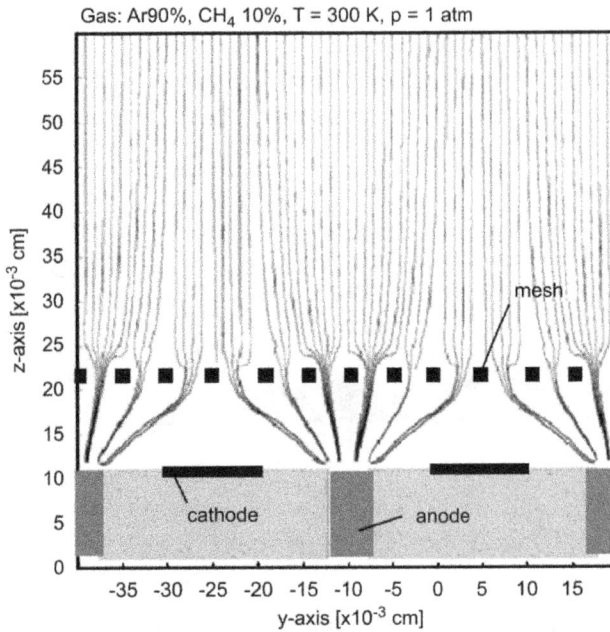

Figure 8.10:   The micro-mesh micro-pixel chamber (M³-PIC).
Source: Ochi *et al.* (2009).

can be realized at moderate costs, making it a suitable candidate for large-area detectors, as the TPCs (Nagayoshi *et al.*, 2003; Kubo *et al.*, 2003).

Addition to the μ-PIC of a micro-mesh a few hundred micron above the structure, in the so-called micro-mesh micro-pixel chamber (M³-PIC), Figure 8.10, permits to pre-amplify the electron charge, reaching overall gains well above 10⁴. The geometry also helps in reducing the backflow of ions from the anodes into the drift volume to about 0.5% for moderate drift field values (Ochi *et al.*, 2009).

As many other MPGDs, the μ-PIC is easily damaged by sparks. To reduce the discharge probability, various methods to coat or replace the metallic cathode with a resistive layer have been developed. An example of a structure combining a DLC cathode and internal embedded electrodes is shown in Figure 8.11; the thin insulating layers between electrodes permit a capacitive signal

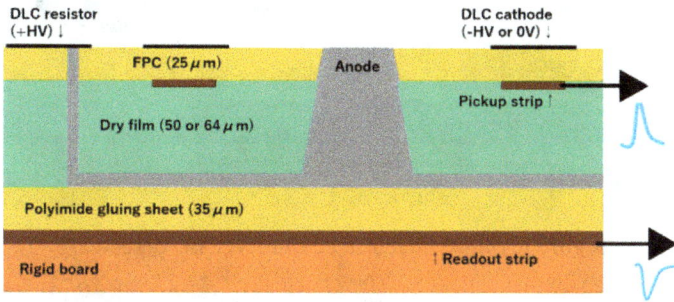

Figure 8.11: Schematic of a multi-layer implementation of a μ-PIC device with DLC resistive cathodes.

Source: Yamane *et al.* (2020).

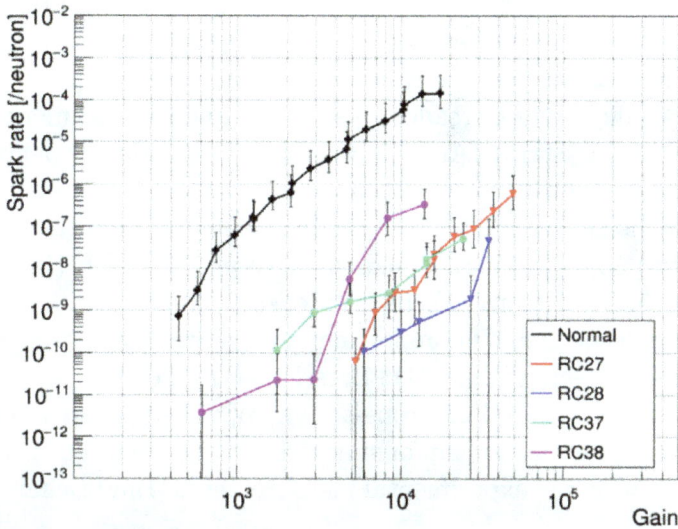

Figure 8.12: Normalized spark rate measured in a high-intensity neutron beam for the normal (highest points) and several models of the resistive DLC μ-PIC detector.

Source: Yamane *et al.* (2020).

readout, avoiding the need for external capacitors. The spark rate of several models of resistive μ-PIC detectors, measured in a neutron beam, is compared in Figure 8.12 with the result obtained with a standard device (Yamane *et al.*, 2020).

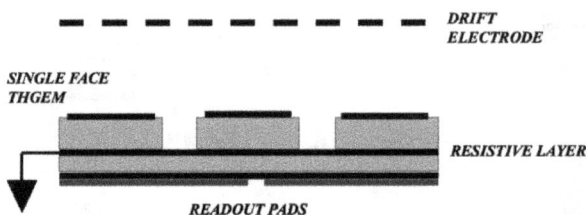

Figure 8.13:   Schematic of the RWELL detector.

8.4   Resistive WELL Devices

Related to the original CAT and WELL structures described in Chapter 3, the resistive WELL (RWELL) detector consists of a single-face thick GEM plate coupled through a thin high-resistivity layer to a segmented readout anode, Figure 8.13; in a modified version named segmented resistive WELL (SRWELL), to reduce the cross talk between readout pads, the anode is segmented with thin conducting strips (Arazi *et al.*, 2014). The rate dependence of gain measured with the various structures, compared to a standard TH-GEM, is shown in Figure 8.14. Implemented on a thicker, high-resistivity support and renamed resistive-plate WELL (RPWELL), Figure 8.15, the detector has been extensively tested in minimum ion-izing particle beams, demonstrating good efficiency and operational robustness (Moleri *et al.*, 2017; Bressler *et al.*, 2020).

The signal formation processes in WELL-type structures has been analysed with the help of simulation programs, showing a good agreement with the experimental measurements (Bhattacharya *et al.*, 2019).

8.5   Micro-Resistive Well

The micro-resistive WELL (μ-RWELL) detector, shown in Figure 8.16, is manufactured on a copper-clad polyimide foil 50 μm thick with the photolithographic techniques developed for GEM detectors to produce a structure with small metallic dots at the bottom of the wells. A high-resistivity layer, typically of 100 MΩ/square, separates the active structure from the readout electrodes. Tested in the

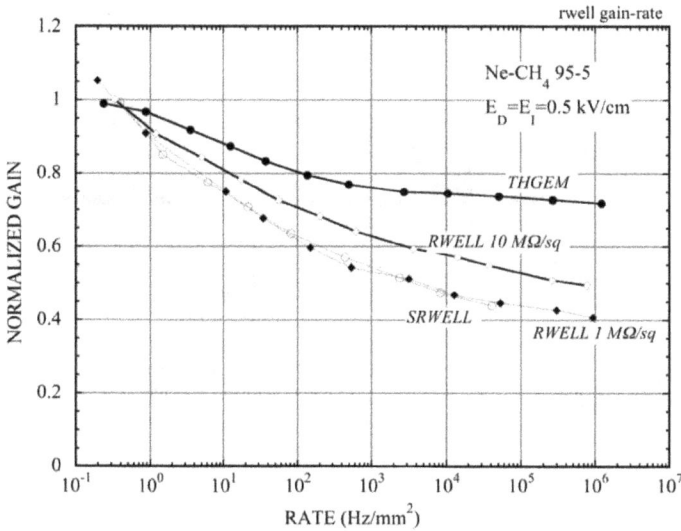

Figure 8.14: Rate dependence of gain for a standard TH-GEM, two RWELL with different surface resistivities and a SRWELL.

Source: Data from Arazi *et al.* (2014).

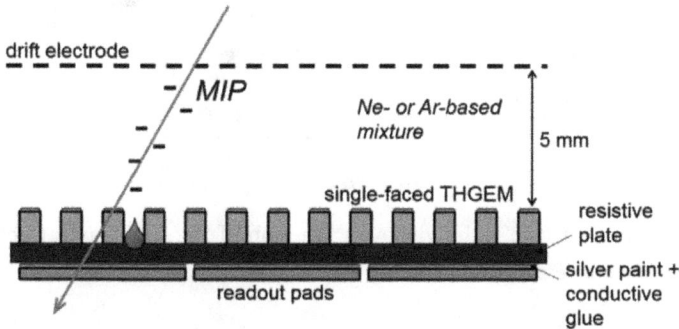

Figure 8.15: Schematic diagram of the RPWELL detector.

Source: Moleri *et al.* (2017).

laboratory with exposure to a soft X-ray generator, the detector has a constant gain up to a flux around $5.10^5 \, \mathrm{Hz\,cm}^{-2}$ without detectable discharges (Bencivenni *et al.*, 2015; Poli Lener *et al.*, 2017).

Further developments using sputtered DLC layers of different thicknesses permit to cover a range of resistivity spanning many

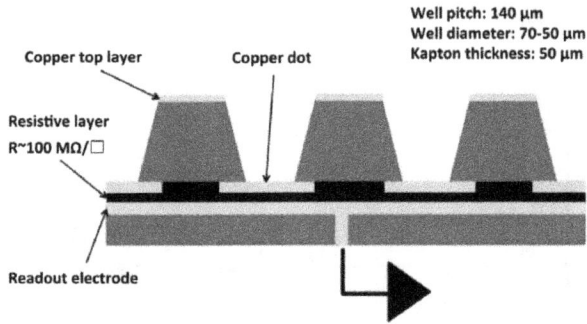

Figure 8.16:   The μ-RWELL detector.
Source: Bencivenni *et al.* (2018).

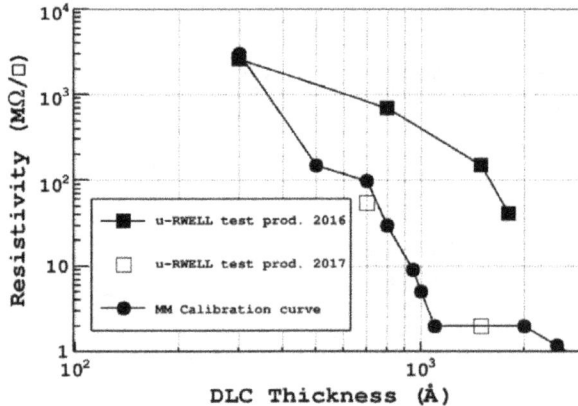

Figure 8.17:   Surface resistivity as a function of the DLC layer thickness.
Source: Bencivenni *et al.* (2018).

orders of magnitude (Figure 8.17); the detector gain dependence from soft X-ray flux is shown in Figure 8.18 for two values of surface resistivity of the anode coating (80 and 100 MΩ/square). Measured at a gain around 3,000 for the larger value of resistivity, Figure 8.19 shows the rate dependence of relative gain for a range of sizes of the source collimator; as expected, a greater drop is observed for wide beams, due to the potential drop induced by larger currents on the resistive layer.

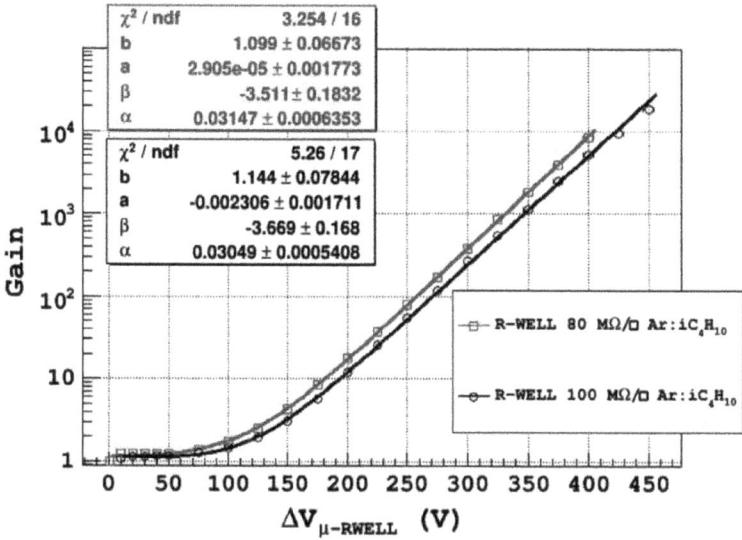

χ^2 / ndf	3.254 / 16
b	1.099 ± 0.06673
a	2.905e-05 ± 0.001773
β	-3.511± 0.1832
α	0.03147 ± 0.0006353

χ^2 / ndf	5.26 / 17
b	1.144 ± 0.07844
a	-0.002306 ± 0.001711
β	-3.669 ± 0.168
α	0.03049 ± 0.0005408

Figure 8.18: μ-RWELL gain as a function of voltage for 80 and 100 MΩ/square DLC layer resistivity.

Source: Poli Lener *et al.* (2017).

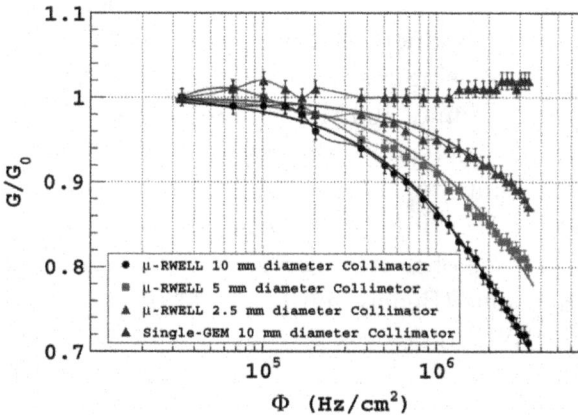

Figure 8.19:   Relative gain versus X-ray flux of the μ-RWELL detector for several values of the collimated beam width. The topmost set is the response of a standard single-GEM device.

Source: Poli Lener *et al.* (2017).

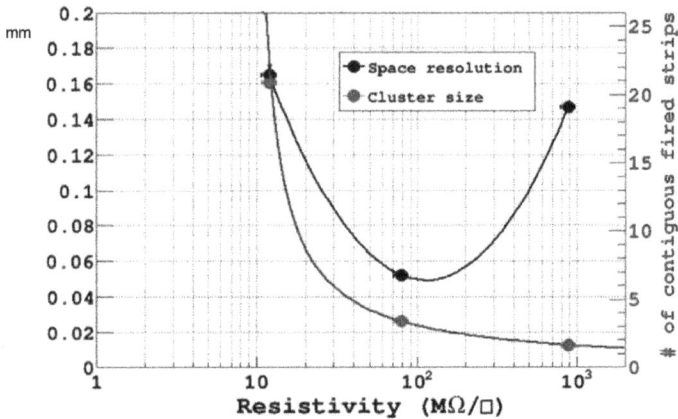

Figure 8.20:   Space resolution (left scale, in mm) and cluster size as a function of DLC resistivity for the μ-WELL.

Source: Bencivenni *et al.* (2018).

The value of the resistive layer affects the lateral spread of the detected charge on the anode, or cluster size, and indirectly the localization accuracy. Measured exposing a set of μ-RWELL chambers to a charged particle beam, the number of readout strips at 400-μm pitch with a detected charge over threshold (cluster size) decreases with increasing resistivity, while the localization accuracy has a minimum around $100\,M\Omega$/square, Figure 8.20 (Bencivenni *et al.*, 2018). The detection efficiency measured for fast particle beams perpendicular to the device is shown in Figure 8.21 as a function of gain. Figure 8.22 is a compilation of measurements of the rate dependence of normalized gain, for several detector design and beam conditions (Bencivenni *et al.*, 2020).

A set of μ-RWELL devices is being considered as tracker for the high-rate upgrade of the LHCb muon detector at CERN (Morello *et al.*, 2019).

8.6   Plasma Diagnostics

The very high rate capability of GEM-based detectors permits to perform two-dimensional imaging of the intense X-ray emission for fusion plasma diagnostics; images of X-ray emission have been

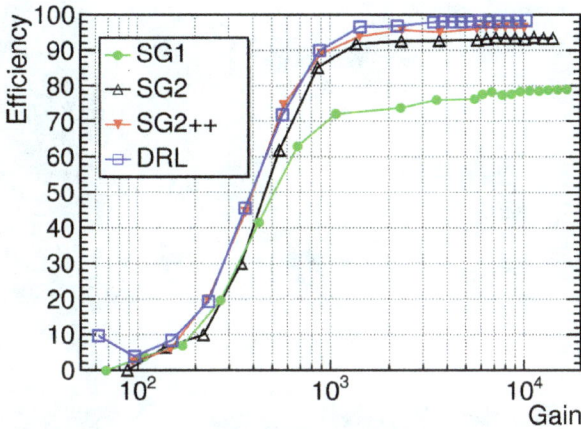

Figure 8.21: μ-RWELL detection efficiency as a function of gain for several detector designs.

Source: Bencivenni *et al.* (2020).

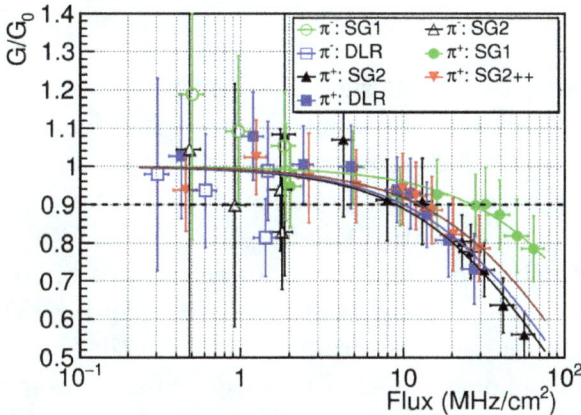

Figure 8.22: μ-RWELL normalized gain as a function of particle flux for several detector designs.

Source: Bencivenni *et al.* (2020).

recorded with a pinhole GEM camera on a matrix of readout pixels, $2\,\text{mm}^2$ each, at rates exceeding $4\,\text{MHz}$ per pixel (Pacella *et al.*, 2001). Further work has extended the use of the device to a mixed photon and neutron field, common around a fusion experimental setup;

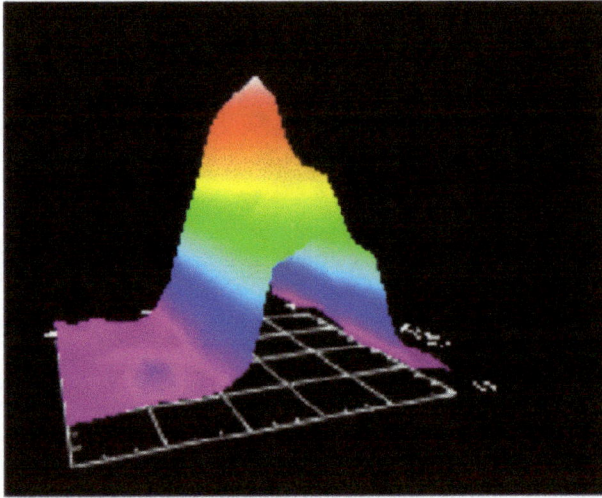

Figure 8.23:   Two-dimensional plot of soft X-rays plasma activity.
Source: Pacella *et al.* (2013).

indeed, inherent to their design, gas detectors have a sensitivity to
neutrons several orders of magnitude lower than for soft X-rays (see
Chapter 9). Figure 8.23 is an example of a two-dimensional image
of the plasma source in the presence of an estimated neutron and
gamma flux above 10^6 MHz (Pacella *et al.*, 2013).

A TGEM pinhole camera with an active area of $10 \times 10 - cm^2$
and 12×12-pixel readout was used to record two-dimensional
images of high-temperature toroidal plasmas at the Korea Institute of
Science and Technology (KAIST). The detector works in the photon-
counting mode, and can acquire up to 60 kframes at 1-kHz sampling
frequency. Figure 8.24 shows examples of tangential images of plasma
shots recorded placing the GEM detector in various positions; the
computed magnetic flux surfaces are superimposed in the images
(Song *et al.*, 2016). A similar device, using a TGEM detector with
one-dimensional projective strip readout has been developed to
monitor the plasma radiation emitted by the Joint European Torus
(JET) (Rzadkiewicz *et al.*, 2013; Chernyshova *et al.*, 2018).

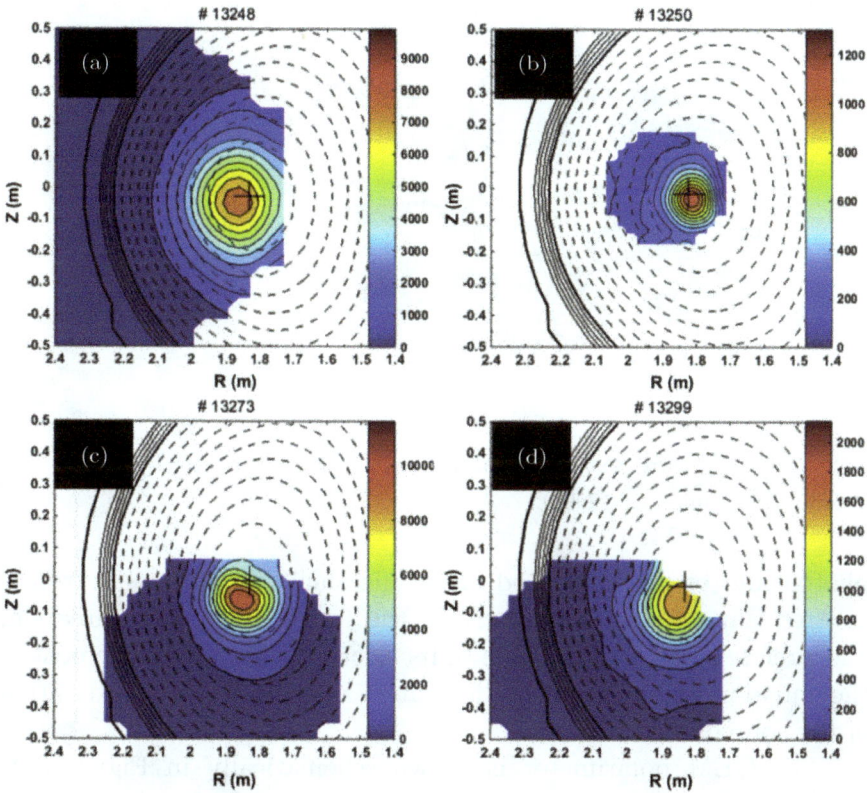

Figure 8.24: Recorded plasma images and computed magnetic flux lines. *Source*: Song *et al.* (2016).

8.7 X-ray Polarimeters

The degree of polarization of X-rays emitted by celestial sources is important information for the investigation of their nature, not widely used in astrophysics due to the inefficiency of existing instrumentation. Micro-pattern detectors, with their high efficiency and tracking capabilities, are a promising tool to measure the polarization of X-rays in the keV energy range. In this region, the interaction of a photon with a gas molecule results in the emission of an electron with an energy corresponding to the difference between that of the incident

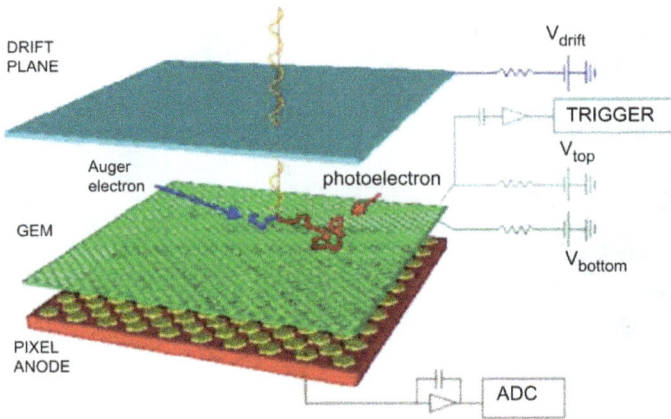

Figure 8.25:   Schematic of the GEM polarimeter.
Source: Bellazzini *et al.* (2003).

photon and of the interested electronic shell. The photoelectron is emitted almost perpendicularly to the incoming photon direction, and preferentially in the plane of its electric field; a measurement of the emission angle provides information on the average polarization of the source.

The GEM polarimeter is shown schematically in Figure 8.25 (Bellazzini *et al.*, 2003). It has a conversion and drift volume where the photon interaction takes place, and a GEM electrode to amplify the released photoelectrons. In the early prototypes, the readout pattern consisted of a matrix of hexagonal pads, at 200-μm pitch, individually connected to charge recording electronics; in a further development, named gas pixel detector (GPD), the discrete readout was replaced by a custom-made solid-state chip, directly collecting and encoding the charge on ~50-μm hexagonal sensors, Figure 8.26. Matching the readout pitch, and to ensure the best position accuracy, the GEM electrodes are realized on thin, metal-coated polymer foils with laser-drilled holes at 50-μm pitch (Bellazzini *et al.*, 2013).

Recorded in a helium-DME gas mixture, optimized for a long photoelectron path and low diffusion, Figure 8.27 shows an example of electron track released by 5.9-keV X-rays and scattering in the gas; the width of the pixels in the image is proportional to recorded

Figure 8.26: Exploded view of the GPD assembly.
Source: Bellazzini *et al.* (2013).

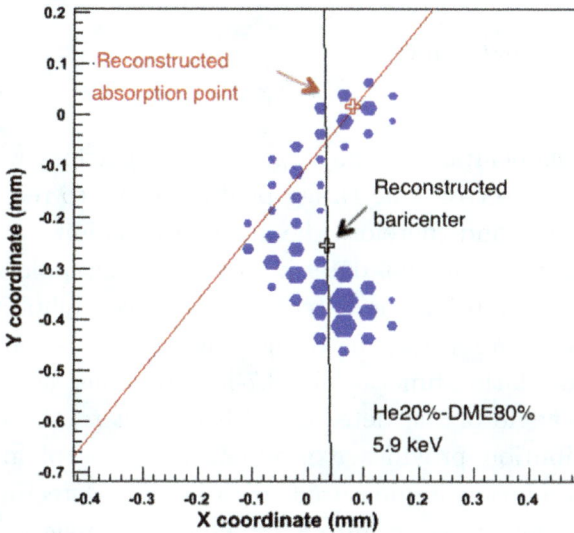

Figure 8.27: Example of a recorded 5.9-keV photoelectron track.
Source: Bellazzini *et al.* (2013).

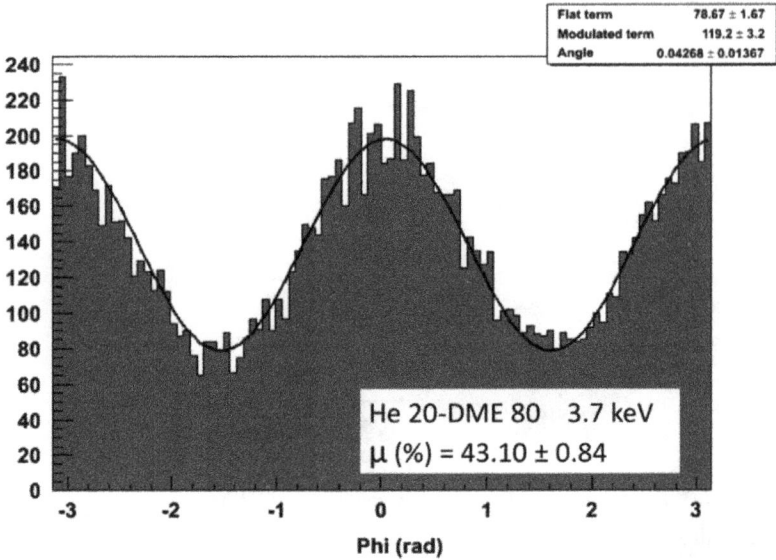

Figure 8.28: Modulation of the reconstructed angles for polarized 3.7-keV X-rays.

Source: Bellazzini *et al.* (2013).

charge. The association of the largest ionization loss with the end of range of the electron (the Bragg peak) permits to reconstruct the interaction point and therefore the angle of emission.

The resolution in the determination of the polarization has been measured exposing the detector to polarized photon beams at the IASF-INAF laboratory in Rome. Figure 8.28 is an example of measured modulation function for 3.7-keV photons, as a function of the rotation angle of the detector with respect to the beam. A fit to the distribution provides a modulation factor of around 43%. The space resolution is measured exposing the detector to a well-collimated X-ray source. Figure 8.29 shows the image of the source recorded in three positions, $300\,\mu m$ apart; the distributions have a FWHM around $80\,\mu m$, including the estimated source width of $30\,\mu m$.

A set of three identical X-ray telescopes with GPDs in the focal plane is foreseen for installation on the NASA Imaging X-ray

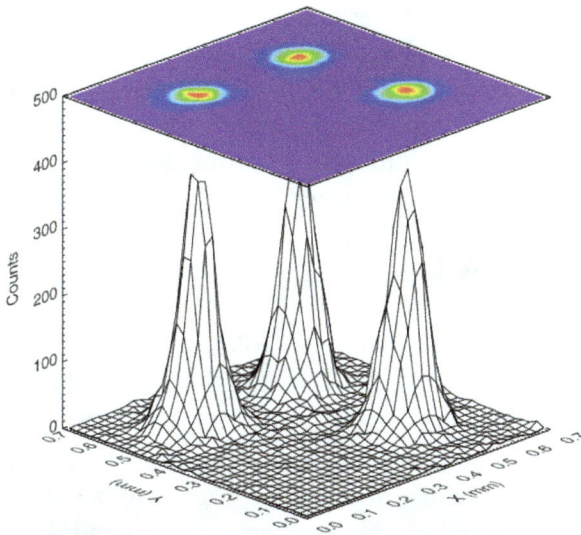

Figure 8.29:   Space distribution measured for three positions of the collimated X-ray source, 300 μm apart.

Source: Li *et al.* (2015).

Polarimeter Explorer (IXPE), planned for launch in 2021 (Weisskopf *et al.*, 2016).

Algorithms for the track reconstruction and angle determination in MPGD-based polarimeters are discussed in Kitaguchi *et al.* (2018).

Conceived to instrument one of the small explorer (SMEX) missions, the polarimeter for relativistic astrophysical X-ray sources (PRAXyS) makes use of a time projection chamber with GEM readout to detect X-rays in the 2–10-keV region. Recording and analysing the detected photoelectron trajectory in the gas, along the lines described above, permits to deduce the average polarization of the source using a dedicated iterative reconstruction algorithm (Iwakiri *et al.*, 2016).

8.8   Hybrid MPGDs

A promising line of development, reaching the ultimate tracking capability of gaseous devices, is to replace the conventional printed circuit board pickup electrode with a solid-state pixel sensor to collect the charge released in an overlying gas layer after amplification by a multiplying electrode integrated on the structure, GEM or MICROMEGAS (van der Graaf *et al.*, 2006; van der Graaf, 2007). Due to the small capacitance of the pixels, the noise on each recording channel is very low, few hundred electrons, facilitating the detection of a few ionization electrons with moderate amplifications. The simultaneous recording of charge and time on each pixel permits to reconstruct the events in a TPC-like device with sub-millimetre resolutions.

Figure 8.30 shows an electron microscopic view of GEMGrid, a hybrid detector realized by post-processing the TIMEPIX silicon chip (Llopart *et al.*, 2007) having $\sim$64,000, 55-μm $\times$ 55-μm pixels, to fabricate a gaseous amplification structure over the solid state sensor. Figure 8.31 is an example of a double-track event recorded with the GEMGrid device; each dot in the figure represents the reconstructed three-dimensional position of a cluster in the ionization trails (Blanco Carballo *et al.*, 2009).

An improvement of the GEMGrid, the GEMpix detector consists of a TGEM amplifier followed by a set of four adjacent MEDIPIX

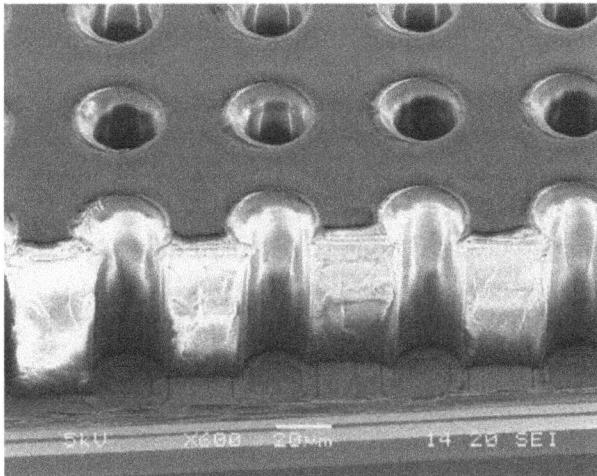

Figure 8.30:   Electron microscopic view of the GEMGrid detector.
Source: Blanco Carballo *et al.* (2009).

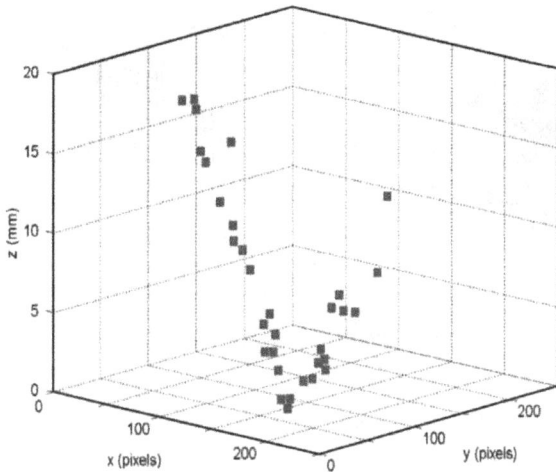

Figure 8.31: A double-track event recorded with the GEMGrid.
Source: Blanco Carballo *et al.* (2009).

silicon sensors (Pacella *et al.*, 2016; George *et al.*, 2015). An outcome of a CERN-based electronics development project, the MEDIPIX C-MOS ASIC is a matrix of 256 × 256 silicon pixels, with an area of 55 × 55 μm^2 each, collecting the electron charge detected and amplified by the gaseous counter. The chip allows single-photon counting at rates up to few MHz per pixel (Llopart *et al.*, 2002).

The high rate, good resolution and large dynamic range of the GEMpix detector are well suited for imaging the intense X-ray emission produced in the study of laser-generated plasmas, such as those produced at the ENEA test facility in Frascati (Italy) (Pacella *et al.*, 2016). The study has been pursued with several targets at the ECLIPSE laser facility in Bordeaux (France) (Claps *et al.*, 2016).

8.9 Fast Timing MPGD

The described MPGD devices share as common feature a moderate time resolution, caused by the diffusion of ionization electrons drifting towards the multiplying electrodes; even in the faster known gases, this can take several tens of nanoseconds per centimetre. This lag can be reduced decreasing the thickness of the drift gap,

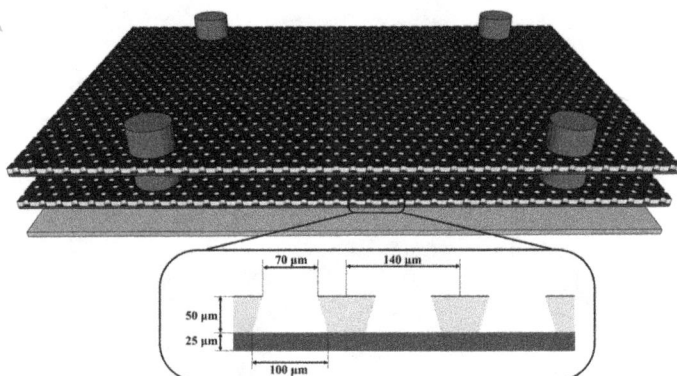

Figure 8.32:   Schematic of the fast timing MPGD.
Source: De Oliveira *et al.* (2015).

resulting however in a loss of efficiency, as fast particles experience around three ionizing collisions per millimetre in gases at atmospheric pressure. In the multi-gap resistive plate counters, time resolution well below 100 ps can be reached adding up the signals of many narrow-gap counters (Akindinov *et al.*, 2000); the devices have, however, moderate rate and localization capabilities.

An alternative using sets of resistive GEMs of a special design, named fast timing micro-pattern (TFM) gas detector, is shown in Figure 8.32 (De Oliveira *et al.*, 2015). The device consists of a cascade of two or more GEMs of the WELL design etched on thin polyimide foils, with arrays of truncated conical holes at a narrow pitch; the perforated foils are coated with a high-resistivity DLC layer. The narrow distance ($250\,\mu m$) between electrodes ensures a fast collection time; signals are transmitted through the resistive electrodes and picked up on the patterned anode at one end. The structure permits in principle to achieve good time resolutions, as far as the proper conditions are met to transmit the signals without losses through all electrodes. Preliminary measurements with a two-layer device achieved a time resolution of 2 nanoseconds (Abbaneo *et al.*, 2017b). Figure 8.33 shows schematically a four-electrode device. A simulation performed with Garfield-Magboltz coupled with a finite element method to compute the field maps demonstrated

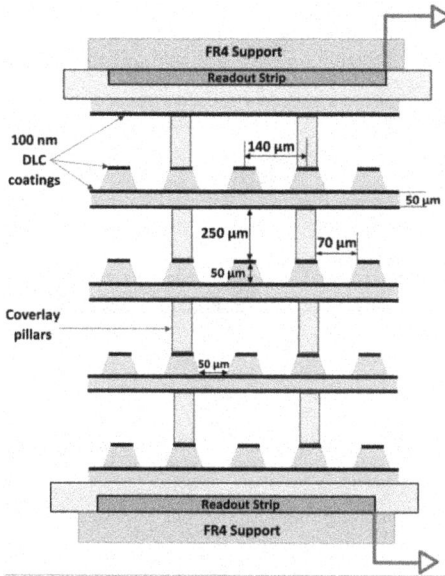

Figure 8.33:   Schematic of a four-layer FTM.
Source: Maghrbi *et al.* (2020).

that a 300-ps time resolution can be achieved with a 16-layer FTM
(Maghrbi *et al.*, 2020).

8.10 PICOSEC Detector

As indicated above, the time resolution of gaseous detectors is limited
by the time spread of the collected amplified charge due to the
primary ionization statistics. Multi-gap devices, such as the described
FTM, improve the time response, adding up signals produced in a
cascade of narrow gaps, but are rather complex to manufacture.
A new approach was proposed in 2015, the so-called PICOSEC
detector, relying on the detection and amplification of photoelectrons
generated by the Cherenkov effect in solid radiators (Papaevangelou
et al., 2018). The device is schematically represented in Figure 8.34
(Bortfeldt *et al.*, 2018). Fast charged particles with momenta above
the Cherenkov threshold of the radiator emit photons, whose UV
component converts to electrons in the thin photosensitive layer

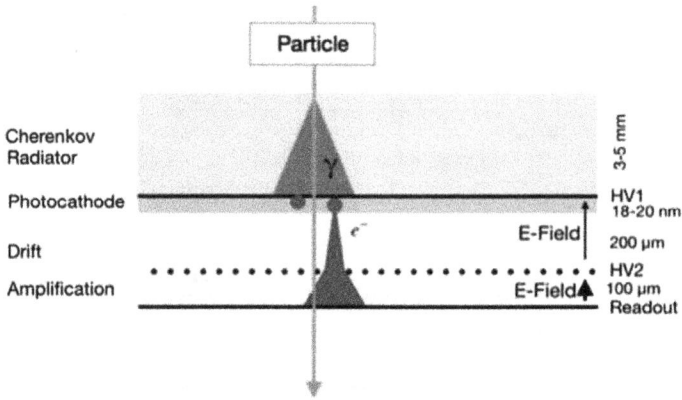

Figure 8.34:   Schematic of the PICOSEC detector.
Source: Bortfeldt *et al.* (2018).

deposited on the entrance face of the gaseous amplifier. A drift region with high electric field helps extract the photoelectrons into the gas and transport the ionization into the main charge amplifier, a MICROMEGAS-like structure; if operated at higher voltages, it can contribute to the charge amplification. With a 3-mm-thick MgF_2 radiator coated with a CsI semi-transparent photocathode deposited on a ~5-nm Cr layer serving as electrode, calculations using the known value of emissivity and quantum efficiency provide an estimated value of ~30 detected photoelectrons for relativistic charged particles.

The timing performance was studied experimentally exposing the detector to fast UV laser pulses, with an intensity that could be reduced to generate single photoelectrons. Figure 8.35 is an example of detected pulses; the first peak is the signal of a reference photodiode providing the time reference of the event.

The single-photon time resolution depends on the drift and anode voltages, and is well below 100 ps (Figure 8.36). Proper fitting algorithms have been developed to extract the best timing information in the presence of noise (Papaevangelou *et al.*, 2018).

The time response for charged particles was measured in a wide range of gas fillings and conditions exposing small-sized prototype

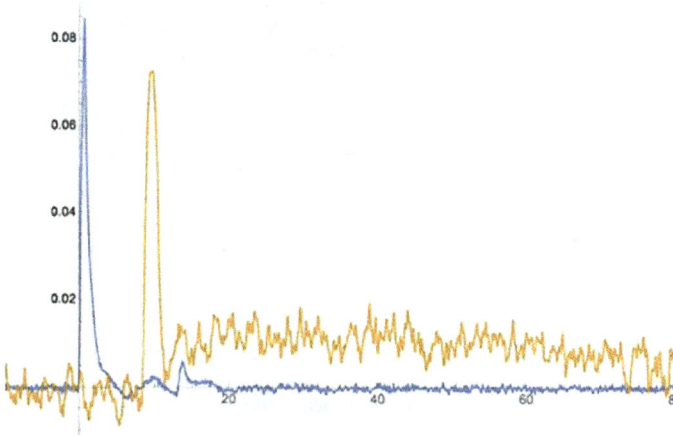

Figure 8.35: Laser-generated PICOSEC single-electron response (right peak). The signal on the left is the pulse provided by a reference photodiode.

Source: Bortfeldt *et al.* (2018).

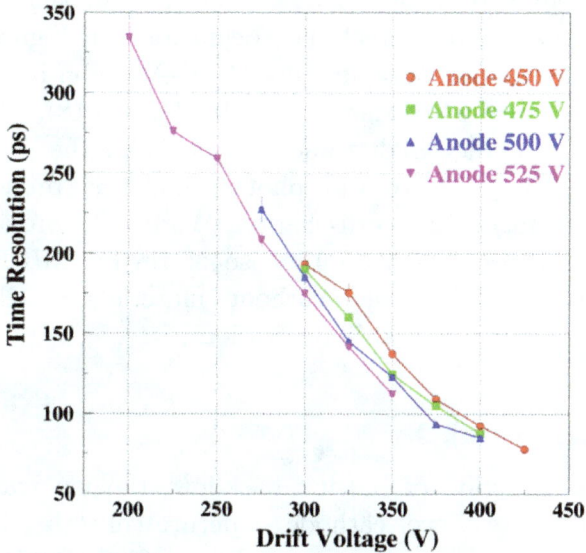

Figure 8.36: PICOSEC time resolution for laser-generated single electrons, as a function of the drift voltage and for several anode voltages.

Source: Bortfeldt *et al.* (2018).

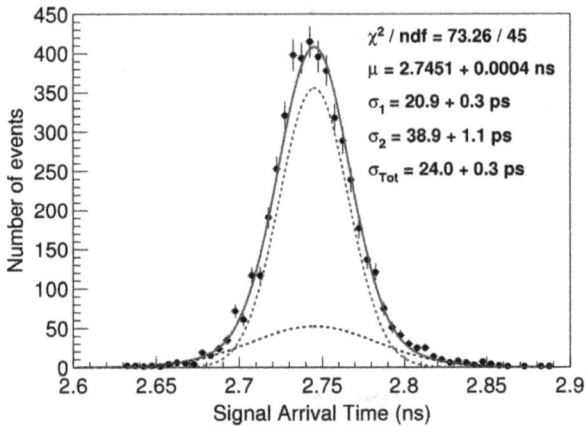

Figure 8.37:   PICOSEC time resolution for a muon beam, with 24 ps rms. *Source*: Bortfeldt *et al.* (2018).

detectors to charged particle beams at CERN; the time reference is obtained from a micro-channel plate beam counter. Figure 8.37 is an example of the results, obtained in a 150-GeV muon beam, showing a time resolution of 24 ps rms (Bortfeldt *et al.*, 2018).

Operating at very high values of the field, the detector, and in particular the delicate CsI photocathode, is prone to suffer permanent damages due to discharges. Work is in progress to find protection schemes, based on the use of restive MICROMEGAS structures and sturdier photocathode materials as diamond-like carbon (DLC) or metallic thin depositions.

8.11   Ion Trap

Targeting a reduction of the ion backflow, the ion trap device is designed with a two-level cathode structure collecting most of the field lines departing from the anode, Figure 8.38 (Bouianov, 2004). Simulation results indicate that an IBF below 2% could be achieved. The work seems, however, not to have been pursued, probably because of manufacturing difficulties and competition from other detectors achieving similar performances.

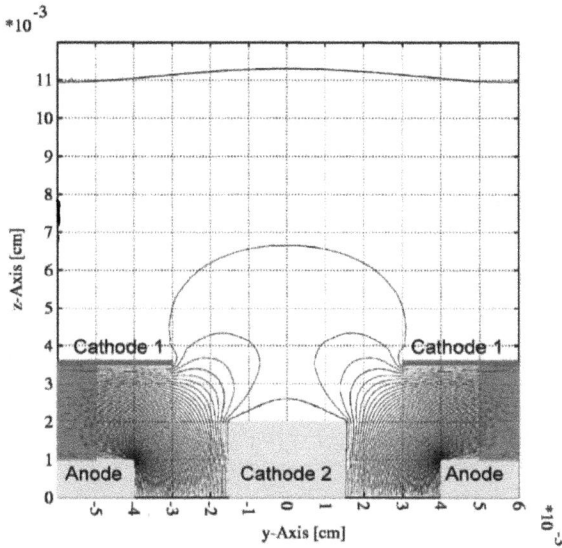

Figure 8.38: Schematic of the ion trap detector.
Source: Bouianov (2004).

8.12 Flame and Smoke Detectors

Ultraviolet flame and smoke detectors operate detecting photons in the wavelength range between 185 and 280 nm, a region in which the radiation from the Sun is fully absorbed by the ozone layer, while the atmosphere at ground level has a good transparency. Gaseous counters filled with the photosensitive vapours and layers developed for the Cherenkov photon detection, described in Chapter 7, are appropriate for this purpose (Peskov and Zichichi, 2007). While commercial flame detectors make use of single-wire counters, recent developments based on MPGDs, and TH-GEM in particular, seem promising in terms of sensitivity, robustness and low cost (Abbrescia *et al.*, 2020).

In Figure 8.39, the relative sensitivities measured with a GEM sealed device at 185 nm for CsI, TMAE and ethyl ferrocene (EF) are compared with the vacuum values for CsI as a function of operating voltage. The loss of sensitivity with the gas fillings is due to the photoelectron backscattering.

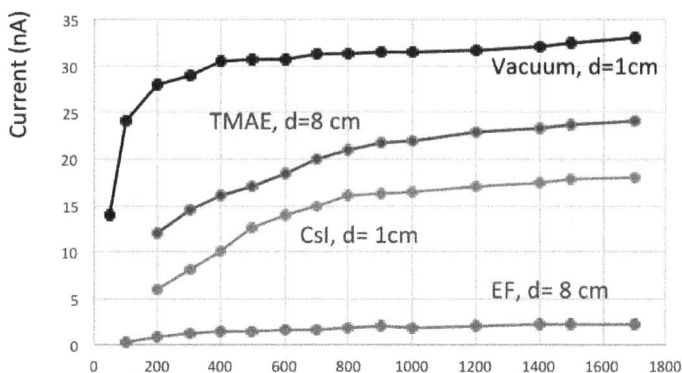

Figure 8.39:   Relative sensitivity of a sealed GEM detector to 185-nm photons in several photosensitive gases an CsI, compared to CsI in vacuum.

Source: Abbrescia *et al.* (2020).

8.13   Negative Ion TPC

The momentum resolution in time projection chambers is limited by the transverse diffusion of electrons, spreading the collected charge and degrading the position determination of the track segments with increasing drift length. An optimal choice of the gas mixture and drift field strength and the presence of a magnetic field reduce the transverse diffusion, that remains however a major dispersion factor; use of high operating pressures, as in the original LBL TPC, improves the performances but requires heavy containment vessels.

Considering that, for a given field value, the diffusion of ions is much lower than for electrons, it was proposed to use as filling a gas with electronegative properties, capturing the ionization electrons soon after production and drifting the negative ions; a suitable high field structure at the anode would then detach and multiply the electrons, providing a detectable signal (Martoff *et al.*, 2000). While many molecules exhibit electronegative properties, some gases have been identified having attachment and detachment probabilities sufficiently large to be used in the Negative Ion TPC (NITPC). One of the preferred gases is carbon disulphide (CS_2), pure or in mixtures, at pressures from a few Torr up to atmospheric (Ohnuki *et al.*, 2001; Martoff *et al.*, 2005). The less chemically aggressive

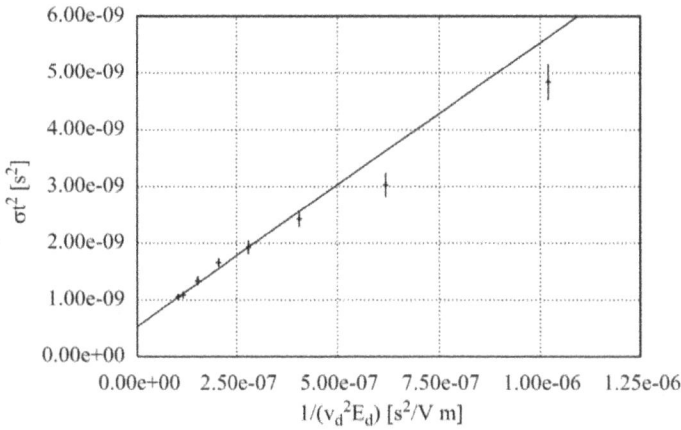

Figure 8.40: Negative ion diffusion as a function of field for nitromethane-carbon dioxide mixtures (20–50 torr).

Source: Martoff *et al.* (2009).

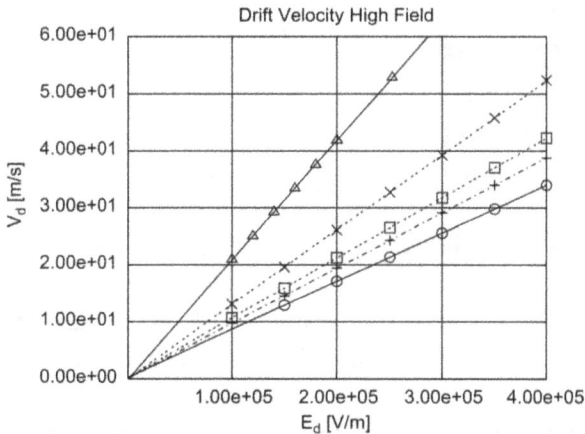

Figure 8.41: Ion drift velocity in various mixtures of nitromethane with CO_2, neon and methane.

Source: Martoff *et al.* (2009).

nitromethane (CH_3NO_2) mixed with noble gases CO_2 and CH_4 has also been successfully tested (Martoff *et al.*, 2009). Figures 8.40 and 8.41, from the last reference, provide the measured values of ion diffusion and drift velocity as a function of field for nitromethane

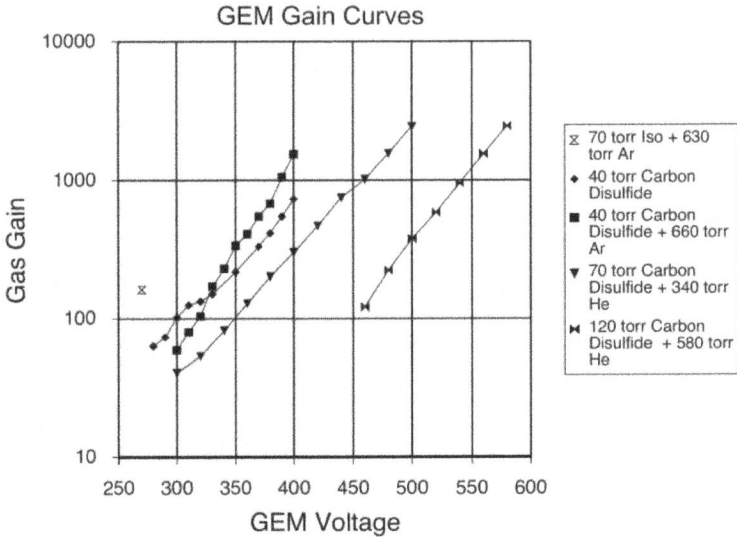

Figure 8.42:   Gain as a function of GEM voltage measured with the NITPC in various mixtures of CS_2 at pressures close to atmospheric.
Source: Miyamoto *et al.* (2004).

mixtures. The drift velocity is about three orders of magnitude lower than for electrons in non-electronegative gases at similar values of field.

The GEM and the thick-GEM (TH-GEM) have been used as end-cap detector in the NITPC, permitting to reach amplification factors above 10^3 (Miyamoto *et al.*, 2004; Phan *et al.*, 2017). Figure 8.42 is an example of gain measured in an NITPC with a GEM amplifier.

The mechanisms of collisional electron detachment and Townsend avalanche development in pure CS_2 have been studied in detail both experimentally and analytically (Dion *et al.*, 2010).

The slow motion of the carrier charges, while easing the realization of the readout electronics, constrains the use of the NITPC for low event rate applications, such as dark matter and WIMP searches (Phan *et al.*, 2017).

Chapter 9

Neutron Detectors

9.1 Introduction

The study of neutron scattering and diffraction is a major tool for the investigation of condensed matter structures, boosted by the commissioning of intense neutron sources; for a review of the field, see for example Gebauer (2004).

Neutrons are detected through their interactions with matter, resulting in the emission of energetic charged particles or gamma rays; the type and probability of the interaction, and consequently the choice of detectors, strongly depend on energy.

Neutron fields are generally named after the range of their energy: cold below ∼25 meV, thermal up to 100 keV and fast above. For a detailed description of the various interaction processes and of classic neutron detectors, see for example Knoll (1989). For neutron energies above a few MeV, the major interaction process is elastic scattering, generally not providing a detectable signal; fast neutron detectors generally use a moderator, a layer of material placed before the active device, whose purpose is to reduce the energy of the field by successive interactions; hydrogenated materials are particularly efficient to this extent. At energies below one MeV, inelastic scattering may occur, leaving the nucleus in an excited state, returning to the ground state through a radiative process that can be detected. At lower energies, thermal and cold neutrons interact with matter through nuclear processes, resulting in the emission of prompt gammas and/or charged prongs: protons, alpha particles, tritons and nuclear fragments; cross sections depend on the material and energy.

A commonly used gaseous detection medium for thermal neutrons is ^{3}He, through the reaction $^3_2$He $+$ n $\rightarrow$ $^3_3$H $+$ p, having a high cross section and resulting in the emission in the gas of a proton and a triton of 573 and 191 keV, respectively, that are easily detected. With their high efficiency, ^{3}He proportional counters are widely used for monitoring and calibration purposes. However, the world shortage of the isotope has oriented the research of alternative gas fillings, as ^{10}B-enriched boron fluoride, or to the use of thin-foil converters with large neutron cross sections: lithium, boron and gadolinium, natural or enriched with their higher cross section isotopes, ^{6}Li, ^{10}B and ^{157}Gd, respectively. As the reaction yields, MeV alpha particles for Li and B, keV conversion electrons for Gd, have a short absorption path in condensed matter, the thickness of the converter is limited and therefore the neutron detection efficiency is smaller than for the gaseous counterpart; this disadvantage is compensated by a reduced sensitivity to gamma rays, a substantial background present around spallation sources.

9.2   MSGC-based Neutron Detectors

Owing to their high efficiency and resolution, micro-pattern devices are well suited as neutron detectors; the micro-strip counter was indeed developed to serve as a focal plane detector for a neutron spectrometer. With a ^{3}He–CF$_4$ gas filling, the MSGC powder diffractometer at ILL operated for many years, with several modifications and improvements, see Section 2.8 (Clergeau *et al.*, 2001). With a similar design, a MSGC-based spectrometer is used for high-resolution neutron imaging at the Japanese J-PARC centre (Fujita *et al.*, 2007). Using a thin-foil Gd converter, and to solve the reliability problem encountered by the ILL detector, Masaoka *et al.* (2003) adopted a double scheme, with a capillary plate pre-amplifier, reminiscent of the multi-GEM structures, to permit the operation of the MSGC at safe voltages, Figure 9.1.

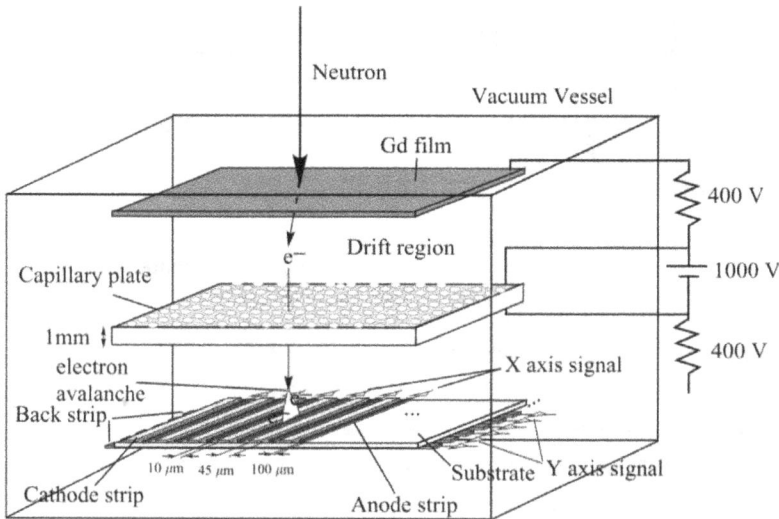

Figure 9.1: Schematic diagram of a neutron detector with Gd thin-film converter and a double amplifying structure.

Source: Masaoka *et al.* (2003).

9.3 GEM-based Devices

With multiple ^{10}B-coated GEMs operated in the transmission mode, the CASCADE detector achieves high detection efficiencies and position resolution, Figure 9.2 (Klein and Schmidt, 2011). The interaction of neutrons with the converters results in the emission of charged prongs, ionizing the gas near the electrodes; electrons are transported through the cascade of foils operated at potentials tuned to ensure transmission without amplification; the last GEM electrode provides the gain needed for detection. Signals are collected on sets of perpendicular strips equipped with fast front-end amplifiers followed by ADC ecordings.

Figure 9.3 shows the measured and computed detection efficiencies as a function of neutron wavelength for two CASCADE prototypes, having three and eight ^{10}B converter layers, and an extrapolation to 20 layers. Figure 9.4 provides the localization accuracy as a function of gas pressure, measured with a three-layer device exposed to a ^{252}Cf collimated neutron source.

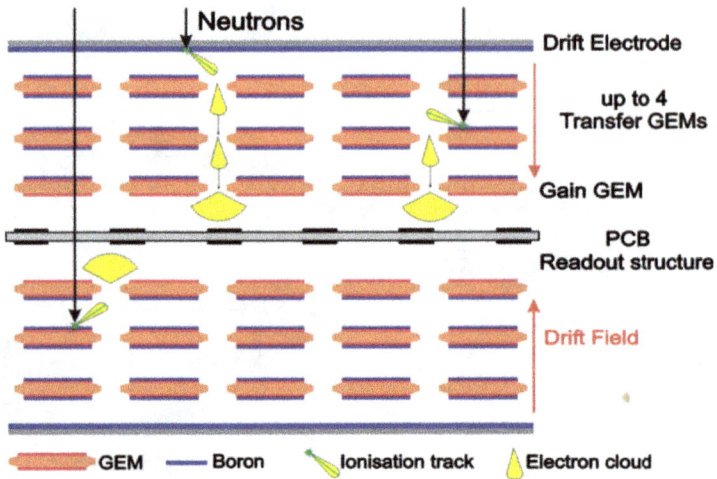

Figure 9.2:   Schematic diagram of the CASCADE neutron detector.
Source: Klein and Schmidt (2011).

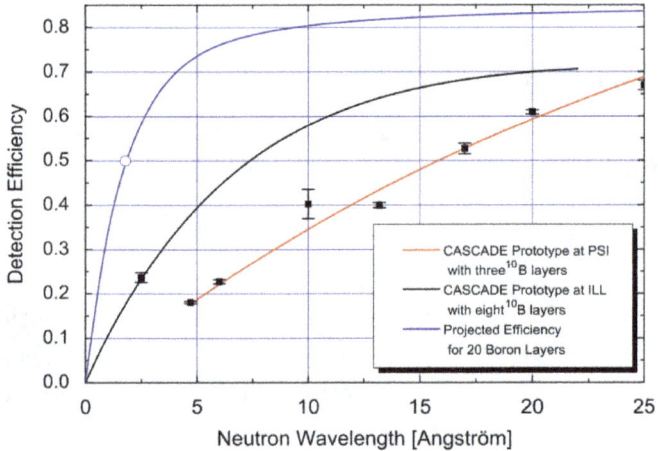

Figure 9.3:   Neutron detection efficiency as a function of wavelength of the CASCADE detector. Data points are measurements for three and eight ^{10}B layer devices; lines are computed values for up to 20 layers.
Source: Klein and Schmidt (2011).

The picture in Figure 9.5 is a neutron radiography of common desk objects recorded exposing the detector to a wide cold neutron beam.

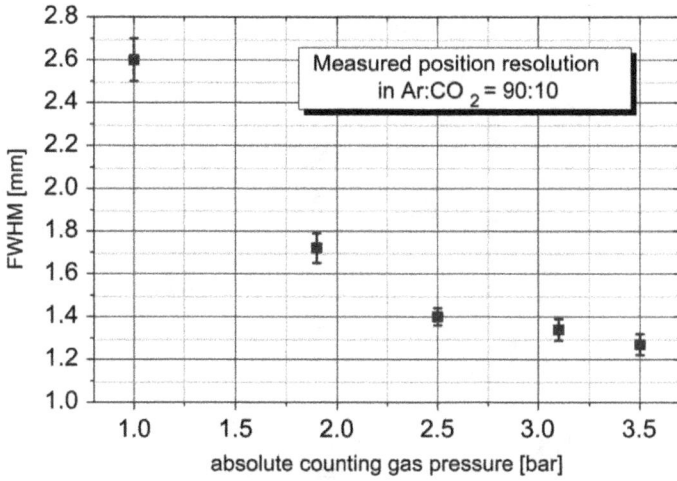

Figure 9.4: Position accuracy measured with the 20×20 cm^2 detector at increasing gas pressures.

Source: Klein and Schmidt (2011).

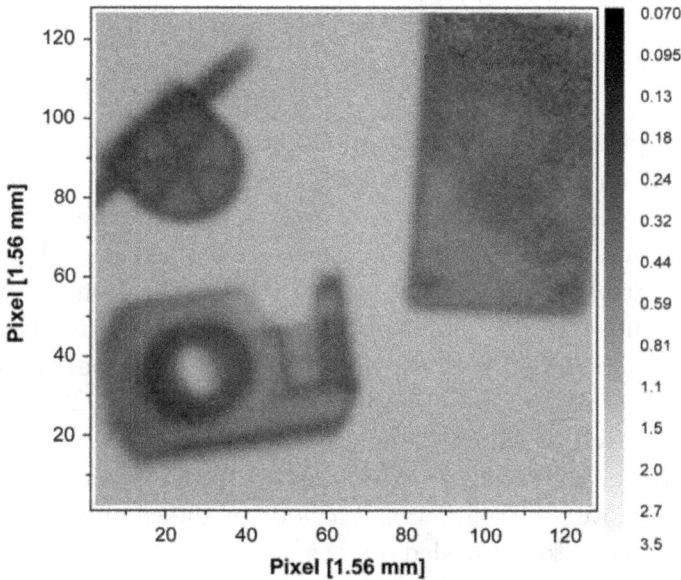

Figure 9.5: Neutron radiography of desk objects recorded with the CASCADE detector.

Source: Klein and Schmidt (2011).

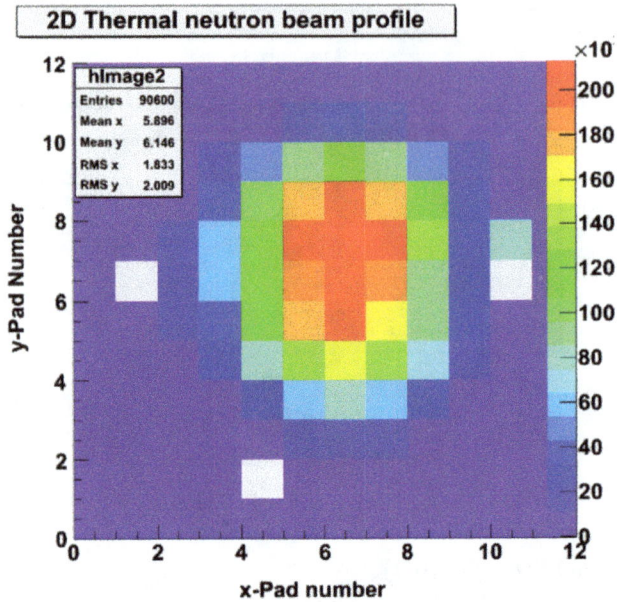

Figure 9.6:   Neutron beam profile recorded at the ISIS spallation source on a matrix of anode pads, 8×8 mm^2 each.
Source: Croci *et al.* (2013).

Developed primarily as a beam monitor for spallation sources, the bGEM has an aluminium cathode coated with a thin ($\sim 1\mu$m) B$_4$C layer acting as thermal neutron converter through the ^{10}B(n, ^{7}Li)α nuclear reaction; ionization electrons are collected and amplified through a triple-GEM (TGEM) structure. Localization is performed recording the amplified charge on a padded anode. Figure 9.6 shows a beam profile recorded with the detector at the ISIS spallation source of Rutherford Appleton Laboratory (UK) (Croci *et al.*, 2013).

A device with a similar design, named nGEM, was used to detect fast neutrons exploiting the protons scattered by a polyethylene layer placed over the aluminium cathode of the chamber. Figure 9.7 shows the detector installed in the VESUVIO beam line at the ISIS spallation source (Croci, Claps *et al.*, 2013). The measured efficiency as a function of neutron energy, Figure 9.8, matches well the values computed with GEANT4 from the known neutron–proton cross sections.

Figure 9.7: The nGEM detector installed in the ISIS beam line.
Source: Croci *et al.* (2013).

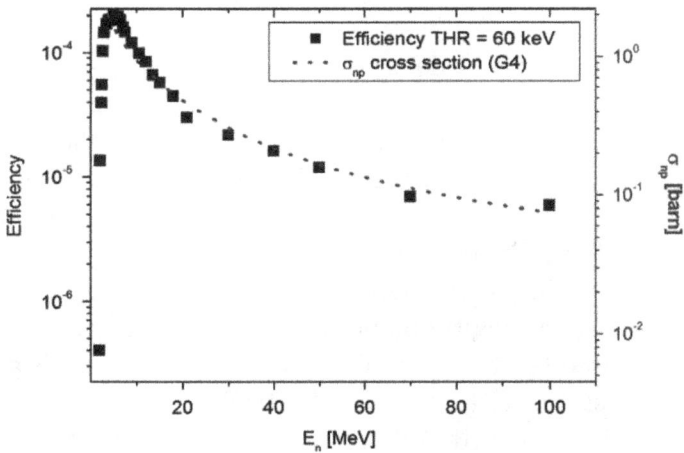

Figure 9.8: Detection efficiency of the nGEM as a function of neutron energy.
Source: Croci *et al.* (2013).

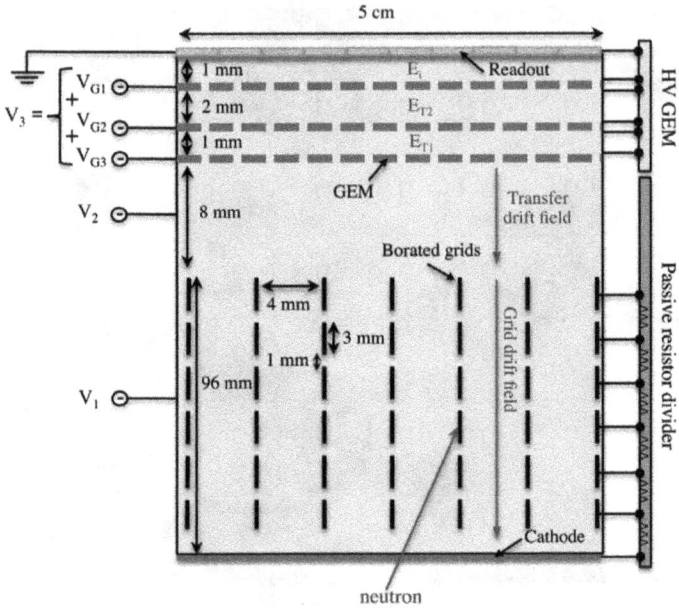

Figure 9.9:   Schematic diagram of the BAND.
Source: Albani *et al.* (2020).

A larger-sized nGEM device is under construction to serve as a fast neutron detector for the close-contact neutron emission surface mapping (CNESM) system, under development at the ITER-NBI facilities in Padova, Italy (Muraro *et al.*, 2019).

A novel GEM-based neutron device, named boron array neutron detector (BAND), Figure 9.9, consists of a set of parallel, boron-coated grids serving as neutron converters. The grids, at graded potentials, create a longitudinal electric field, drifting the ionization electrons released by the neutron interaction yields towards a TGEM amplification structure (Albani *et al.*, 2020).

Developed for neutron detection and imaging at the China Spallation Neutron Source, a multi-layer detector includes 10 GEM electrodes, eight of which are coated with enriched ^{10}B layers in two stacks mounted back-to-back, Figure 9.10; the computed detection efficiency is compared to measured values in Figure 9.11 (Zhou *et al.*, 2020).

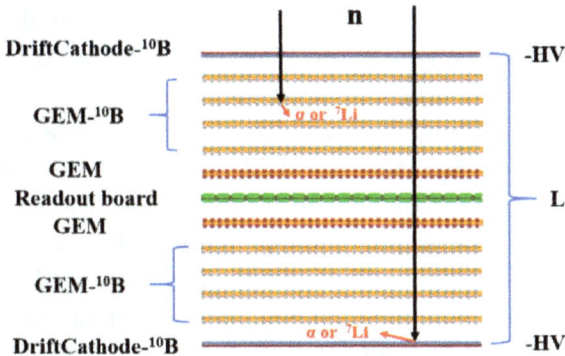

Figure 9.10: Multi-GEM neutron detector developed for the China Spallation Neutron Source.

Source: Zhou *et al.* (2020).

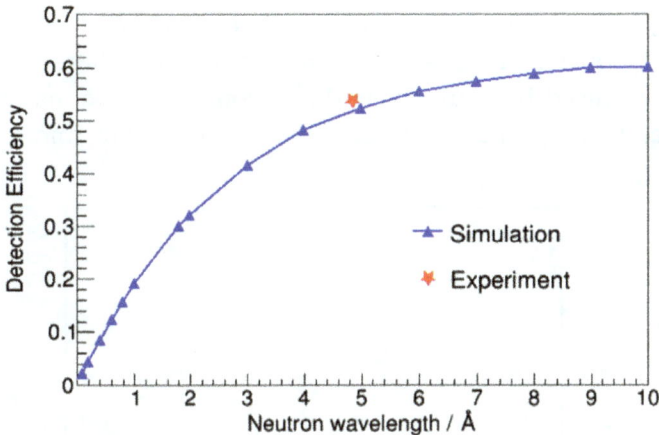

Figure 9.11: Neutron detection efficiency as a function of wavelength.

Source: Zhou *et al.* (2020).

9.4 MICROMEGAS-based Devices

A MICROMEGAS detector with ^{6}Li and ^{10}B thin converter layers deposited on the cathode, tested in a neutron beam at the Centre d'Etudes Nucléaires de Bordeaux-Gradignan (CENBG), served as the prototype for the neutron time-of-flight (nTOF) facility at CERN aimed at the study of the space distribution of the neutron beam as a function of its energy (Andriamonje *et al.*, 2002).

An intense neutron beam is produced by a spallation process induced by 20-GeV proton interactions with a target. With a ^{6}Li-enriched lithium fluoride layer deposited on the cathode and hydrocarbon gas filling, the detector covers a wide region of sensitivity, from a few keV to several MeV. Charged prongs generated by the neutron interactions ionize the gas in the drift region between the cathode and the mesh, and are amplified in the high field of MICROMEGAS (Pancin *et al.*, 2004). The measured reaction rate, a product of the detection efficiency and neutron flux, shows the transition from events generated by the LiF converter to those by the gas recoils, Figure 9.12.

Optimized for the detection of neutrons in a high-γ background, a MICROMEGAS-based device named detector MICROMEGAS for neutrons (DEMIN), Figure 9.13, is designed to operate as a monitor at deuterium–tritium ignition facilities (Houry *et al.*, 2006). A 2-mm-thick polypropylene layer serves as a converter for charged particles; protons produced by neutrons and electrons from gamma rays enter the device through a 2-μm-thick aluminium window and ionize the

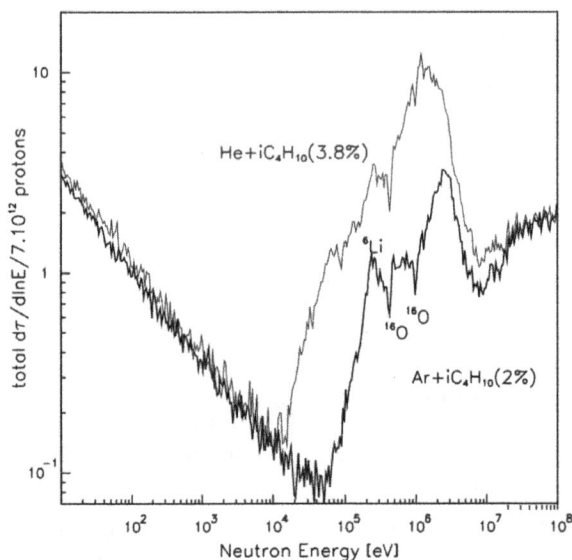

Figure 9.12:   Measured reaction rate in the nTOF detector.
Source: Pancin *et al.* (2004).

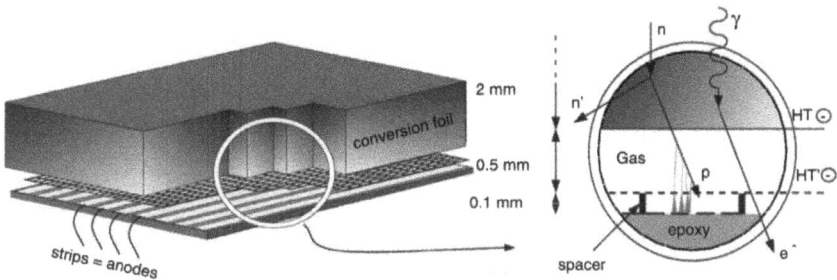

Figure 9.13: Schematic diagram of the MICROMEGAS-based DEMIN detector. *Source*: Houry *et al.* (2006).

gas in the drift region. Charge multiplication in the high field of MICROMEGAS enhances the signal. The detector, with a very fast response and low sensitivity to photons, is optimized for use around high-intensity neutron sources.

Developed to study the neutron flux in the presence of strong X-ray and gamma ray background, the PICCOLO detector, Figure 9.14, consists of a cylindrical vessel with a matrix of converters mounted on the cathode of a MICROMEGAS. Charged prongs emitted by the converters ionize the gas in the drift region, and the released electrons are collected and amplified by the multiplier. A choice of different materials for deposits permits to detect neutrons in a wide energy domain: ^{10}B via the $(n\alpha)$ process, ^{232}Th and ^{235}U through a fission reaction. One sector is left without converter for calibration purposes. The detector has been tested extensively around a reactor, as part of the program for nuclear waste disposal. Figure 9.15 shows the good linearity of the dependence of the detector current from the reactor power (Pancin *et al.*, 2008).

A MICROMEGAS-based TPC has been built and operated to perform precision neutron-induced cross-section measurements at the neutron-induced fission fragment tracking experiment (NIFFTE). The fission chamber is shown in Figure 9.16, and the correlation between the measured track length and energy loss resulting from an exposure to an actinide source is given in Figure 9.17 (Snyder *et al.*, 2018). The work includes a detailed comparison of performances of the detector for various gas fillings, with particular emphasis on the discharge probability.

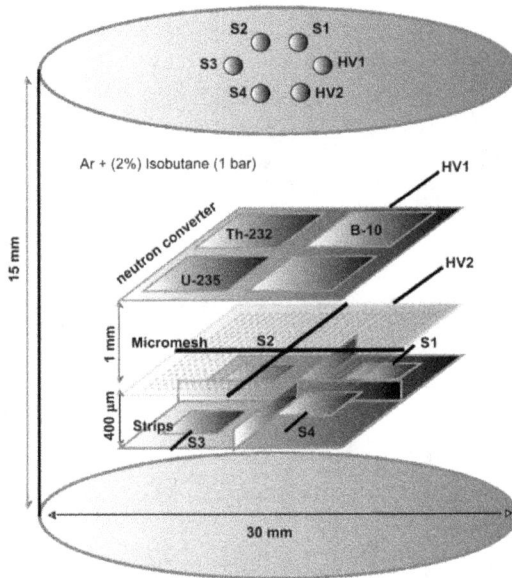

Figure 9.14:   Schematic diagram of the PICCOLO detector with multiple targets and a MICROMEGAS amplifier.

Source: Pancin *et al.* (2008).

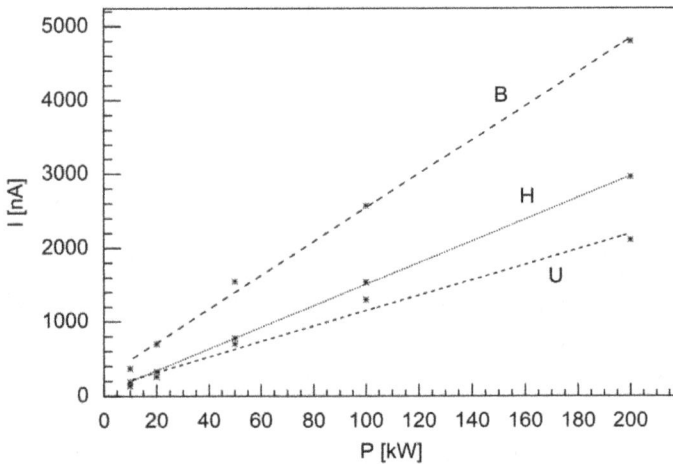

Figure 9.15:   Dependence of the PICCOLO measured current from the reactor power for three converters.

Source: Pancin *et al.* (2008).

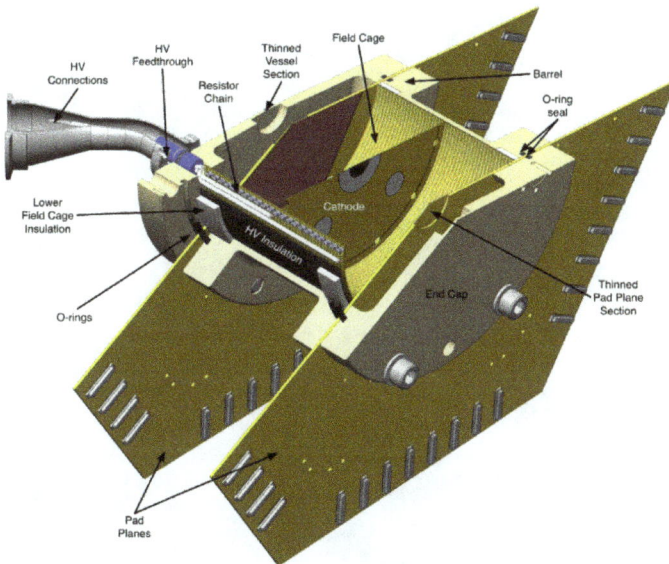

Figure 9.16: Schematic diagram of the NIFFTE reaction chamber.
Source: Snyder *et al.* (2018).

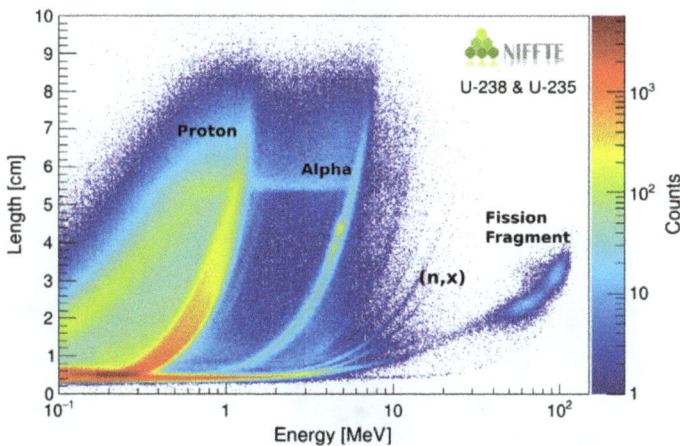

Figure 9.17: Track length versus energy measured with the NIFFTE fission
TPC chamber on exposure to an actinide source. The gas filling is a mixture
of argon–isobutane 95–5 at 550 torr.
Source: Snyder *et al.* (2018).

Chapter 10

Cryogenic and Dual-Phase Detectors

10.1 Introduction

Gaseous counters operated near atmospheric pressures do not have sufficiently high mass to permit detection of rare events, such as those investigated in neutrino astrophysics and dark matter searches. Increasing size and pressure of the sensitive volume helps, but involves serious technical difficulties; the use of liquefied noble gases as detection media is more appropriate. Large liquid argon and xenon TPC-like systems are widely used worldwide; they operate in the ionization mode, setting a limit in the minimum amount of charge needed to provide a detectable signal.

Charge multiplication, observed in liquid xenon (Derenzo *et al.*, 1974) and argon (Kim *et al.*, 2004), is generally considered not secure enough to be exploitable in large devices, making it difficult to detect low ionization yields. A solution has been found with the development of dual-phase detectors, where the electron charges released in the liquid are extracted by an electric field into an overlying gaseous sheet, instrumented with wire or micro-pattern counters and used to perform charge multiplication and localization in the gas phase.

For a review of noble gas properties, cryogenic and dual-phase detectors, see Aprile *et al.* (2006), Buzulutskov, (2012), Chepel and Aráujo (2013) and González-Díaz *et al.* (2018).

10.2   Dual-Phase Detectors

Dual-phase detectors exploit two fundamental features: a field-assisted electron extraction from the liquid, and a high gain operation of gaseous devices at cryogenic temperatures. Both processes have been extensively studied in the course of development of the new generations of micro-pattern devices. For argon, two extraction time constants are observed: a fast component, 1 ns or less, and slower delayed emissions with millisecond time constants caused by electron trapping under the liquid surface. Compared to argon, in liquid xenon, the larger surface barrier requires higher fields for extraction (Chepel and Aráujo, 2013).

Figure 10.1 shows schematically the setup used to study the electron extraction process (Bondar *et al.*, 2006). The multi-GEM electrodes are powered through a resistive high-voltage divider; the extraction field is defined by the potential difference between the

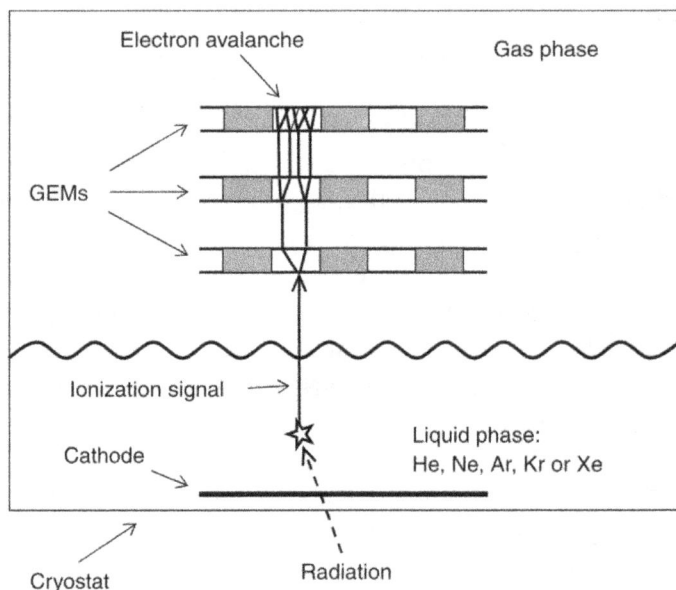

Figure 10.1:   Schematic diagram of the dual-phase cryogenic detector with a triple-GEM (TGEM) extraction and amplification structure.
Source: Bondar *et al.* (2006).

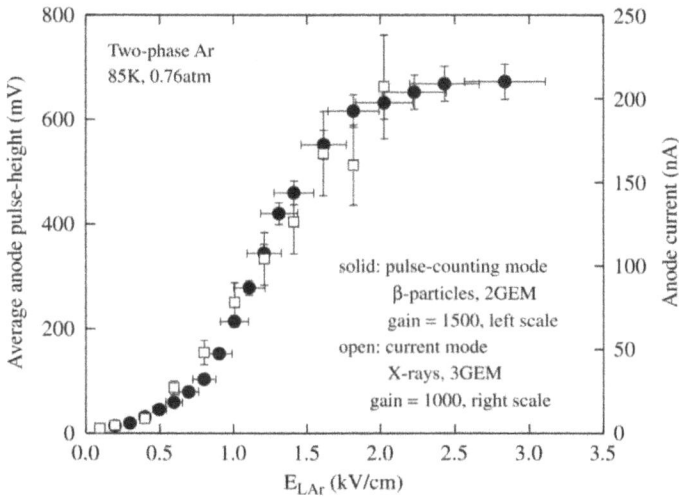

Figure 10.2: Anode current and average pulse height recorded with the dual-phase argon detector as a function of extraction field.

Source: Bondar *et al.* (2006).

lowermost side of the last GEM and the bottom of the vessel, acting as the cathode. The current and average pulse heights measured on the first electrode of the cascade in various conditions as a function of extraction field are given in Figure 10.2; at saturation, the electron extraction efficiency is close to 100% (Bondar *et al.*, 2006). As seen in Figure 10.3, while the maximum gas gain is limited to a few hundred at cryogenic temperatures in xenon and krypton, larger gains can be reached in argon.

A double large electron multiplier (LEM) has been used to instrument an argon dual-phase TPC detector; geometry and operation of the multiplier were described in Section 5.4. The prototype detector has a 10-cm-thick drift volume; all materials for the assembly are selected to permit operation at very low temperature, avoiding outgassing and thermal deformations (Badertscher *et al.*, 2009).

Based on the two-dimensional GEM scheme (Bressan *et al.*, 1999), the readout system collects and records in time slots the charge on strips along the X and Y directions, thus permitting full three-dimensional reconstruction of tracks.

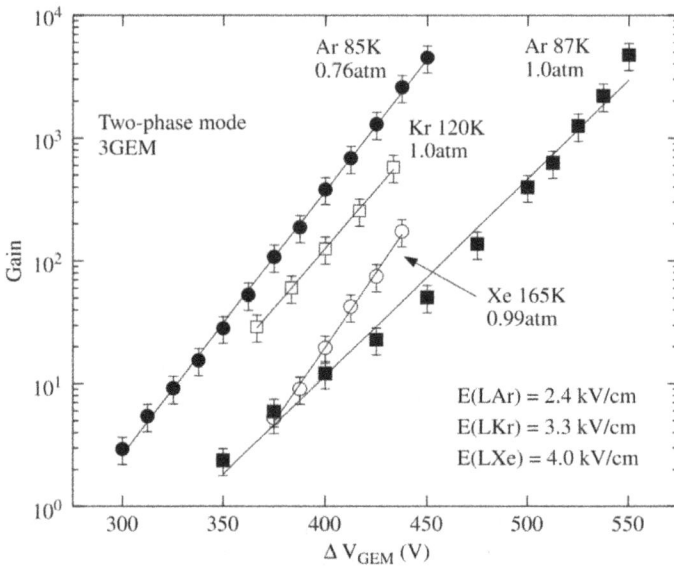

Figure 10.3: Dual-phase TGEM gain as a function of voltage applied in noble gases at cryogenic temperatures.

Source: Bondar *et al.* (2006).

In further work, the double LEM was replaced with a single structure, capable of providing the gain needed for detection. A detailed analysis of gain, operating stability and performances as a function of LEM geometry is given in Cantini *et al.* (2015). Figure 10.4 shows the 21-cm drift LEM-TPC prototype before insertion in a liquid argon cryostat. Figure 10.5 is an example of a cosmic ray event recorded with the device (Badertscher *et al.*, 2011).

The liquid argon volume was progressively extended to 200 L, with a drift length of 60 cm and an active area of $40 \times 76\,\mathrm{cm}^2$ (Badertscher *et al.*, 2013). The experience acquired with the prototypes served as the basis for the realization of a larger device, a $3 \times 1 \times 1\mathrm{m}^3$ (4.2 tonnes) dual-phase LA LEM-TPC, successfully tested at CERN. Electrons released by ionization in the sensitive volume drift in the liquid are extracted into the gas phase and amplified by the LEM structure; the amplified charge is detected and recorded on sets of perpendicular strips on the anode plane,

Figure 10.4:   The prototype 21-cm drift dual phase LEM-TPC before insertion in the cryostat.

Source: Badertscher *et al.* (2011).

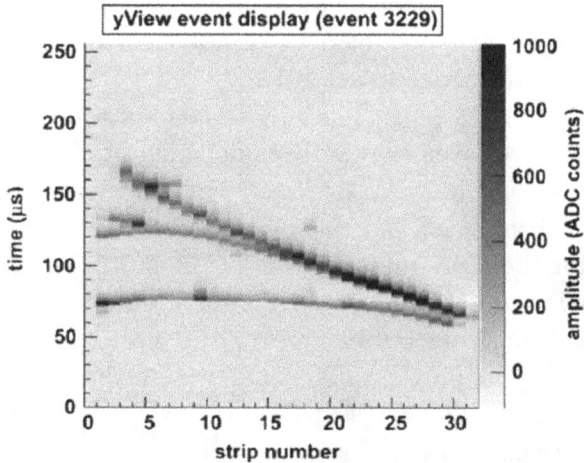

Figure 10.5:   Cosmic ray event recorded with the 21-cm LA LEM-TPC.

Source: Badertscher *et al.* (2011).

with 3.125-mm pitch, providing a two-dimensional view of the events. Five photomultipliers, coated with wavelength shifters and mounted underneath the TPC field cage, detect the primary argon scintillation light, providing the time of the interaction, as well as the secondary scintillation from the gas phase; a set of cryogenic cameras is used to monitor the liquid level.

Small and medium-sized MICROMEGAS, manufactured with the bulk technology have been tested successfully at high pressures and cryogenic temperatures to assess their performances in view of applications with liquid argon TPCs (Delbart *et al.*, 2011).

10.3   Bubble-assisted MPGD

While very successful, dual-phase detectors necessitate rather stringent requirements on the control and stability of the liquid-gas interface. A possible alternative is to immerse into the liquid a micro-pattern structure (GEM, Thick-GEM, LEM) and detect the secondary scintillation photons emitted by electrons in high fields (Breskin, 2013). Initial observations with such structures in liquid xenon, providing yields larger than expected, led to the conjecture that the electroluminescence was in fact generated within thin gas layers or bubbles trapped below the holes of the microstructures (Arazi *et al.*, 2015; Erdal *et al.*, 2015).

With a TH-GEM or a standard GEM electrode coated with a CsI photosensitive layer, the authors demonstrated direct detection of the primary scintillation UV photons, in addition to the ionization of electrons (Erdal *et al.*, 2017). Figure 10.6 shows schematically the setup used for the early measurements. The multiplying structure is immersed in liquid noble gas; a grid of resistive wires is used to heat the liquid and generate bubbles below the electrode. A radioactive source emits alpha particles, ionizing the medium, and inducing primary (S1) and secondary scintillations (S1' and S2) generated by the photoelectrons extracted from the CsI layer and by the ionization electrons accelerated in the gas bubble. Figure 10.7 is an example of signals detected by the photomultiplier, with the primary and delayed bubble-induced pulses.

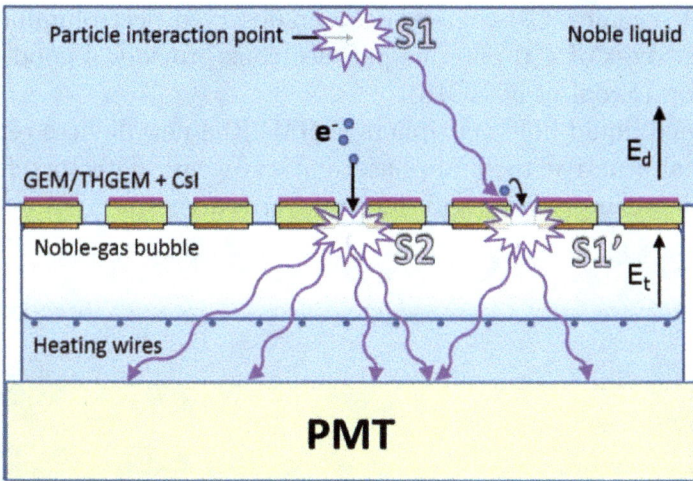

Figure 10.6:   Schematic diagram of the bubble-assisted detector, immersed in liquid xenon.

Source: Breskin *et al.* (2019).

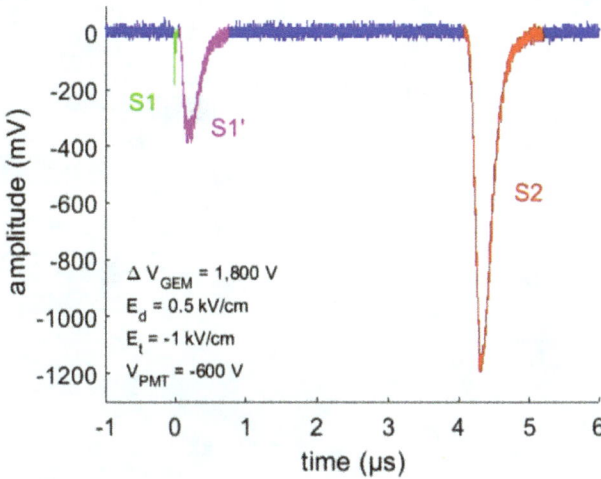

Figure 10.7:   Primary and bubble-assisted secondary scintillation signals.

Source: Breskin *et al.* (2019).

With a conical CsI-coated GEM, an observed electroluminescence yield in excess of 400 photons per electrons provides a good energy resolution (Erdal *et al.*, 2018).

Named liquid-hole multipliers (LHM), the new devices represent a valid alternative to dual-phase and cryogenic detectors for rare events, neutrino physics and dark matter searches.

Chapter 11

Optical Imaging Chambers

11.1 Introduction

The majority of the detectors described in previous chapters exploit the processes of collisional charge multiplication in high electric fields to increase the primary ionization charge and achieve electronic detection. Photons are, however, abundantly emitted both by the primary interactions between radiation and matter and as an outcome of inelastic electron-molecule collisions in high electric fields, in a process named scintillation or luminescence. For rarefied noble gases, the photon emission energy corresponds to the atomic excitation lines; at higher pressures, collisions between excited and neutral atoms may result in the formation of short-lived molecules, named dimers, having characteristic emission bands at lower energies. For a comprehensive analysis of radioluminescence processes in rare gases, see Salete Leite (1980).

For molecules, the photon luminescence emission spans over a wide range of wavelengths, depending on the nature, physical conditions of the medium and applied electric fields; in mixtures, the insurgence of a multitude of energy transfer processes complicates the issue even further. For a review, see for example Chapter 5 in the author's book on gaseous radiation detectors (Sauli, 2014).

Primary and field-enhanced photon emission has been exploited in the gas scintillating proportional counters; owing to the absence of the dispersions intrinsic in charge multiplication processes, these devices are inherently capable of providing energy resolutions close to the statistical limit. Policarpo (1977) is a good review on the operating principles and performances of gas scintillation counters.

11.2   Early Imaging Chambers

While the main photon emission wavelength is in the far and vacuum ultraviolet regions for most gases, the copious emission of photons at wavelengths close to the visible of some molecules has suggested to perform direct optical detection of radiation using photography or television cameras. As low photoionization threshold vapours studied in the framework of Cherenkov ring imaging, TMAE and TEA (see Chapter) are attractive for imaging, with their secondary emission spectra centred on 460 and 260 nm, respectively, Figure 11.1, that can be detected directly or using an appropriate wavelength shifter medium in front of the optical recording device. An example of minimum ionizing tracks recorded with a TPC-like optical detector

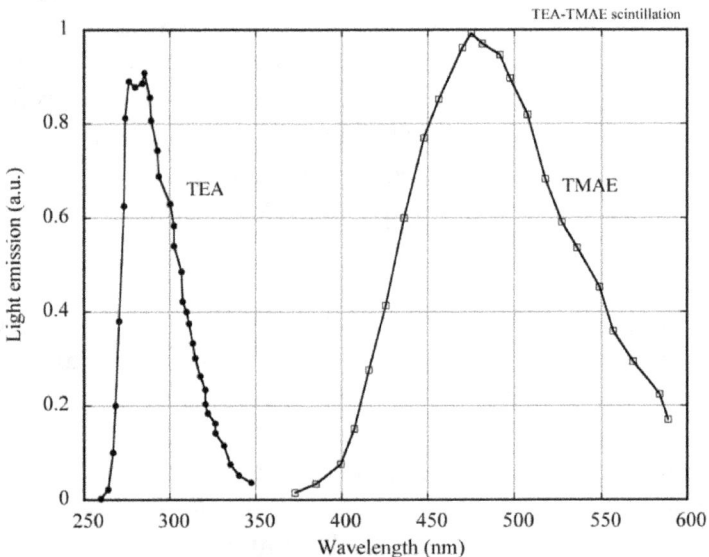

Figure 11.1:   Secondary photon emission spectra in TEA and TMAE. *Source*: Data from Suzuki *et al.* (1987) and Charpak *et al.* (1988).

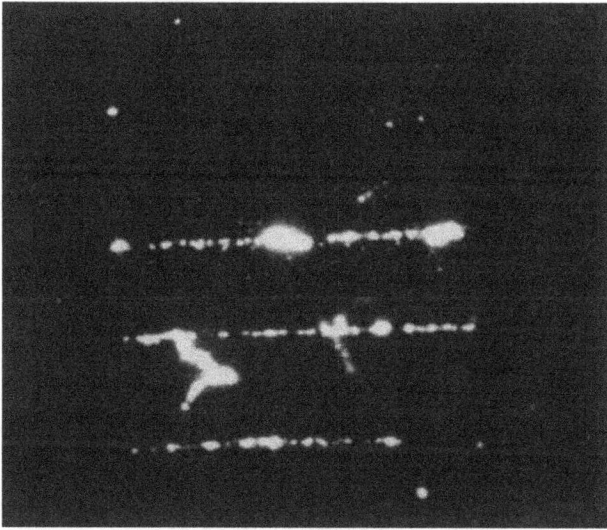

Figure 11.2: Beam tracks recoded with the optical TPC; gas filling Ar–CH$_4$–TEA.
Source: Charpak *et al.* (1988).

is shown in Figure 11.2 (Charpak *et al.*, 1988); ionized trails released in the detector's sensitive volume are drifted into a parallel plate avalanche multiplier, located at the focal plane of the optical recording system.

Visual imaging chambers have been used in applications where the modest acquisition rates, constrained by the characteristics of the recording system, are not a limiting factor, such as autoradiography and radiochromatography (Dominik *et al.*, 1989), as well dosimetry (Titt *et al.*, 1998) and imaging of nuclear decays (Miernik *et al.*, 2007).

The introduction in the late nineties of the gas electron multiplier (GEM), and the observation in some gases of a copious photon emission at near-visible wavelengths, both primary and field-induced, concurrently with the availability of fast solid-state cameras, opened novel possibilities for the use of imaging chambers in many applied fields (Sauli, 2018).

11.3   Carbon Tetrafluoride Scintillation

Carbon tetrafluoride (CF_4) offers several advantages when used as a component in the filling of gaseous counters: good quenching of secondary processes, permitting to reach high gains; fast electron drift velocity; low cross section for neutron background as compared to methane. It suffers however from two disadvantages: high electron capture probability, and chemical reactivity induced by the release in the avalanche process of fluorine, directly reacting with the detector materials or, combined with water vapours, creating the very aggressive hydrofluoric acid (HF) (Alfonsi *et al.*, 2005). The precautions and results achieved operating GEM chambers with a thorough moisture control in mixtures including CF_4 were discussed in Section 5.11.

The photon emission of CF_4, generated both by primary and by field-induced collisions, extends from the vacuum ultraviolet to the visible; it is attributed to the decay of excited states $(CF_{3+})^*$ and $(CF_{4+})^*$ formed in the avalanches. The process has been investigated in gaseous detectors at low to moderate pressures exposing a parallel plate counter to various radiation sources and recording the photon yield with photomultipliers and selected optical filters (Pansky *et al.*, 1995).

The secondary, field-induced scintillation of carbon tetrafluoride in mixtures with noble gases has been studied in the course of the development of radiation imaging systems based on GEM devices. Figure 11.3 shows the emission spectra recorded with a monochromator at a charge gain around 170 in two gas mixtures at atmospheric pressures, and Figure 11.4 provides the number of photons per electron emitted above 400 nm in various gas mixtures (Fraga *et al.*, 2003).

The emission spectra depend on the gas pressure. Figure 11.5 is an example of the primary scintillation yield measured in pure CF_4 exposed to a ^{241}Am α-particle source between one and five bars; the spectrum has two distinct peaks around 300 and 650 nm. In the VUV region (220–500 nm), the photon yield is about 2,000 per MeV at one bar, decreasing to ~600 at five bars, while the emission in the

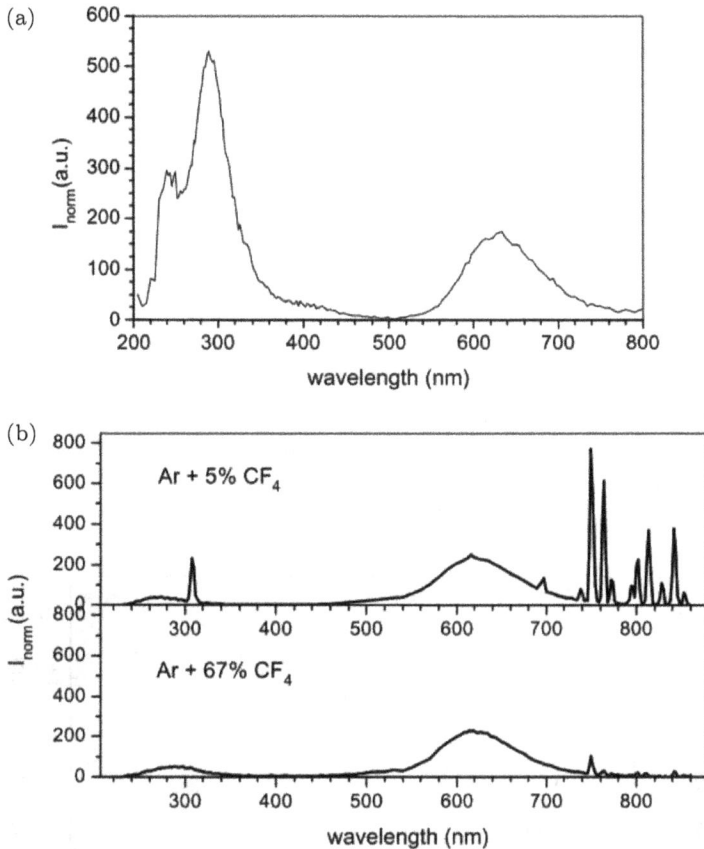

Figure 11.3: Secondary scintillation spectra in He–CF$_4$ (60–40) (a) and in two Ar–CF$_4$ mixtures (b).

Source: Fraga *et al.* (2003).

visible increases to 6,000 photons per MeV between 550 and 750 nm (Morozov *et al.*, 2010).

At high pressures, the strength of the applied electric field affects the primary scintillation in the UV and visible regions, as seen in Figure 11.6, while there is almost no effect at one bar (Morozov *et al.*, 2011). The scintillation yield can be strongly reduced by the presence of contaminations in the gas, namely, due to atmospheric oxygen or nitrogen (Margato *et al.*, 2011).

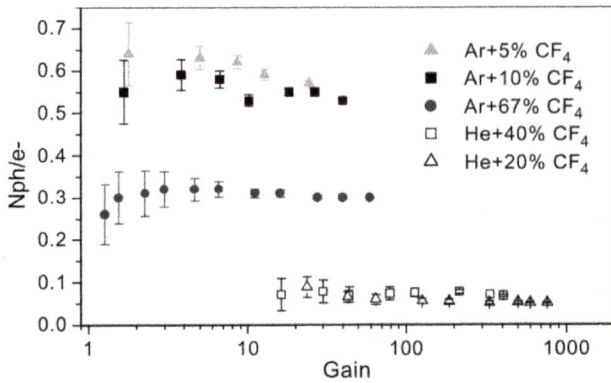

Figure 11.4:   Normalized number of scintillation photons per electron as a function of GEM gain in Ar–CF$_4$ and He–CF$_4$.

Source: Fraga *et al.* (2003).

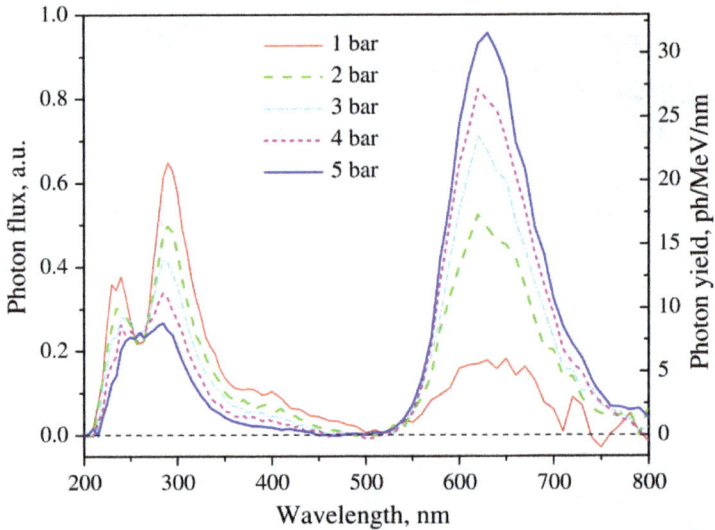

Figure 11.5:   CF$_4$ primary scintillation under α-particles irradiation at increasing pressures.

Source: Morozov *et al.* (2010).

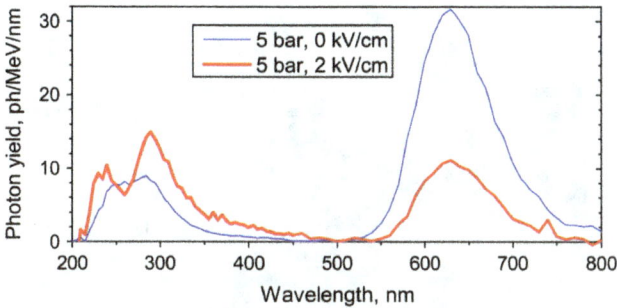

Figure 11.6:   Field dependence of the CF_4 scintillation spectrum at five bars. *Source*: Morozov *et al.* (2011).

11.4   MPGD-based Optical Imaging

Detectors using one or more GEMs are particularly well suited for optical imaging applications, since they permit to reach high gains in most gases and can be directly viewed through a glass window with a solid-state camera. The last GEM in the cascade can be operated in the collection mode, not requiring an anode; alternatively, the anode can be a mesh with good optical transparency, or the window can be coated by a transparent resistive layer such as indium tin oxide (ITO) to collect the electron charge.

While in early measurement, mixtures of argon-CO_2 were used to reach high gains (Fraga *et al.*, 2001), in later studies, the copious scintillation of CF_4 has been exploited in TPC-like structures, achieving better image quality in the detection of neutrons (Fraga *et al.*, 2002) and charged prongs, (Margato *et al.*, 2004). Figure 11.7 shows an image of overlapping neutron interactions in a triple-GEM (TGEM) detector filled with a ^{3}He–CF_4 mixture, recorded with a CCD camera; in many events, the two tracks resulting from the interaction (a proton and a triton) can be identified.

A small-sized TPC, read out with a TGEM and a CCD camera, was used to record α-particles emitted by an internal ^{241}Am source; integration of the light yield along the tracks permits to reconstruct the characteristic end-of-range energy loss profile (Bragg peak). By combining the optical image of the projected track with an analysis of

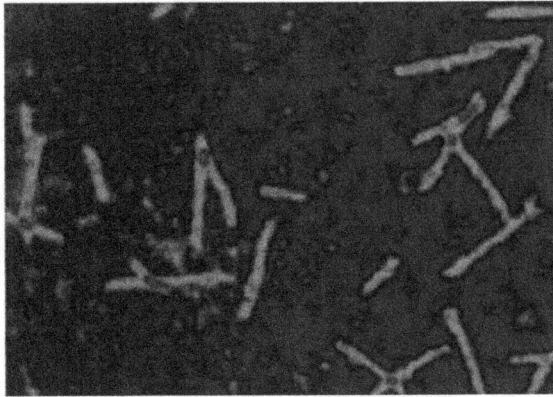

Figure 11.7: Neutron interactions in ^{3}He–CF$_4$ recorded with the optical GEM.
Source: Fraga *et al.* (2002).

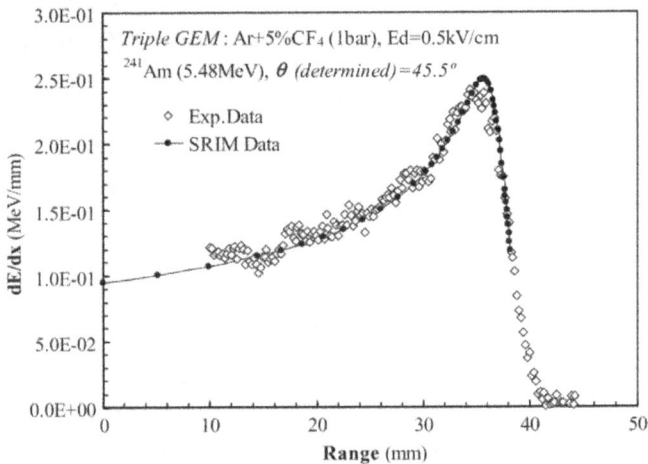

Figure 11.8: Bragg peak and angle determination for 5-MeV α-particles recorded with the optical GEM–TPC.
Source: Margato *et al.* (2004).

the signals detected by a photomultiplier, the authors could deduce the emission angle, Figure 11.8 (Margato *et al.*, 2004).

Well suited for high-resolution imaging, solid state cameras set an intrinsic limitation to the event acquisition rate, up to one kHz for advanced devices. An alternative recording system making use of a set of photomultipliers reading the scintillation signals of a multi-GEM detector permits in principle to achieve better rate performances; the track coordinates are deduced from the relative amplitude of the time-sampled signals exploiting the Anger camera principle. With small-sized ($10 \times 10 \times 5 \, \text{cm}^3$) double-GEM detectors and four PM, Fetal *et al.* (2007) achieved the reconstruction of α-particles with $\sim 2°$ angular resolution.

The tracking and background subtraction capabilities of a low-pressure GEM-based TPC have been analysed in view of applications in dark matter experiments (Phan *et al.*, 2016). A systematic investigation of light yield and localization properties for single and multiple TH-GEM is described by Rubin *et al.* (2013).

Designed by CERN's GDD group to perform a systematic search on imaging detectors, the setup shown in Figure 11.9 combines a TGEM, optically read out with a CCD camera, with a photomultiplier detecting the primary scintillation induced by the tracks and providing the time of the interaction, thus permitting a full three-dimensional reconstruction of the events, Figure 11.10 (Brunbauer *et al.*, 2018).

Figure 11.11 is an example of double α-particle tracks recorded exposing the detector to an internal ^{220}Rn source; together with the primary, a secondaryα is emitted by the subsequent decay of ^{216}Po to ^{212}Pb. Integration of the light intensity along the top track provides the Bragg energy loss profile (Brunbauer *et al.*, 2018).

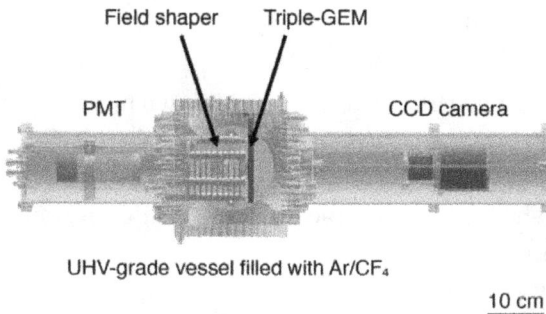

Figure 11.9: GEM–TPC prototype detector.
Source: Brunbauer *et al.* (2018).

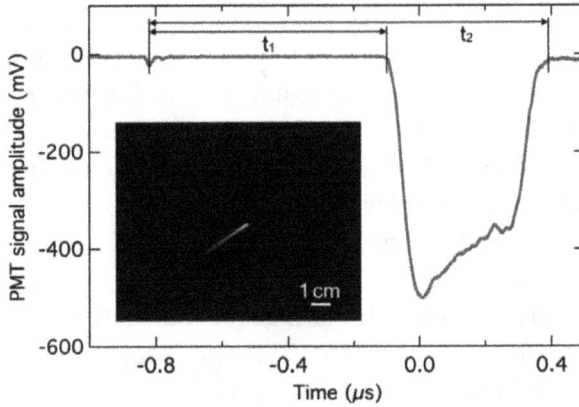

Figure 11.10:   Primary and secondary scintillation of α-particles detected with the GEM–TPC.

Source: Brunbauer *et al.* (2018).

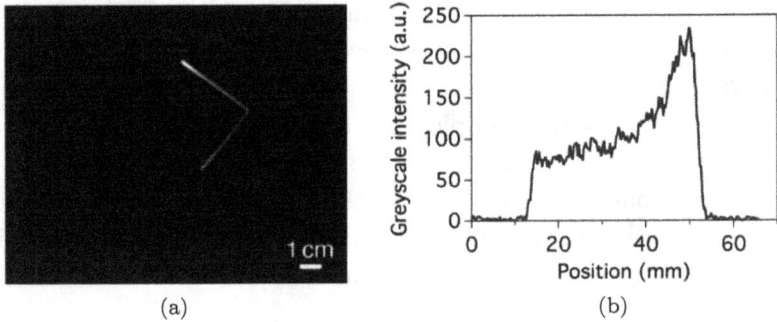

(a)   (b)

Figure 11.11:   Double-αtrack event from the decay of ^{220}Rn (a); integrated light emission along the top track (b).

Source: Brunbauer (2018).

The high gains that can be reached with the chamber permit to detect low-energy losses, down to single electrons. Recorded with the detector exposed to a high-energy charged particles beam, Figure 11.12 shows an example of a hadronic shower, with several delta rays clearly visible.

In a study on detectors aimed at dark matter searches, several prototype GEM–TPC with He–CF$_4$ have been tested with sources and in an electron beam, demonstrating high efficiency and good

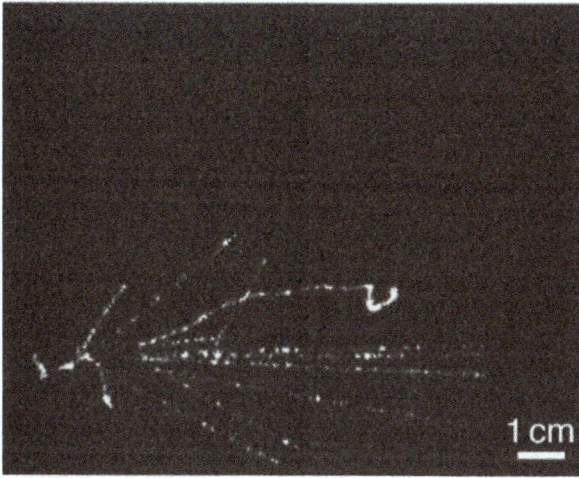

Figure 11.12:   Hadronic shower recorded with the optical GEM–TPC.
Source: Brunbauer (2018).

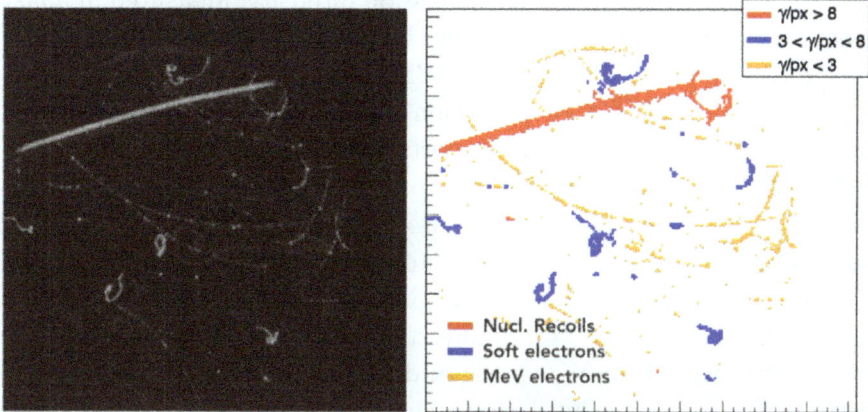

Figure 11.13:   Recorded and reconstructed tracks from an Am–Be source with the LEMOn GEM–TPC detector operated in He–CF$_4$.
Source: Pinci *et al.* (2019).

space resolution (Mazzitelli *et al.*, 2018; Cavoto *et al.*, 2020). Figure 11.13 is an example of tracks from by an Am–Be source recorded with a 20-cm drift prototype (LEMOn) in a magnetic field parallel to the drift direction; the different types of tracks can be

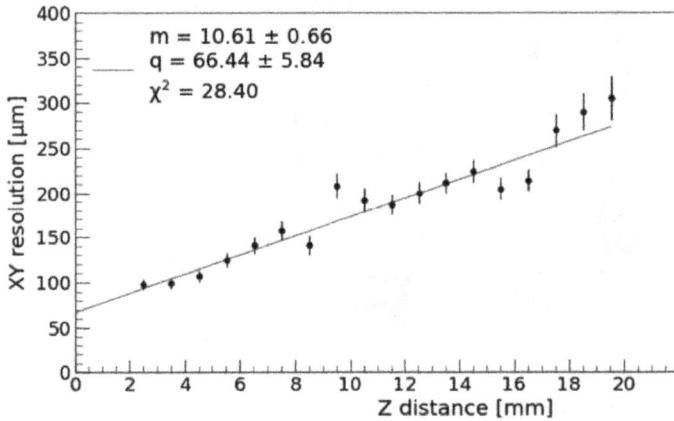

Figure 11.14:   Space resolution for fast electron tracks as a function of drift distance.

Source: Pinci *et al.* (2019).

identified according to the light emission intensity. Measured in a fast electron beam, Figure 11.14 shows the space resolution as a function of the track's distance from the GEM (Pinci *et al.*, 2019).

In their standard construction, making use of thick printed circuit boards as the main supports for the multiplying electrode, MICROMEGAS are unsuited to optical imaging. The development of a device manufactured on a glass plate permits to record images from a MICROMEGAS-based detector (Brunbauer *et al.*, 2020). To ensure stable operation, the glass substrate is coated with a thin indium-tin oxide (ITO) layer, electrically conductive with a resistivity around 15 W/square and transparent in the wavelength region of the CF_4 scintillation. Measurements with a small-sized prototype confirm the good energy and optical imaging resolutions of the detector. Figure 11.15 compares the images obtained in similar operating conditions with electronic signal recording and optical devices making use of GEM and MICROMEGAS; owing to the absence of diffusion between cascaded GEMs, the image resolution is better for the latter detector. The left image also shows a recurrent problem met with electronic recording, the artefacts generated by dead channels, absent with optical recordings.

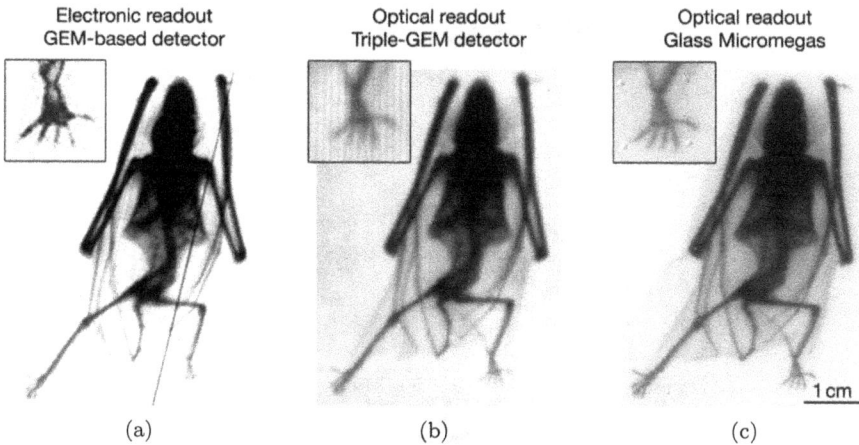

Figure 11.15: Soft X-ray radiography of a bat recorded with electronic readout (a), and optical readout GEM (b) and MICROMEGAS (c).

Source: Brunbauer *et al.* (2020).

11.5 Optical GEM Detector Applications

Beam dose monitoring is an essential tool in hadrotherapy, and is normally ensured by the use of ionization chambers, deriving the beam intensity and position from the current recorded on a matrix of charge-collecting electrodes covering the beam area. Limited by the number of electronic channels and interconnection, the detectors have modest position accuracy (Braccini *et al.*, 2015).

The GEM-based beam dosimeter shown in Figure 11.16 permits to achieve sub-millimetre position accuracies, while ensuring a large dynamic range, thanks to the controlled charge gain of the detectors. To minimize the amount of material in the beam, the image is reflected off-beam by a mirror to the recording CCD camera (Fetal *et al.*, 2003).

Dose response, linearity and stability have been extensively studied exposing the detector to clinical proton and carbon beams (Seravalli *et al.*, 2008, 2009).

At therapeutic beam intensities (typically 10^8–10^9 ions s^{-1}), a single GEM operated at low gains is generally sufficient to provide well-contrasted images, thus minimizing the amount of material in

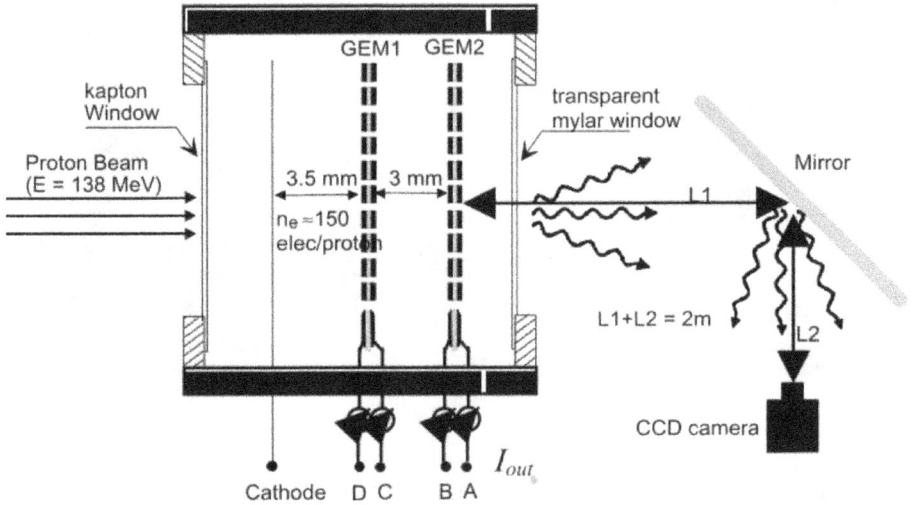

Figure 11.16:   GEM-based beam dosimeter.

Source: Fetal *et al.* (2003).

the beam and possible distortions due to space charge (Brunbauer *et al.*, 2018).

A scintillating GEM imager has also been developed for monitoring the high-energy X-ray flux in intensity-modulated radiation therapy (IMRT) (Simon *et al.*, 2005).

The high detection efficiency for low-energy X-rays, submillimetre position accuracy and simplicity of the setup permit to realize good quality transmission radiographies. Figure 11.17 is the X-ray radiography at 8 keV of a small drone. Within the limits of the data acquisition rate of the solid state camera, the images can be time resolved, permitting to realize a tomography of moving objects (Brunbauer *et al.*, 2018).

Aiming at applications in material and cultural heritage studies, the setup shown in Figure 11.18 has been used to perform the analysis of the fluorescence yield of composite samples. A wide band X-ray beam is used to hit a target, and a scintillating TGEM amplifies the charge released in a drift space by fluorescence photons emitted by the excited sample. Recording of the scintillation spectra position and

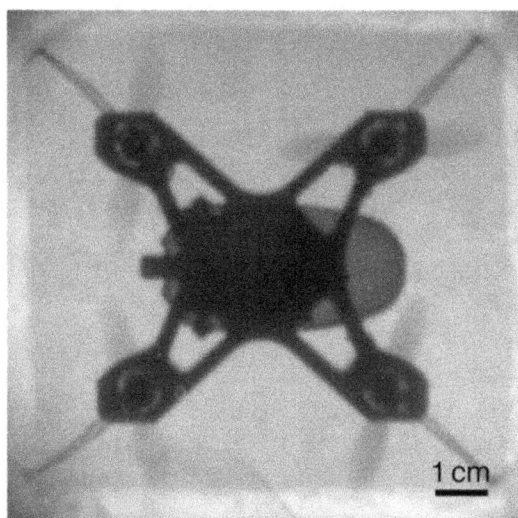

Figure 11.17: 8-keV radiography of a small drone.
Source: Brunbauer (2018).

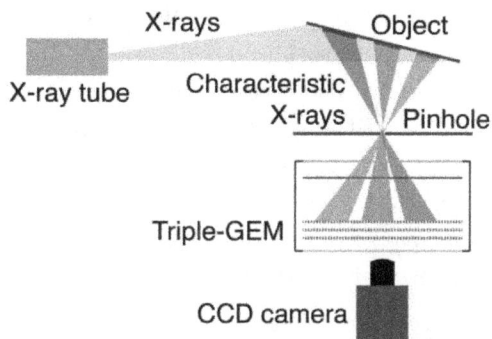

Figure 11.18: Schematic diagram of the fluorescence analysis setup using a scintillating GEM detector with optical recording.
Source: Brunbauer (2018).

amplitude permits to perform a coarse position-dependent, energy-resolved analysis of the sample (see also Section 8.2). An example is shown in Figure 11.19, realized imaging a target composed of several materials (Brunbauer *et al.*, 2018).

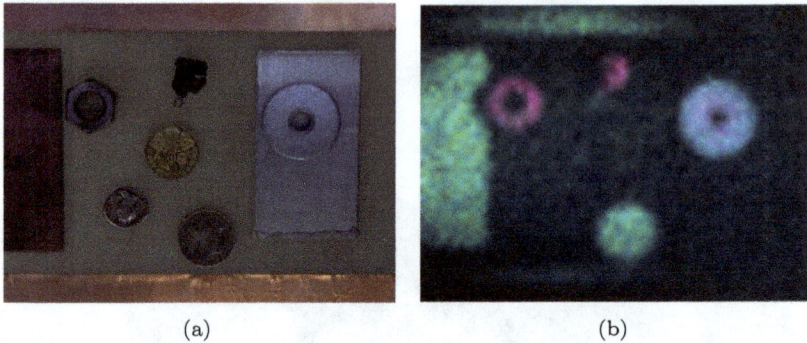

(a)   (b)

Figure 11.19:   A composite target (a) and the energy-resolved image of the fluorescence yield (b).
Source: Brunbauer (2018).

A limitation in the application of gaseous counters for detection and localization of X-rays results from the increasing absorption length with energy. For point sources, as those used for fluorescence analysis, crystal diffraction studies and similar analyses, this sets a limit on the maximum conversion thickness, hence on the efficiency, to limit the dispersion due to the parallax error for inclined photon directions. A possible solution is to design a detector with radial electric field geometry in the conversion gap; this was achieved long ago with the so-called spherical chamber (Charpak *et al.*, 1977). More recently, a technology was developed to build a self-supporting GEM electrode with spherical geometry (Duarte Pinto *et al.*, 2011). Rather complex to fabricate, while very efficient, these devices had limited applications.

An intrinsically simpler way to realize a structure with radial electric field is to manufacture GEM electrodes with radial segmentation, with each segment powered at voltages suitably graded to build up a quasi-spherical field in the overlaying drift gap. Facing segments on the anode side of the foil are powered with the same gradient, increased by a potential difference ensuring charge multiplication in the holes (Sauli, 1999). An upper electrode, segmented with similar geometry, constitutes the entrance window for the X-rays. This so-called planispherical GEM chamber is relatively simple to

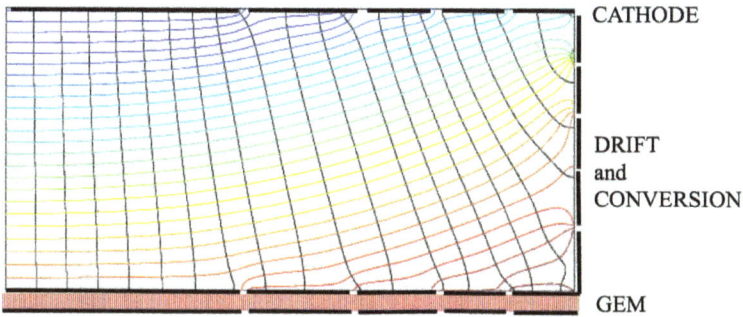

CATHODE

DRIFT
and
CONVERSION

GEM

Figure 11.20: Computed electric field in the planispherical chamber.

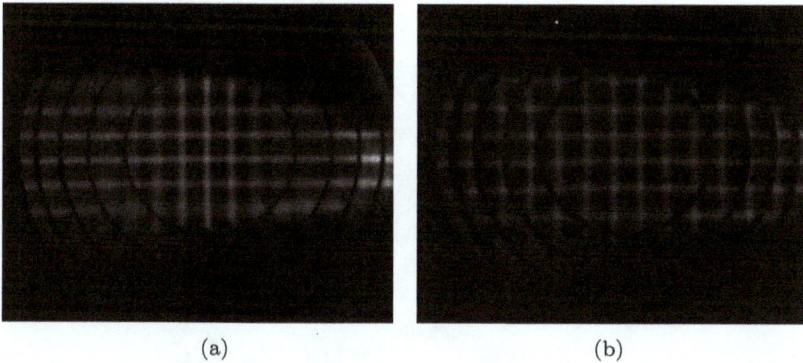

(a)   (b)

Figure 11.21: X-ray fluorescence images of a metal grid with a parallel (a) and radial electric field (b), recorded with the planispherical chamber.

Source: Brunbauer *et al.* (2017).

manufacture, and with an appropriate choice of the applied potentials permits to easily change the point of convergence of the radial filed lines, or focal length; see Figure 11.20. The effectiveness of the detector in removing the parallax error is demonstrated by the comparison of two fluorescence images of a metallic grid with a conventional parallel and a radially shaped drift field, Figure 11.21 (Brunbauer *et al.*, 2017).

Chapter 12

Future Challenges and Prospects

The development in recent years of innovative gaseous detectors has been largely motivated by the demanding requirements of particle physics experimentation: high radiation fluxes, better position and multi-track resolution, survivability in hostile environments. The variety of conception of the new generations of devices partly fulfils the needs of existing experiments; further improvements are certainly crucial to allow operation at higher rates and particle multiplicities. The long-established competition with sensors exploiting solid state technologies, silicon micro-strips and pixel detectors has been a motivating factor for many innovative developments of gas-based devices, their lower intrinsic resolution rewarded by lesser costs and easier manufacturing, permitting to realize large experimental structures. While antagonist, gaseous detectors have largely profited from the implementation of high-density sensing electronics used for the readout of silicon devices, compensating with high gains the lower ionization yield. However, their intrinsic resolution remains limited by statistical dispersions due the distance between ionization clusters, electron collection time and diffusion in the gas. Use of fast gases and high drift fields can improve the time response, which remains however limited to several nanoseconds, marginal to disentangle events produced by high-luminosity colliders.

Several ways have been devised to overcome these limitations. The simplest, increasing the gas pressure, is possible in some cases, but entails a considerable complication of the apparatus. The poor performance of a single device can be compensated adding up signals generated in stacks of identical thin elements, each providing better timing albeit with lower efficiency: this is the approach of the multi-gap resistive plate chambers, and of their modern embodiment, the Fast Timing MPGDs (Maghrbi *et al.*, 2020). For high-energy charged particles, detecting the photons emitted in a radiator with a fast gaseous counter, the PICOSEC detector permits to reach time resolutions nearing 25 ps (White, 2018). Being very promising, these innovative devices are, at the time of writing, in the early development stage, and doubts subsist on their long-term survivability, hindered by the sporadic insurgence of damaging discharges.

A potentially far-reaching solution has been explored replacing the counter's gas filling with a layer of porous material, serving as higher density ionization medium and electron multiplier in the high electric field of the interstices (Lorikyan *et al.*, 2007). While reasonable efficiency has been demonstrated in the detection of ionizing particles, the tendency of the insulating material to retain the charges released and the need to operate in near vacuum seriously limit the range of applications. A possible alternative was proposed by the writer in a stimulating paper entitled "Six Concepts in Search of an Author" (Sauli, 2019): use as detection medium a low-density foam made with electron-conducting glass, manufactured with the technology used to make the silica aerogel slabs used as radiators in Cherenkov counters. Designed with a resistivity high enough to sustain an applied voltage, and saturated with a noble gas, such a medium might allow charge multiplication in the interstices with detection of the signals on external electrodes. This structure is still a figment of imagination.

Exploiting avalanche multiplication to increase the electron signal charge, gaseous counters suffer from the simultaneous production of positive ions, slowly receding to the cathodes and affecting strength and direction of the electric field. Reducing the perturbing effects of this positive space charge was one of the motivations to develop the new families of micro-pattern devices. Using multiple cascaded gas electron multiplier electrodes, the positive ion backflow can be reduced to a few percent of the collected charge; this is still large enough to perturb the detector operation, particularly in large-volume drift chambers. While ions and electrons follow the same field lines in opposite directions, the smaller transverse diffusion of ions can be exploited, suitably offsetting the holes in the GEM electrodes, to unbalance their transparency to the two signs of charge. The electrodes can be designed to optimize their transparency and permit blocking of the unwanted species (Sauli *et al.*, 2006). A potentially far reaching approach has been explored, using as gating electrode a thin monolayer of graphene, supposedly transparent to electrons but blocking the larger ions thanks to its nanometre-scale structure. Preliminary promising results have been reported (Thuiner *et al.*, 2015); however, while graphene is one of the strongest known materials, handling and framing the molecular monolayers is technically very challenging.

Another field where gaseous counters can find possible applications is the search for rare events: dark matter, WIMPS, axions, high-energy neutrinos and the like. Widely used, large systems using liquefied noble gases as detection medium are normally equipped with photomultiplier sensors to record the scintillation and sets of electrodes to collect the ionization charge. As charge gain cannot be safely obtained in liquids, dual-phase devices are designed to extract the ionization electrons from the liquid into a gas layer where multiplication is achieved using a micro-pattern detector. Requiring thorough control of the liquid level, gas pressure and temperature, the dual-phase devices are rather difficult to operate. An original alternative is the so-called bubble-assisted liquid hole multiplier: the micro-pattern detector, immersed in the liquid, is placed near a mesh where a thin gas layer is built up by local heaters (Erdal *et al.*, 2018). Ionization electrons drifted into the gas layer or created by conversions of scintillation photons on a photosensitive layer on the MPGD are amplified and detected. Still in its infancy, this approach holds promise to largely enhance the signals created in the liquid by the rare events under study.

The integration of gaseous multipliers with high-density silicon pixel readout chips permits to combine the excellent tracking properties of a time projection chamber with the granularity of the solid state readout (Van Der Graaf, 2007). Achieving single electron sensitivity, these hybrid systems hold the promise to provide the best multi-particle resolution for complex events. The technological effort to permit covering square metres of active detection area, as required for particle physics experiments, remains a great challenge (Ligtenberg *et al.*, 2020).

As an alternative, with similar resolutions and capable of covering large sensitive volumes, detection of the fluorescence photons emitted in the primary interactions and by the multiplying electrons in MPGD structures permits to image complex events; the copious scintillation at visible wavelengths observed in gas mixtures containing carbon tetrafluoride permits the use of commercial solid state cameras and recording hardware (Margato *et al.*, 2004). Combining the two-dimensional projection recorded by the imager and the timing information provided by photomultipliers, a full three-dimensional reconstruction of the vents can be performed (Brunbauer *et al.*, 2018). Limited to frame rates below a kilohertz with commercial devices, optical recording systems are in continuing evolution and hold the promise to permit increasingly high acquisition rates. Intrinsically simpler to operate than complex electronic systems, they find applications in as astrophysics, dark matter searches, medical diagnostics and cultural heritage; some were illustrated in the previous chapters.

As a final remark, the author recollects that with the emergence of the silicon micro-strip detectors in the early 1990s, a frequent comment was that they would soon replace all the gas-based systems popular at the time. The variety of detectors illustrated in the book and the speculative devices described in this chapter certainly contradict that allegation!

Further Readings on Gaseous Detectors

S.C. Curran and J.D. Craggs: *Counting Tubes Theory and Applications. Butterworth*, London (1949).

S.A. Korff: *Electron and Nuclear Counters*. Van Nostrand New York (1955).

G. Charpak: Evolution of the automatic spark chambers, *Ann. Rev. Nucl. Sci.* 20 (1970) 195.

P. Rice-Evans: Spark, *Streamer, Proportional and Drift Chambers*. Richelieu, London (1974).

F. Sauli: From bubble chambers to electronic systems: 25 years of evolution in particle detectors at CERN (1979-2004). *Phys. Reports*, 403–404 (1974) 471.

G. Knoll: *Radiation Detection and Measurements*. Wiley & Sons, New York (1989).

W. Blum, and G. Rolandi: *Particle Detection with Drift Chambers*. Springer-Verlag, Berlin (1993).

C. Grupen: *Particle Detectors*, Cambridge University Press, Cambridge (1996).

F. Sauli and A. Sharma, Micropattern gaseous detectors. *Ann. Rev. Nucl. Part. Sci.* 49 (1999) 341.

E. Nappi and V. Peskov: *Imaging Gaseous Detectors and their Applications*. Wiley VCH (2013).

J. Van Der Marel: *Microstrip & Microgap Chambers: A New Generation of Gaseous Radiation Detectors*. Delft University Press (1997).

F. Sauli: *Gaseous Radiation Detectors: Fundamentals and Applications.* Cambridge University Press, Cambridge (2014).

T. Francke and V. Peskov: *Innovative Applications and Developments of Micropattern Gaseous Detectors.* IGI Global (2014).

V. Peskov and T. Francke: *Position-Sensitive Gaseous Photomultipliers: Research and Applications.* IGI Global (2016).

References

Abbaneo, D. *et al.* (2013) 'GEM based detector for future upgrade of the CMS forward muon system', *Nucl. Instr. Meth. Phys. Res. A*, 718, pp. 383–386.

Abbaneo, D. *et al.* (2017a) 'Overview of large area triple-GEM detectors for the CMS forward muon upgrade', *Nucl. Instr. Meth. Phys. Res. A*, 845, pp. 298–303.

Abbaneo, D. *et al.* (2017b) 'R&D on a new type of micropattern gaseous detector: The Fast Timing Micropattern detector', *Nucl. Instr. Meth. Phys. Res. A*, 845, pp. 313–317.

Abbon, P. *et al.* (2001) 'MICROMEGAS, a microstrip detector for Compass', *Nucl. Instr. Meth. Phys. Res. A*, 461, pp. 29–32.

Abbon, P. *et al.* (2007) 'The COMPASS experiment at CERN', *Nucl. Instr. Meth. Phys. Res. A*, 577, pp. 455–518.

Abbon, P. *et al.* (2011) 'Particle identification with COMPASS RICH-1', *Nucl. Instr. Meth. Phys. Res. A*, 631, pp. 26–39.

Abbrescia, M. *et al.* (2020) 'Systematic studies and optimization of super sensitivity gaseous detectors of sparks, open flames and smoke', arXiv:2003.06941v1, pp. 1–24.

Abe, K. *et al.* (2011) 'The T2K experiment', *Nucl. Instr. Meth. Phys. Res. A*, 659, pp. 106–135.

Abgrall, N. *et al.* (2011) 'Time projection chambers for the T2K near detectors', *Nucl. Instr. Meth. Phys. Res. A*, 637, pp. 25–46.

Abi Akl, M. *et al.* (2016) 'Uniformity studies in large area triple-GEM based detectors', *Nucl. Instr. Meth. Phys. Res. A*, 832, pp. 1–7.

Ableev, V. *et al.* (2004) 'TPG development', *Nucl. Instr. Meth. Phys. Res. A*, 518, pp. 113–116.

Acconcia, T. V. *et al.* (2015) 'VHMPID RICH prototype using pressurized C4F8O radiator gas and VUV photon detector', *Nucl. Instr. Meth. Phys. Res. A*, 767, pp. 50–60.

Acker, A. *et al.* (2020) 'The CLAS12 MICROMEGAS Vertex Tracker', *Nucl. Instr. Meth. Phys. Res. A*, 957, p. 163423.

Adak, R. P. *et al.* (2017) 'Performance of a large size triple GEM detector at high particle rate for the CBM Experiment at FAIR', *Nucl. Instr. Meth. Phys. Res. A*, 846, pp. 29–35.

Adams, M. *et al.* (1983) 'π/K/p Identification with a Large-Aperture Ring-Imaging Cherenkov Counter', *Nucl. Instr. Meth.*, 217, pp. 237–243.

Adeva, B. *et al.* (2003) 'DIRAC: A high resolution spectrometer for pionium detection', *Nucl. Instr. Meth. Phys. Res. A*, 515, pp. 467–496.

Adeva, B. *et al.* (2016) 'Upgraded DIRAC spectrometer at CERN PS for the investigation of $\pi\pi$ and πK atoms', *Nucl. Instr. Meth. Phys. Res. A*, 839, pp. 52–85.

Agarwala, J. *et al.* (2018) 'Novel MPGD based detectors of single photons in COMPASS RICH-1', *Nucl. Instr. Meth. Phys. Res. A*, 912, pp. 158–162.

Agarwala, J. *et al.* (2019) 'The MPGD-based photon detectors for the upgrade of COMPASS RICH-1 and beyond', *Nucl. Instr. Meth. Phys. Res. A*, 936, pp. 416–419.

Aggarwal, M. M. *et al.* (2018) 'Particle identification studies with a full-size 4-GEM prototype for the ALICE TPC upgrade', *Nucl. Instr. Meth. Phys. Res. A*, 903, pp. 215–223.

Aidala, C. *et al.* (2003) 'A Hadron Blind Detector for PHENIX', *Nucl. Instr. Meth. Phys. Res. A*, 502, pp. 200–204.

Aiola, S. *et al.* (2016) 'Combination of two Gas Electron Multipliers and a MICROMEGAS as gain elements for a time projection chamber', *Nucl. Instr. Meth. Phys. Res. A*, 834, pp. 149–157.

Akindinov, A. *et al.* (2000) 'The multigap resistive plate chamber as a time-of-flight detector', *Nucl. Instr. Meth. A*, 456, p. 16.

Albani, G. *et al.* (2020) 'High-rate measurements of the novel BAND-GEM technology for thermal neutron detection at spallation sources', *Nucl. Instr. Meth. Phys. Res. A*, 957, p. 163389.

Albrecht, A. *et al.* (2005) 'Status and characterization of COMPASS RICH-1', *Nucl. Instr. Meth. Phys. Res. A*, 553, p. 215.

Albrecht, E. *et al.* (2003) 'COMPASS RICH-1', *Nucl. Instr. Meth. Phys. Res. A*, 502, pp. 112–116.

Alexeev, M. *et al.* (2008) 'Micropattern gaseous photon detectors for cherenkov imaging counters', *IEEE Nucl. Sci. Symp. Conf. Rec.*, p. 1335.

Alexeev, M. *et al.* (2010) 'Micropattern gaseous photon detectors for Cherenkov imaging counters', *Nucl. Instr. Meth. Phys. Res. A*, 623, pp. 129–131.

Alexeev, M. *et al.* (2012) 'Detection of single photons with THickGEM-based counters', *Nucl. Instr. Meth. Phys. Res. A*, 695, pp. 159–162.

Alexeev, M. *et al.* (2017) 'The MPGD-based photon detectors for the upgrade of COMPASS RICH-1', *Nucl. Instr. Meth. Phys. Res. A*, 876, pp. 96–100.

Alexopoulos, T. *et al.* (2010) 'Development of large size MICROMEGAS detector for the upgrade of the ATLAS Muon system', *Nucl. Instr. Meth. Phys. Res. A*, 617, pp. 161–165.

Alexopoulos, T. *et al.* (2011) 'A spark-resistant bulk-MICROMEGAS chamber for high-rate applications', *Nucl. Instr. Meth. Phys. Res. A*, 640, pp. 110–118.

Alexopoulos, T. *et al.* (2019) 'Performance studies of resistive-strip bulk MICROMEGAS detectors in view of the ATLAS New Small Wheel upgrade', *Nucl. Instr. Meth. Phys. Res. A*, 937, pp. 125–140.

Alexopoulos, T. *et al.* (2020) 'Construction techniques and performances of a full-size prototype MICROMEGAS chamber for the ATLAS muon spectrometer upgrade', *Nucl. Instr. Meth. Phys. Res. A*, 955, p. 162086.

Alfonsi, M. *et al.* (2004) 'High-rate particle triggering with triple-GEM detector', *Nucl. Instr. Meth. Phys. Res. A*, 518, p. 106.

Alfonsi, M. *et al.* (2005) 'Studies of etching effects on triple-GEM detectors operated with CF4-based gas mixtures', *IEEE Trans. Nucl. Sci.*, NS-52(6), pp. 2872–2878.

Alfonsi, M. *et al.* (2006) 'The LHCb triple-GEM detector for the inner region of the first station of the muon system: Construction and module-0 performance', *IEEE Trans. Nucl. Sci.*, NS-53, pp. 322–325.

Alfonsi, M. *et al.* (2007) 'Status of triple GEM muon chambers for the LHCb experiment', *Nucl. Instr. Meth. Phys. Res. A*, 581, pp. 283–286.

Alfonsi, M. *et al.* (2010) 'Activity of CERN and LNF groups on large area GEM detectors', *Nucl. Instr. Meth. Phys. Res. A*, 617, pp. 151–154.

Alfonsi, M. *et al.* (2012) 'Simulation of the dielectric charging-up effect in a GEM detector', *Nucl. Instr. Meth. Phys. Res. A*, 671, pp. 6–9.

ALICE Collaboration (2014) *Technical Design Report for the Alice Time Projection Chamber.* CERN.

Alkhazov, G. D. (1969) 'Mean value and variance of gas amplification in proportional counters', *Nucl. Instr. Meth.*, 75, pp. 161–162.

Alkhazov, G. D. (1970) 'Statistics of electron avalanches and ultimate resolution of proportional counters', *Nucl. Instr. Meth.*, 89, pp. 155–165.

Alme, J. *et al.* (2010) 'The ALICE TPC, a large 3-dimensional tracking device with fast readout for ultra-high multiplicity events', *Nucl. Instr. Meth. Phys. Res. A*, 622, p. 316.

Altunbas, C. *et al.* (2002) 'Construction, test and commissioning of the triple-gem tracking detector for compass', *Nucl. Instr. Meth. Phys. Res. A*, 490, pp. 177–203.

Altunbas, M. C. *et al.* (2003) 'Aging measurements with the gas electron multiplier (GEM)', *Nucl. Instr. Meth. Phys. Res. A*, 515, pp. 249–254.

Alunni, L. *et al.* (1994) 'Performance of MSGC on electronically and ionically conductive substrata in various operational conditions', *Nucl. Instr. Meth. Phys. Res. A*, 348, pp. 344–350.

Amaldi, U. *et al.* (2011) 'Construction, test and operation of a proton range radiography system radiography system', *Nucl. Instr. Meth. Phys. Res. A*, 629, pp. 337–344.

Amaro, F. *et al.* (2010) 'The Thick-COBRA: a new gaseous electron multiplier for radiation detectors', *JINST*, 5, p. P10002.

Amoroso, A. *et al.* (2016) 'A cylindrical GEM detector with analog readout for the BESIII experiment', *Nucl. Instr. Meth. Phys. Res. A*, 824, pp. 515–517.

Anderson, D. F. (1980) 'A xenon gas scintillation proportional counter coupled to a photoionization detector', *Nucl. Instr. Meth.*, 178, pp. 125–130.

Anderson, W. *et al.* (2011) 'Design, construction, operation and performance of a Hadron Blind Detector for the PHENIX experiment', *Nucl. Instr. Meth. Phys. Res. A*, 646, pp. 35–58.

Andriamonje, S. *et al.* (2002) 'Experimental studies of a MICROMEGAS neutron detector', *Nucl. Instr. Meth. Phys. Res. A*, 481, pp. 120–129.

Andriamonje, S. *et al.* (2004) 'A MICROMEGAS detector for the CAST experiment', *Nucl. Instr. Meth. Phys. Res. A*, 518, pp. 252–255.

Andriamonje, S. *et al.* (2010) 'Development and performance of Microbulk MICROMEGAS detectors', *JINST*, 5, p. P02001.

Angelini, F., Bellazzini, R., Brez, A., Lomtadze, T., *et al.* (1993) 'A thin, large area microstrip gas chamber with strip and pad readout', *Nucl. Instr. Meth. Phys. Res. A*, 336, pp. 106–115.

Angelini, F., Bellazzini, R., Brez, A., Massai, M. M., *et al.* (1993) 'The micro-gap chamber', *Nucl. Instr. Meth. Phys. Res. A*, 335, pp. 69–77.

Angelini, F. *et al.* (1996) 'Operation of MSGCs with gold strips built on surface-treated thin glass', *Nucl. Instr. Meth. Phys. Res. A*, 382, p. 461.

Ansoft (2019) 'MAXWELL', http://www.ansoft.com/.

Antchev, G. *et al.* (2010) 'The TOTEM detector at LHC', *Nucl. Instr. Meth. Phys. Res. A*, 617, pp. 62–66.

Anulli, F. *et al.* (2007) 'A triple GEM gamma camera for medical application', *Nucl. Instr. Meth. Phys. Res. A*, 572, pp. 266–267.

Anvar, S. *et al.* (2009) 'Large bulk MICROMEGAS detectors for TPC applications', *Nucl. Instr. Meth. Phys. Res. A*, 602, pp. 415–420.

Aprile, E. *et al.* (2006) *Noble Gas Detectors.* Wiley-VCH.

Arazi, L. *et al.* (2014) 'Laboratory studies of THGEM-based WELL structures with resistive anode', *JINST*, 9, p. P04011.

Arazi, L. *et al.* (2015) 'Liquid hole multipliers: bubble-assisted electroluminescence in liquid xenon Related content', *JINST*, 10, p. P08015.

Arogancia, D. C. *et al.* (2009) 'Study in a beam test of the resolution of a MICROMEGAS TPC with standard readout pads', *Nucl. Instr. Meth. Phys. Res. A*, 602, pp. 403–414.

Arora, R. *et al.* (2012) 'A large GEM-TPC prototype detector for Panda', *Physics Procedia*, 37, pp. 491–498.

Attié, D. (2008) 'TPC Review', *Nucl. Instr. Meth. Phys. Res. A*, 598, pp. 89–93.

Attié, D. *et al.* (2017) 'A time projection chamber with GEM-based readout', *Nucl. Instr. Meth. Phys. Res. A*, 856, pp. 109–118.

Attié, D. *et al.* (2020) 'Performances of a resistive MICROMEGAS module for the time projection chambers of the T2K near detector upgrade', *Nucl. Instr. Meth. Phys. Res. A*, 957, p. 163286.

Aune, S. *et al.* (2009) 'MICROMEGAS tracker project for CLAS12', *Nucl. Instr. Meth. Phys. Res. A*, 604, pp. 53–55.

Azevedo, C. D. R. *et al.* (2010) 'Towards THGEM UV photon detectors for RICH on single photon detection efficiency in Ne/CH4 and Ne/CF4', *JINST*, 5, p. P01002.

Azevedo, C. D. R. *et al.* (2015) 'Position resolution limits in pure noble gaseous detectors for X-ray energies from 1 to 60 keV', *Physics Letters B*, 741, pp. 272–275.

Babichev, E. A. *et al.* (1991) 'Digital radiographic scanning installation with multiwire proportional chamber', *Nucl. Instr. Meth. Phys. Res. A*, 310, pp. 449–454.

Babichev, E. A. *et al.* (1998) 'Photon counting and integrating analog gaseous detectors for digital scanning radiography', *Nucl. Instr. Meth. Phys. Res. A*, 419, pp. 290–294.

Bachmann, S. *et al.* (1998) 'Test beam results of closed MSGC modules for the CMS forward tracker', *Nucl. Instr. Meth. Phys. Res. A*, 409, pp. 6–8.

Bachmann, S. *et al.* (1999) 'Charge amplification and transfer processes in the gas electron multiplier', *Nucl. Instr. Meth. Phys. Res. A*, 438, pp. 376–408.

Bachmann, S., Bressan, A., *et al.* (2002) 'Discharge studies and prevention in the gas electron multiplier (GEM)', *Nucl. Instr. Meth. Phys. Res. A*, 479, pp. 294–308.

Bachmann, S., Kappler, S., *et al.* (2002) 'High rate X-ray imaging using multi-GEM detectors with a novel readout design', *Nucl. Instr. Meth. Phys. Res. A*, 478, pp. 104–108.

Badertscher, A. *et al.* (2009) 'Operation of a double-phase pure argon large electron multiplier time projection chamber: Comparison of single and double phase operation', *Nucl. Instr. Meth. Phys. Res. A*, 617, pp. 188–192.

Badertscher, A. *et al.* (2011) 'First operation of a double phase LAr Large Electron Multiplier Time Projection Chamber with a 2D projective readout anode', *Nucl. Instr. Meth. Phys. Res. A*, 641, pp. 48–57.

Badertscher, A. *et al.* (2013) 'First operation and performance of a 200 lt double phase LAr LEM-TPC with a 40×76 cm^2 readout', *JINST*, 8, p. P04012.

Bagaturia, Y. *et al.* (2002) 'Studies of aging and HV break down problems during development and operation of MSGC and GEM detectors for the inner tracking system of HERA-B', *Nucl. Instr. Meth. Phys. Res. A*, 490, pp. 223–242.

Bagliesi, M. G. *et al.* (2010) 'The TOTEM T2 telescope based on triple-GEM chambers', *Nucl. Instr. Meth. Phys. Res. A*, 617, pp. 134–137.

Bal, A. and Dubey, A. K. (2020) *Simulation of Triple GEM, with animations, RD51 Miniweek CERN 10–13 February 2020.*

Ball, M., Eckstein, K. and Gunji, T. (2014) 'Ion backflow studies for the ALICE TPC upgrade with GEMs', *JINST*, 9, p. C04025.

Balla, A. *et al.* (2013) 'Construction and test of the cylindrical-GEM detectors for the KLOE-2 Inner Tracker', *Nucl. Instr. Meth. Phys. Res. A*, 732, pp. 221–224.

Baranov, D. *et al.* (2017) 'GEM tracking system of the BM@N experiment', *JINST*, 12, p. C06041.

Barbosa, F. *et al.* (2019) 'A new transition radiation detector based on GEM technology', *Nucl. Instr. Meth. Phys. Res. A*, 942, p. 162356.

Baron, P. *et al.* (2008) 'AFTER, an ASIC for the readout of the large T2K time projection chambers', *IEEE Trans. Nucl. Sci.*, NS-55, p. 1744.

Barouch, G. *et al.* (1999) 'Development of a fast gaseous detector: "MICROMEGAS"', *Nucl. Instr. Meth. Phys. Res. A*, 423, pp. 32–48.

Barr, A., Bachmann, S., *et al.* (1998) 'Construction, test and operation in a high intensity beam of a small system of micro-strip gas chambers', *Nucl. Instr. Meth. Phys. Res. A*, 403, pp. 31–56.

Barr, A., Boimska, B., *et al.* (1998) 'Operation of high rate microstrip gas chambers', *Nucl. Phys. B (Proc. Suppl.)*, 61B, pp. 264–269.

Barthe, S. *et al.* (1998) 'Micro-Gap Chambers with self-aligned geometry', *Nucl. Instr. Meth. Phys. Res. A*, 414, pp. 283–288.

Bartol, F. *et al.* (1996) 'The C.A.T. Pixel Proportional Gas Counter Detector', *J. Phys. III France*, 6, pp. 337–347.

Bateman, J. E. *et al.* (2002) 'The pin pixel detector-X-ray imaging', *Nucl. Instr. Meth. Phys. Res. A*, 482, pp. 707–714.

Bay, A. *et al.* (2002) 'Study of sparking in MICROMEGAS chambers', *Nucl. Instr. Meth. Phys. Res. A*, 488, pp. 162–174.

Beckers, T. *et al.* (1994) 'Optimization of microstrip gas chamber design and operating conditions', *Nucl. Instr. Meth. Phys. Res. A*, 346, pp. 95–101.

Beirle, S. *et al.* (1999) 'Carbon coated gas electron multipliers', *Nucl. Instr. Meth. Phys. Res. A*, 423, pp. 297–302.

Bellazzini, R. (1996) 'The micro-gap chamber: new developments', *Nucl. Instr. Meth. Phys. Res. A*, 384, pp. 192–195.

Bellazzini, R. *et al.* (1998) 'Technique for the characterization of discharges in micro-strip gas chambers', *Nucl. Instr. Meth. Phys. Res. A*, 398, pp. 426–428.

Bellazzini, R. *et al.* (1999a) 'The micro-groove detector', *Nucl. Instr. Meth. Phys. Res. A*, 424, pp. 444–458.

Bellazzini, R. *et al.* (1999b) 'The WELL detector', *Nucl. Instr. Meth. Phys. Res. A*, 423, pp. 125–134.

Bellazzini, R. *et al.* (2001) 'The CMS micro-strip gas chamber project-development of a high-resolution tracking detector for harsh radiation environments', *Nucl. Instr. Meth. Phys. Res. A*, 457, pp. 22–42.

Bellazzini, R. *et al.* (2003) 'A photoelectric polarimeter based on a Micropattern Gas Detector for X-ray astronomy', *Nucl. Instr. Meth. Phys. Res. A*, 510, pp. 176–184.

Bellazzini, R. *et al.* (2007) 'Imaging with the invisible light', *Nucl. Instr. Meth. Phys. Res. A*, 581, pp. 246–253.

Bellazzini, R. *et al.* (2013) 'Photoelectric X-ray polarimetry with gas pixel detectors', *Nucl. Instr. Meth. Phys. Res. A*, 720, pp. 173–177.

Bellazzini, R. and Spandre, G. (1995) 'The microgap chamber: a new detector for the next generation of high energy, high rate experiments', *Nucl. Instr. Meth. Phys. Res. A*, 368, pp. 259–264.

Belloni, F. *et al.* (2012) 'Neutron beam imaging with an XY-MICROMEGAS detector at n_TOF at CERN', *Phys. Scripta*, 150, p. 014004.

Bencivenni, G. *et al.* (2002) 'A triple GEM detector with pad readout for high rate charged particle triggering', *Nucl. Instr. Meth. Phys. Res. A*, 488, pp. 493–502.

Bencivenni, G. *et al.* (2015) 'The micro-Resistive WELL detector: a compact spark-protected single amplification-stage MPGD', *JINST*, 10, p. P02008.

Bencivenni, G. *et al.* (2018) 'Performance of μ-RWELL detector vs resistivity of the resistive stage', *Nucl. Instr. Meth. Phys. Res. A*, 886, pp. 36–39.

Bencivenni, G. *et al.* (2020) 'High space resolution μ-RWELL for high particle rate', *Nucl. Instr. Meth. Phys. Res. A*, 958, p. 162050.

Benlloch, J., Bressan, A., Bttner, C., *et al.* (1998) 'Development of the Gas Electron Multiplier (GEM)', *IEEE Trans. Nucl. Sci.*, NS-45, pp. 234–243.

Benlloch, J., Bressan, A., Capens, M., *et al.* (1998) 'Further developments and beam tests of the gas electron multiplier (GEM)', *Nucl. Instr. Meth. Phys. Res. A*, 419, pp. 410–417.

van den Berg, F. D. *et al.* (1998) 'Design and characteristics of microgap counters on silicon', *Nucl. Instr. Meth. Phys. Res. A*, 409, pp. 90–94.

Berger, H. *et al.* (1995) 'Recent results on the properties of CsI photocathodes', *Nucl. Instr. Meth. Phys. Res.*, 360, pp. 411–415.

Berger, M. *et al.* (2017) 'A large ungated TPC with GEM amplification', *Nucl. Instr. Meth. Phys. Res. A*, 869, pp. 180–204.

Bernard, D. (2013) 'HARPO-A gaseous TPC for high angular resolution gamma-ray astronomy and polarimetry from the MeV to the TeV', *Nucl. Instr. Meth. Phys. Res. A*, 718, pp. 395–399.

Bernet, C. *et al.* (2005) 'The 40×40 cm^2 gaseous microstrip detector MICROMEGAS for the high-luminosity COMPASS experiment at CERN', *Nucl. Instr. Meth. Phys. Res. A*, 536, pp. 61–69.

Bhattacharya, D. S. *et al.* (2017) 'Experimental and numerical simulation of a TPC like set up for the measurement of ion backflow', *Nucl. Instr. Meth. Phys. Res. A*, 861, pp. 64–70.

Bhattacharya, P. *et al.* (2013) 'Comparison of bulk MICROMEGAS with different amplification gaps', *Nucl. Instr. Meth. Phys. Res. A*, 732, pp. 208–212.

Bhattacharya, P. *et al.* (2015) 'The effect of spacers on the performance of MICROMEGAS detectors: A numerical investigation', *Nucl. Instr. Meth. Phys. Res. A*, 793, pp. 41–48.

Bhattacharya, P. *et al.* (2017) '3D simulation of electron and ion transmission of GEM-based detectors', *Nucl. Instr. Meth. Phys. Res. A*, 870, pp. 64–72.

Bhattacharya, P., Moleri, L. and Bressler, S. (2019) 'Signal formation in THGEM-like detectors', *Nucl. Instr. Meth. Phys. Res. A*, 916, pp. 125–132.

Biagi, S. F. *et al.* (1997) 'Further experimental results of gas microdot detectors', *Nucl. Instr. Meth. Phys. Res. A*, 392, pp. 131–134.

Biagi, S. F. (1999) 'Monte Carlo simulation of electron drift and diffusion in counting gases under the influence of electric and magnetic fields', *Nucl. Instr. Meth. Phys. Res. A*, 421, pp. 234–240.

Biagi, S. F., Duxbury, D. and Gabathuler, E. (1998) 'Experimental results from a microdot detector overcoated with a semiconducting layer', *Nucl. Instr. Meth. Phys. Res. A*, 419, pp. 438–443.

Biagi, S. F. and Jones, T. J. (1995) 'The microdot gas avalanche chamber: an investigation of new geometries', *Nucl. Instr. Meth. Phys. Res. A*, 361, pp. 72–76.

Bianco, M. *et al.* (2016) 'Construction of two large-size four-plane MICROMEGAS detectors', *Nucl. Instr. Meth. Phys. Res. A*, 814, pp. 117–130.

Biglietti, M. *et al.* (2016) 'Development of a novel micro pattern gaseous detector for cosmic ray muon tomography', *Nucl. Instr. Meth. Phys. Res. A*, 824, pp. 220–222.

Bishai, M. *et al.* (1997) 'Micro strip gas chambers overcoated with carbon, hydrogenated amorphous silicon, and glass films', *Nucl. Instr. Meth. Phys. Res. A*, 400, pp. 233–242.

Blake, T. *et al.* (2015) 'Quenching the scintillation in CF 4 Cherenkov gas radiator', *Nucl. Instr. Meth. Phys. Res. A*, 791, pp. 27–31.

Blanco Carballo, V. M. *et al.* (2009) 'GEMGrid: a wafer post-processed GEM-like radiation detector', *Nucl. Instr. Meth. Phys. Res. A*, 608, pp. 86–91.

Bleeker, J. A. M. *et al.* (1980) 'A parallel grid imaging proportional counter optimized for detection of low brightness stellar XUV-sources', *IEE Trans. Nucl. Sci.*, NS-27, p. 176.

Bohm, J. *et al.* (1995) 'High rate operation and lifetime studies with micro-strip gas chambers', *Nucl. Instr. Meth. Phys. Res. A*, 360, pp. 34–41.

Bhmer, F. V. *et al.* (2013) 'Simulation of space-charge effects in an ungated GEM-based TPC', *Nucl. Instr. Meth. Phys. Res. A*, 719, pp. 101–108.

Bhmer, F. V. *et al.* (2014) 'First measurement of dE/dx with a GEM-based TPC', *Nucl. Instr. Meth. Phys. Res. A*, 737, pp. 214–221.

Boimska, B. *et al.* (1998) 'Progress with diamond over-coated microstrip gas chambers', *Nucl. Instr. Meth. Phys. Res. A*, 404, pp. 57–70.

Bondar, A. *et al.* (2003) 'Study of ion feedback in multi-GEM structures', *Nucl. Instr. Meth. Phys. Res. A*, 496, pp. 325–332.

Bondar, A. *et al.* (2006) 'Two-phase argon and xenon avalanche detectors based on Gas Electron Multipliers', *Nucl. Instr. Meth. Phys. Res. A*, 556, pp. 273–280.

Bortfeldt, J. *et al.* (2013) 'High-resolution MICROMEGAS telescope for pion- and muon-tracking', *Nucl. Instr. Meth. Phys. Res. A*, 718, pp. 406–408.

Bortfeldt, J. *et al.* (2017) 'Low material budget floating strip MICROMEGAS for ion transmission radiography', *Nucl. Instr. Meth. Phys. Res.*, 845, pp. 210–214.

Bortfeldt, J. *et al.* (2018) 'PICOSEC: Charged particle timing at sub-25 picosecond precision with a MICROMEGAS based detector', *Nucl. Instr. Meth. Phys. Res. A*, 903, pp. 317–325.

Bosch, R. E. *et al.* (2003) 'ALTRO chip: A 16-channel A/D converter and digital processor for gas detectors', *IEE Trans. Nucl. Sci.*, NS-50, p. 2460.

Bouclier, R. *et al.* (1992) 'High flux operation of microstrip gas chambers on glass and plastic supports', *Nucl. Instr. Meth. Phys. Res. A*, 323, pp. 240–246.

Bouclier, R. *et al.* (1993) 'Performance of gas microstrip chambers on glass substrata with electronic conductivity', *Nucl. Instr. Meth. Phys. Res. A*, 332, pp. 100–106.

Bouclier, R. *et al.* (1994) 'Ageing studies with microstrip gas chambers', *Nucl. Instr. Meth. Phys. Res. A*, 348, pp. 109–118.

Bouclier, R. *et al.* (1994) 'Development of microstrip gas chambers on substrata with electronic conductivity', *IEEE Trans. Nucl. Sci.*, NS-41, p. 821.

Bouclier, R., Capens, M., Garabatos, C., *et al.* (1995) 'On some factors affecting discharge conditions in micro-strip gas chambers', *Nucl. Instr. Meth. Phys. Res. A*, 365, pp. 65–69.

Bouclier, R., Capens, M., Evans, J., *et al.* (1995) 'Optimization of design and beam test of microstrip gas chambers', *Nucl. Instr. Meth. Phys. Res. A*, 367, pp. 163–167.

Bouclier, R. *et al.* (1996) 'High rate operation of micro-strip gas chambers on diamond-coated glass', *Nucl. Instr. Meth. Phys. Res. A*, pp. 328–331.

Bouclier, R. *et al.* (1997) 'New observations with the gas electron multiplier (GEM)', *Nucl. Instr. Meth. Phys. Res. A*, 396, pp. 50–66.

Bouclier, R., Gaudaen, J. and Sauli, F. (1991) 'The coated cathode conductive layer chamber', *Nucl. Instr. Meth. Phys. Res. A*, 310, pp. 74–77.

Boudjemline, K. *et al.* (2007) 'Spatial resolution of a GEM readout TPC using the charge dispersion signal', *Nucl. Instr. Meth. Phys. Res. A*, 574, pp. 22–27.

Bouianov, O. (2004) 'The ion trap: a new approach to gaseous microstructure detectors', *Nucl. Instr. Meth. Phys. Res. A*, 526, pp. 413–419.

Bouteille, S. *et al.* (2016) 'A MICROMEGAS-based telescope for muon tomography: The WatTo experiment', *Nucl. Instr. Meth. Phys. Res. A*, 834, pp. 223–228.

Bouvet, D. *et al.* (2000) 'Results on the radiation hardness of small gap chambers', *Nucl. Instr. Meth. Phys. Res. A*, 454, pp. 359–363.

Braccini, S. *et al.* (2015) 'Segmented ionization chambers for beam monitoring in hadrontherapy', *Modern Phys. Lett. A*, 30, p. 1540026.

Breskin, A. *et al.* (1974) 'Further results on the operation of high-accuracy drift chambers', *Nucl. Instr. Meth.*, 119, pp. 9–28.

Breskin, A. *et al.* (1975) 'Recent observations and measurements with high-accuracy drift chambers', *Nucl. Instr. Meth.*, 124, pp. 189–214.

Breskin, A. *et al.* (1994) 'Secondary electron emission gaseous detectors for fast X-ray imaging', *Nucl. Instr. Meth. Phys. Res. A*, 353, pp. 302–306.

Breskin, A. (1996) 'CsI UV photocathodes: History and mystery', *Nucl. Instr. Meth. Phys. Res. A*, 371, pp. 116–136.

Breskin, A. *et al.* (2000) 'Advances in gas avalanche photomultipliers', *Nucl. Instr. Meth. Phys. Res. A*, 442, pp. 58–67.

Breskin, A. *et al.* (2002) 'Sealed GEM photomultiplier with a CsI photocathode: ion feedback and ageing', *Nucl. Instr. Meth. Phys. Res. A*, 478, pp. 225–229.

Breskin, A. *et al.* (2009) 'A concise review on THGEM detectors', *Nucl. Instr. Meth. Phys. Res. A*, 598, pp. 107–111.

Breskin, A. *et al.* (2010) 'Progress in gaseous photomultipliers for the visible spectral range', *Nucl. Instr. Meth. Phys. Res. A*, 623, pp. 318–320.

Breskin, A. (2013) 'Liquid Hole-Multipliers: A potential concept for large single-phase noble-liquid TPCs of rare events', *J. Phys.: Conf. Ser.*, 460, p. 012020.

Breskin, A. *et al.* (2019) 'Liquid hole multiplier (LHM)-local dual-phase TPC: potential tools for future noble-liquid detectors', in *MPGD 2019 La Rochelle*.

Breskin, A., Buzulutskov, A. and Chechik, R. (2002) 'GEM photomultiplier operation in CF 4', *Nucl. Instr. Meth. Phys. Res. A*, 483, pp. 670–675.

Bressan, A., Labb, J. C., *et al.* (1999) 'Beam tests of the gas electron multiplier', *Nucl. Instr. Meth. Phys. Res. A*, 425, pp. 262–276.

Bressan, A., Hoch, M., *et al.* (1999) 'High rate behavior and discharge limits in micro-pattern detectors', *Nucl. Instr. Meth. Phys. Res. A*, 424, pp. 321–342.

Bressan, A., De Oliveira, R., *et al.* (1999) 'Two-dimensional readout of GEM detectors', *Nucl. Instr. Meth. Phys. Res. A*, 425, pp. 254–261.

Bressler, S. *et al.* (2020) 'Novel resistive-plate WELL sampling element for (S)DHCAL', *Nucl. Instr. Meth. Phys. Res. A*, 958, p. 162861.

Brunbauer, F. (2018) *Dissertation Anwendungen von Eigenschaften szintillierender Gase in optisch ausgelesenen GEM-basierten Detektoren.* Technische Universitat Vienna.

Brunbauer, F. *et al.* (2018) 'Radiation imaging with optically read out GEM-based detectors', *JINST*, 13, p. T02006.

Brunbauer, F. M. *et al.* (2018) 'Live event reconstruction in an optically read out GEM-based TPC', *Nucl. Instr. Meth. Phys. Res. A*, 886, pp. 24–29.

Brunbauer, F. M. *et al.* (2020) 'Radiation imaging with glass MICROMEGAS', *Nucl. Instr. Meth. Phys. Res. A*, 955, p. 163320.

Brunbauer, F., Ropelewski, L. and Sauli, F. (2017) 'The planispherical chamber: a Parallax-free X-ray gaseous detector for imaging applications', *Nucl. Instr. Meth. Phys. Res. A*, 875, pp. 16–20.

Bucciantonio, M., Amaldi, U., Kieffer, R., Sauli, F., *et al.* (2013) 'Development of a fast proton range radiography system for quality assurance in hadrontherapy', *Nucl. Instr. Meth. Phys. Res. A*, 732, pp. 564–567.

Bucciantonio, M., Amaldi, U., Kieffer, R., Malakhov, N., *et al.* (2013) 'Fast readout of GEM detectors for medical imaging', *Nucl. Instr. Meth. Phys. Res. A*, 718, pp. 160–163.

Bucciantonio, M. and Sauli, F. (2015) 'Proton computed tomography', *Modern Phys. Lett. A*, 30(17), p. 1540024.

Budtz-Jrgensen, C. (1992) 'Features of the micro-strip proportional counter technology', *Rev. Sci. Instrum.*, 63, p. 648.

Buzulutskov, A. *et al.* (1999) 'GEM operation in pure noble gases and the avalanche confinement', *Nucl. Instr. Meth. Phys. Res. A*, 433, pp. 471–475.

Buzulutskov, A. *et al.* (2000) 'Further studies of the GEM photomultiplier', *Nucl. Instr. Meth. Phys. Res. A*, 442, pp. 68–73.

Buzulutskov, A. (2012) 'Advances in Cryogenic avalanche detectors', *JINST*, 7, p. C02025.

Byrne, J. (1969) 'Statistics of Electron Avalanches in the Proportional Counter', *Nucl. Instr. Meth.*, pp. 291–296.

Cantini, C. *et al.* (2015) 'Performance study of the effective gain of the double phase liquid Argon LEM Time Projection Chamber', *JINST*, 10, p. P03017.

Capeans, M. (2003) 'Aging and materials: lessons for detectors and gas systems', *Nucl. Instr. Meth. Phys. Res. A*, 515, pp. 73–88.

Capeans, M. *et al.* (2003) *Construction of GEM detectors, CERN Technical Note TA1/00-03.*

Cardini, A., Bencivenni, G. and De Simone, P. (2012) 'The operational experience of the triple-GEM detectors of the lhcb muon system: summary of 2 years of data taking', *IEEE 2003 Nucl. Sci. Symp. Conf. Records*, N13-5.

Carnegie, R. *et al.* (2005) 'Resolution studies of cosmic-ray tracks in a TPC with GEM readout', *Nucl. Instr. Meth. Phys. Res. A*, 538, pp. 372–383.

Carver, J. H. and Mitchell, P. (1964) 'Ionization chambers for the vacuum ultra-violet', *J. Sci. Instrum.*, 41, p. 555.

Causa, F. *et al.* (2015) 'A phenomenological model to study the energy discrimination potential of GEM detectors in the X-ray range', *Nucl. Instr. Meth. Phys. Res. A*, 799, pp. 75–80.

Cavoto, G. *et al.* (2020) 'Micro pattern gas detector optical readout for directional dark matter searches', *Nucl. Instr. Meth. Phys. Res. A*, 958, p. 162400.

Chaplier, G. *et al.* (1999) 'Preliminary results of the experimental and simulated intrinsic properties of the Compteur A Trou (CAT) detector: behavior with synchrotron radiation', *Nucl. Instr. Meth. Phys. Res. A*, 426, pp. 339–355.

Chaplier, G., Lemonnier, M. and Megtert, S. (2000) 'Derivation of a simple analytical expression of the gain of pure argon filled CAT proportional counters', *Nucl. Instr. Meth. Phys. Res. A*, 440, pp. 466–470.

Charpak, G. *et al.* (1968) 'The use of Multiwire Proportional Counters to select and localize charged particles', *Nucl. Instr. Meth.*, 62, pp. 262–268.

Charpak, G. *et al.* (1977) 'Some properties of spherical drift chambers', *Nucl. Instr. Meth.*, 141, pp. 449–455.

Charpak, G. *et al.* (1988) 'Studies of light emission by continuously sensitive avalanche chambers', *Nucl. Instr. Meth. Phys. Res. A*, 269, pp. 142–148.

Charpak, G. *et al.* (2002) 'MICROMEGAS, a multipurpose gaseous detector', *Nucl. Instr. Meth. Phys. Res. A*, 478, pp. 26–36.

Charpak, G. C., Policarpo, A. and Sauli, F. (1980) 'The photo-ionization proportional scintillation chamber', *IEEE Trans. Nucl. Sci.*, NS-27, pp. 212–215.

Charpak, G., Peskov, V. and Sauli, F. (1992) 'Preliminary results of the study of gaseous detectors with solid photocathodes sensitive in the spectral region from 105 to 300 nm', *Nucl. Instr. Meth. Phys. Res. A*, 323, pp. 445–451.

Charpak, G., Rahm, D. and Steinert, H. (1970) 'Some developments in the operation of Multiwire Proportional Chambers', *Nucl. Instr. Meth.*, 80, pp. 13–34.

Charpak, G. and Sauli, F. (1978) 'The multistep avalanche chamber: a new high rate, high accuracy gaseous detector', *Phys. Letters*, 78 B(4), pp. 523–528.

Chechik, R. *et al.* (2004) 'Thick GEM-like hole multipliers: properties and possible applications', *Nucl. Instr. Meth. Phys. Res. A*, 535, pp. 303–308.

Chefdeville, M. *et al.* (2016) 'Resistive MICROMEGAS for sampling calorimetry, a study of charge-up effects', *Nucl. Instr. Meth. Phys. Res. A*, 824, pp. 510–511.

Chepel, V. and Arujo, H. (2013) 'Liquid noble gas detectors for low energy particle physics', *JINST*, 8, p. R04001.

Chernyshova, M. *et al.* (2018) 'Measuring issues in the GEM detector system for fusion plasma imaging', *JINST*, 13, p. C08001.

Cho, H. S. *et al.* (1997) 'Performance of microgap gas chambers fabricated with selected anode metals', *IEEE Trans. Nucl. Sci.*, NS-44, p. 747.

Cho, H. S. *et al.* (1998) 'Performance and spark damage studies of microgap gas chambers with various anode strip metals', *Nucl. Physics*, 61B, p. 258.

Chorowicz, V. *et al.* (1997) 'Performance of a small gap chamber', *Nucl. Instr. Meth. Phys. Res. A*, 401, pp. 238–242.

Cicognani, G., Guerard, B. and Oed, A. (1997) 'Performance of MSGC on diamond-coated glass', *Nucl. Instr. Meth. Phys. Res. A*, 392, pp. 115–119.

Claps, G. *et al.* (2016) 'The GEMpix detector as new soft X-rays diagnostic tool for laser produced plasmas', *Rev. Sci. Instrum.*, 87, p. 103505.

Clergeau, J.-F. *et al.* (1997) 'Performance of three variants of micro-gap chambers', *Nucl. Instr. Meth. Phys. Res. A*, 392, pp. 140–144.

Clergeau, J.-F. *et al.* (1997) 'Simulation of MSGC signals and application to the front-end electronics for LHC experiments', *Nucl. Instr. Meth. Phys. Res. A*, 392, pp. 109–114.

Clergeau, J. F. *et al.* (2001) 'Operation of sealed microstrip gas chambers at the ILL', *Nucl. Instr. Meth. Phys. Res. A*, 471, pp. 60–68.

Colas, P., Giomataris, I. and Lepeltier, V. (2004) 'Ion backflow in the MICROMEGAS TPC for the future linear collider', *Nucl. Instr. Meth. Phys. Res. A*, 535, pp. 226–230.

Comby, G. *et al.* (1980) 'Detecteur multipointes focalisation cathodique', *Nucl. Instr. Meth.*, 174, pp. 77–92.

Conceio, A. S. *et al.* (2010) 'Ion back-flow suppression in GEM-MIGAS', *IEE Trans. Nucl. Sci.*, NS-57, pp. 3753–3759.

Corradi, G., Murtas, F. and Tagnani, D. (2007) 'A novel High-Voltage System for a triple GEM detector', *Nucl. Instr. Meth. Phys. Res. A*, 572, pp. 96–97.

Correia, P. *et al.* (2014) 'A dynamic method for charging-up calculations: the case of GEM', *JINST*, 9, p. P07025.

Cozza, D. *et al.* (2003) 'The CSI-based RICH detector array for the identification of high momentum particles in ALICE', *Nucl. Instr. Meth. Phys. Res. A*, 502, pp. 101–107.

Croci, G., Alfonsi, M., *et al.* (2013) 'Discharge probability measurement of a Triple GEM detector irradiated with neutrons', *Nucl. Instr. Meth. Phys. Res. A*, 712, pp. 108–112.

Croci, G., Claps, G., Caniello, R., *et al.* (2013) 'GEM-based thermal neutron beam monitors for spallation sources', *Nucl. Instr. Meth. Phys. Res. A*, 732, pp. 217–220.

Croci, G., Claps, G., Cavenago, M., *et al.* (2013) 'nGEM fast neutron detectors for beam diagnostics', *Nucl. Instr. Meth. Phys. Res. A*, 720, pp. 144–148.

Curran, S. C. and Craggs, J. D. (1949) *Counting Tubes Theory and Applications.* London: Butterworth.

Dalla Torre, S. (2011) 'Status and perspectives of gaseous photon detectors', *Nucl. Instr. Meth. Phys. Res. A*, 639, pp. 111–116.

Dalla Torre, S. (2020) 'Gaseous counters with CsI photocathodes: The COMPASS RICH', *Nucl. Instr. Meth. Phys. Res. A*, 970, p. 163768.

Dangendorf, V. *et al.* (1990) 'A gas-filled UV-photon detector with CsI photocathode for the detection of Xe light', *Nucl. Instr. Meth. Phys. Res. A*, 289, pp. 322–324.

Das, S. (2016) 'Study of gain variation as a function of physical parameters of GEM foil', *Nucl. Instr. Meth. Phys. Res. A*, 824, pp. 518–520.

Dehmelt, K. (2015) 'Performance of a quintuple-gem based RICH detector prototype', arXiv:submit/1158752.

Deines-Jones, P. *et al.* (2002) 'Large-area imaging micro-well detectors for high-energy astrophysics', *Nucl. Instr. Meth. Phys. Res. A*, 478, pp. 130–134.

Deisting, A. *et al.* (2019) 'Secondary discharge studies in single- and multi-GEM structures', *Nucl. Instr. Meth. Phys. Res. A*, 937, pp. 168–180.

Delbart, A. *et al.* (2001) 'New developments of MICROMEGAS detector', *Nucl. Instr. Meth. Phys. Res. A*, 461, pp. 84–87.

Delbart, A. *et al.* (2002) 'Performance of MICROMEGAS with preamplification at high intensity hadron beams', *Nucl. Instr. Meth. Phys. Res. A*, 478, pp. 205–209.

Delbart, A. (2010) 'Production and calibration of 9 m^2 of bulk-MICROMEGAS detectors for the readout of the ND280/TPCs of the T2K experiment', *Nucl. Instr. Meth. Phys. Res. A*, 623, pp. 105–107.

Delbart, A. *et al.* (2011) 'MICROMEGAS for charge readout of double phase Liquid Argon TPCs', *J. Phys.: Conf. Ser.*, 308, p. 012017.

Derenzo, S. E. *et al.* (1974) 'Electron avalanche in liquid xenon', *Phys. Rev A*, 9(6), pp. 2582–2591.

Derr, J. *et al.* (2000) 'Fast signals and single electron detection with a MICROMEGAS photodetector', *Nucl. Instr. Meth. Phys. Res. A*, 449, pp. 314–321.

Derr, J. *et al.* (2001) 'Spatial resolution in MICROMEGAS detectors', *Nucl. Instr. Meth. Phys. Res. A*, 459, pp. 523–531.

Derr, J. and Giomataris, J. (2002) 'Recent experimental results with MICROMEGAS', *Nucl. Instr. Meth. Phys. Res. A*, 477, pp. 23–28.

Despre, P. *et al.* (2005) 'Evaluation of a full-scale gas microstrip detector for low-dose X-ray imaging', *Nucl. Instr. Meth. Phys. Res. A*, 536, pp. 52–60.

Diener, R. (2012) 'Development of a TPC for an ILC detector', *Phys. Procedia*, 37, pp. 456–463.

Dion, M. P., Martoff, C. J. and Hosack, M. (2010) 'On the mechanism of Townsend avalanche for negative molecular ions', *Astroparticle Physics*, 33, pp. 216–220.

Divani Veis, N., Ehret, A., *et al.* (2018) 'Implementation of the PANDA planar-GEM tracking detector in Monte Carlo simulations', *Nucl. Instr. Meth. Phys. Res. A*, 880, pp. 201–209.

Divani Veis, N., Firoozabadi, M. M., *et al.* (2018) 'Performance studies of the PANDA planar GEM-tracking detector in physics simulations', *Nucl. Instr. Meth. Phys. Res. A*, 884, pp. 150–156.

Dixit, M. *et al.* (1994) 'Gas Microstrip Detectors on Resistive Plastic Substrates', *Nucl. Instr. Meth. Phys. Res. A*, 348, p. 365.

Dixit, M. S. *et al.* (2004) 'Position sensing from charge dispersion in micro-pattern gas detectors with a resistive anode', *Nucl. Instr. Meth. Phys. Res. A*, 518, pp. 721–727.

Dominik, W. *et al.* (1989) 'A gaseous detector for high-accuracy autoradiography of radioactive compounds with optical readout of avalanche positions', *Nucl. Instr. Meth. Phys. Res. A*, 278, pp. 779–787.

Duarte Pinto, S. *et al.* (2008) 'A large area GEM detector', *IEEE Nucl. Sci. Symp. Conf. Rec.*, pp. 1426–1432.

Duarte Pinto, S. *et al.* (2011) 'First results of spherical GEMs', *IEEE Nucl. Sci. Symp. Conf. Rec.*, arXiv:1011.

Erdal, E. *et al.* (2015) 'Direct observation of bubble-assisted electroluminescence in liquid xenon', *JINST*, 10, p. P11002.

Erdal, E. *et al.* (2017) 'First demonstration of VUV-photon detection in liquid xenon with THGEM and GEM-based Liquid Hole Multipliers', *Nucl. Instr. Meth. Phys. Res. A*, 845, pp. 218–221.

Erdal, E. *et al.* (2018) 'Recent advances in bubble-assisted liquid hole-multipliers in liquid xenon', *JINST*, 13, p. P12008.

Everaerts, P. (2006) 'Rate Capabilty and Ion Feedback in GEM Detectors', *Doctoral Thesis at Gent University.*

Fabbietti, L. *et al.* (2003) 'Photon detection efficiency in the CsI based HADES RICH', *Nucl. Instr. Meth. Phys. Res. A*, 502, pp. 256–260.

Fallavollita, F. (2019) 'Aging phenomena and discharge probability studies of the triple-GEM detectors for future upgrades of the CMS muon high rate region at the HL-LHC', *Nucl. Instr. Meth. Phys. Res. A*, 936, pp. 427–429.

Fang, R. *et al.* (1995) 'Charge accumulation at the interface between two dielectrics and gas gain variation of microstrip gas chambers', *Nucl. Instr. Meth. Phys. Res. A*, 365, pp. 59–64.

Fenker, H. *et al.* (2008) 'BoNus: Development and use of a radial TPC using cylindrical GEMs', *Nucl. Instr. Meth. Phys. Res. A*, 592, pp. 273–286.

Ferretti, A., De Nardo, L. and Lombardi, M. (2008) 'Preliminary study of a leak microstructure detector as a new single-electron counter for STARTRACK experiment', *Nucl. Instr. Meth. Phys. Res. A*, 599, pp. 215–220.

Fetal, S. *et al.* (2003) 'Dose imaging in radiotherapy with an Ar-CF 4 filled scintillating GEM', *Nucl. Instr. Meth. Phys. Res. A*, 513, pp. 42–46.

Fetal, S. T. G. *et al.* (2007) 'Towards a PMT based optical readout GEM TPC-First results', *Nucl. Instr. Meth. Phys. Res. A*, 581, pp. 201–205.

Fonte, P. *et al.* (1999) 'Single-electron pulse-height spectra in thin-gap parallel-plate chambers', *Nucl. Instr. Meth. Phys. Res. A*, 433, pp. 513–517.

Fraenkel, Z. *et al.* (2005) 'A hadron blind detector for the PHENIX experiment at RHIC', *Nucl. Instr. Meth. Phys. Res. A*, 546, pp. 466–480.

Fraga, F. A. F. *et al.* (2001) 'Optical readout of GEMs', *Nucl. Instr. Meth. Phys. Res. A*, 471, pp. 125–130.

Fraga, F. A. F. *et al.* (2002) 'CCD readout of GEM-based neutron detectors', *Nucl. Instr. Meth. Phys. Res. A*, 478, pp. 357–361.

Fraga, M. *et al.* (2003) 'The GEM scintillation in He-CF4, Ar-CF4, Ar-TEA and Xe-TEA mixtures', *Nucl. Instr. Meth. Phys. Res. A*, 504, pp. 88–92.

Friese, J. *et al.* (1999) 'Enhanced quantum efficiency for CsI grown on a graphite-based substrate coating', *Nucl. Instr. Meth. A*, 438, p. 86.

Fujita, K. *et al.* (2007) 'A high-resolution two-dimensional 3He neutron MSGC with pads for neutron scattering experiments', *Nucl. Instr. Meth. Phys. Res. A*, 580, pp. 1027–1030.

Fujiwara, T., Mitsuya, Y. and Takahashi, H. (2018) 'Radiation imaging with glass gas electron multipliers (G-GEMs)', *Nucl. Instr. Meth. Phys. Res. A*, 878, pp. 40–49.

Galan, J. *et al.* (2012) 'Aging studies of MICROMEGAS prototypes for the HL-LHC', *JINST*, 7, p. C01041.

Galn, J. *et al.* (2013) 'Characterization and simulation of resistive-MPGDs with resistive strip and layer topologies', *Nucl. Instr. Meth. Phys. Res. A*, 732, pp. 229–232.

Galavanov, A. *et al.* (2019) 'Performance of the BM@N GEM/CSC tracking system at the Nuclotron beam', *EPJ Web of Conf*, 204, p. 07009.

Gallas, A. (2005) 'Performance of the high momentum particle identification CsI-RICH for ALICE at CERN-LHC', *Nucl. Instr. Meth. Phys. Res. A*, 553, pp. 345–350.

Gasik, P. (2016) 'Discharge propagation studies with 1-, 2- and 4-GEM structures in Ar- and Ne-based mixtures', in *RD51 Coll. meeting CERN, April 13–17, 2016*.

Gasik, P. (2017) 'Building a large-area GEM-based readout chamber for the upgrade of the ALICE TPC', *Nucl. Instr. Meth. Phys. Res. A*, 845, pp. 222–225.

Gasik, P. *et al.* (2017) 'Charge density as a driving factor of discharge formation in GEM-based detectors', *Nucl. Instr. Meth. Phys. Res. A*, 870, pp. 116–122.

Gauci, J. L., Gatt, E. and Casha, O. (2020) 'Preparing the ALICE-HMPID RICH for the high-luminosity LHC period 2021–2023', *Nucl. Instr. Meth. Phys. Res. A*, 952, p. 161798.

Gebauer, B. (2004) 'Towards detectors for next generation spallation neutron sources', *Nucl. Instr. Meth. Phys. Res. A*, 535, pp. 65–78.

Geerebaert, Y. *et al.* (2017) 'Measurement of 1.7-74 MeV polarised γ rays with the HARPO TPC', *Nucl. Instr. Meth. Phys. Res. A*, 845, p. 228.

George, S. P. *et al.* (2015) 'Particle tracking with a Timepix based triple GEM detector', *JINST*, 10, p. P11003.

Giomataris, I. *et al.* (2006) 'MICROMEGAS in a bulk', *Nucl. Instr. Meth. Phys. Res. A*, 560, pp. 405–408.

Giomataris, Y. *et al.* (1996) 'MICROMEGAS: a high-granularity position-sensitive gaseous detector for high particle-flux environments', *Nucl. Instr. Meth. Phys. Res. A*, 376, pp. 29–35.

Giomataris, Y. (1998) 'Development and prospects of the new gaseous detector: "MICROMEGAS"', *Nucl. Instr. Meth. Phys. Res. A*, 419, pp. 239–250.

Giomataris, Y. and Charpak, G. (1991) 'A hadron-blind detector', *Nucl. Instr. Meth. Phys. Res. A*, 310, pp. 589–695.

Giordanengo, S. *et al.* (2013) 'Design and characterization of the beam monitor detectors of the Italian National Center of Oncological Hadron-therapy (CNAO)', *Nucl. Instr. Meth. Phys. Res. A*, 698, pp. 202–207.

Gnanvo, K. *et al.* (2015) 'Large size GEM for Super Bigbite Spectrometer (SBS) polarimeter for Hall A 12 GeV program at JLab', *Nucl. Instr. Meth. Phys. Res. A*, 782, pp. 77–86.

Gnanvo, K. *et al.* (2016) 'Performance in test beam of a large-area and light-weight GEM detector with 2D stereo-angle (U-V) strip readout', *Nucl. Instr. Meth. Phys. Res. A*, 808, pp. 83–92.

Gonzlez-Daz, D., Monrabal, F. and Murphy, S. (2018) 'Gaseous and dual-phase time projection chambers for imaging rare processes', *Nucl. Instr. Meth. Phys. Res. A*, 878, pp. 200–255.

van der Graaf, H. *et al.* (2006) 'Recent GridPix results: An integrated MICROMEGAS grid and an ageing test of a MICROMEGAS chamber', *Nucl. Instr. Meth. Phys. Res. A*, 566, pp. 62–65.

van der Graaf, H. (2007) 'GridPix: An integrated readout system for gaseous detectors with a pixel chip as anode', *Nucl. Instr. Meth. Phys. Res. A*, 580, pp. 1023–1026.

Gros, P. *et al.* (2013) 'Blocking positive ion backflow using a GEM gate: experiment and simulations', *JINST*, 8, p. C11023.

Gutierrez, R. M., Cerquera, E. A. and Manana, G. (2012) 'MPGD for breast cancer prevention: a high resolution and low dose radiation medical imaging', *JINST*, 7, p. C07007.

Hanson, K. M. *et al.* (1981) 'Computed tomography using proton energy loss', *Phys. Med. Biol.*, 26, p. 965.

Harrach, D. V and Specht, H. J. (1979) 'A square meter position sensitive parallel plate detector for heavy ions', *Nucl. Instr. Meth.*, 164, pp. 477–490.

Hattori, K. *et al.* (2007) 'Gamma-ray imaging with a large micro-TPC and a scintillation camera', *Nucl. Instr. Meth. Phys. Res. A*, 581, pp. 517–521.

Hauer, P. *et al.* (2019) 'Measurements of the charging-up effects in gas electron multipiers', arXiv:1911.01848.

Hendrix, J. (1984) 'The fine grid detector: a parallel electrode position sensitive detector', *IEEE Trans. Nucl. Sci.*, NS-31, p. 281.

Hesser, J. E. and Dressler, K. (1967) 'Radiative lifetimes of ultraviolet emission systems excited in BF3, CF4 and SiF4', *J. Chem. Phys.*, 47, p. 1621.

Hildn, T. *et al.* (2014) 'Optical quality assurance of GEM foils', *Nucl. Instr. Meth. Phys. Res. A*, 770, pp. 113–122.

Hoedlmoser, H. *et al.* (2006) 'Production technique and quality evaluation of CsI photocathodes for the ALICE/HMPID detector', *Nucl. Instr. Meth. Phys. Res. A*, 566, pp. 338–350.

Hoedlmoser, H. *et al.* (2007) 'Long term performance and ageing of CsI photocathodes for the ALICE/HMPID detector', *Nucl. Instr. Meth. Phys. Res. A*, 574, pp. 28–38.

Hollenstein, C. (1994) 'Thin film diamond deposition by plasmas, a review', in *1st Int. Workshop on Electronics and Detector Cooling WELDEC*. Lausanne, Suisse.

Holroyd, R. A. *et al.* (1987) 'Measurement of the absorption length and absolute quantum efficiency of TMAE and TEA from Threshold to 120 nm', *Nucl. Instr. Meth. Phys. Res. A*, 261, pp. 440–444.

Hott, T. (1998) 'MSGC development for the inner tracker of HERA-B', *Nucl. Instr. Meth. Phys. Res. A*, 408, p. 258.

Houry, M. *et al.* (2006) 'DEMIN: A neutron spectrometer, MICROMEGAS-type, for inertial confinement fusion experiments', *Nucl. Instr. Meth. Phys. Res. A*, 557, pp. 648–656.

Iakovidis, G. (2013) 'The MICROMEGAS project for the ATLAS upgrade', *JINST*, 8, p. C12007.

Inuzuka, M. *et al.* (2004) 'Gas Electron Multiplier produced with the plasma etching method', *Nucl. Instr. Meth. Phys. Res. A*, 525, pp. 529–534.

Iwakiri, W. B. *et al.* (2016) 'Performance of the PRAXyS X-ray polarimeter', *Nucl. Instr. Meth. Phys. Res. A*, 838, pp. 89–95.

Janssen, M. E. *et al.* (2006) 'R&D studies ongoing at DESY on a time projection chamber for a detector at the International Linear Collider', *Nucl. Instr. Meth. Phys. Res. A*, 566, pp. 75–79.

Jeanneau, F., Kebbiri, M. and Lepeltier, V. (2010) 'Ion back-flow gating in a MICROMEGAS device', *Nucl. Instr. Meth. Phys. Res. A*, 623, pp. 94–96.

Jibaly, M. *et al.* (1989) 'The aging of wire chambers filled with dimethyl ether: wire and construction materials and freon impurities', *Nucl. Instr. Meth. Phys. Res. A*, 283, pp. 692–701.

Kadyk, J. (1991) 'Wire chamber ageing', *Nucl. Instr. Meth. Phys. Res. A*, 300, pp. 436–479.

Kalliokoski, M. *et al.* (2012) 'Optical scanning system for quality control of GEM-foils', *Nucl. Instr. Meth. Phys. Res. A*, 664, pp. 223–230.

Kaminski, J. *et al.* (2004) 'Development and studies of a time projection chamber with GEMs', *Nucl. Instr. Meth. Phys. Res. A*, 535, pp. 201–205.

Kanno, K. *et al.* (2016) 'Development of a hadron blind detector using a finely segmented pad readout', *Nucl. Instr. Meth. Phys. Res. A*, 819, pp. 20–24.

Kappler, S. *et al.* (2004) 'A GEM-TPC prototype with low-noise highly integrated front-end electronics for linear collider studies', *IEEE Trans. Nucl. Sci.*, 51(3), pp. 1039–1043.

Karlen, D., Poffenberger, P. and Rosenbaum, G. (2005) 'TPC performance in magnetic fields with GEM and pad readout', *Nucl. Instr. Meth. Phys. Res. A*, 555, pp. 80–92.

Ketzer, B. *et al.* (2002) 'Triple GEM tracking detectors for COMPASS', *IEEE Trans. Nucl. Sci.*, NS-49(5), pp. 2403–2410.

Ketzer, B. *et al.* (2004) 'Performance of triple GEM tracking detectors in the COMPASS experiment', *Nucl. Instr. Meth. Phys. Res. A*, 535, pp. 314–318.

Ketzer, B. (2013) 'A time projection chamber for high-rate experiments: Towards an upgrade of the ALICE TPC', *Nucl. Instr. Meth. Phys. Res. A*, 732, pp. 237–240.

Killenberg, M. *et al.* (2004) 'Charge transfer and charge broadening of GEM structures in high magnetic fields', *Nucl. Instr. Meth. Phys. Res. A*, 530, pp. 251–257.

Kim, J. G. *et al.* (2004) 'Electron avalanches in liquid argon mixtures', *Nucl. Instr. Meth. Phys. Res. A*, 534, pp. 376–396.

Kitaguchi, T. *et al.* (2018) 'An optimized photoelectron track reconstruction method for photoelectric X-ray polarimeters', *Nucl. Instr. Meth. Phys. Res. A*, 880, pp. 188–193.

Klein, M. and Schmidt, C. J. (2011) 'CASCADE, neutron detectors for highest count rates in combination with ASIC/FPGA based readout electronics', *Nucl. Inst. Meth. Phys. Res. A*, 628, pp. 9–18.

Knoll, G. (1989) *Radiation Detection and Measurements*. New York: Wiley & Sons.

Kobayashi, M. *et al.* (2007) 'Performance of MPGD-based TPC prototypes for the linear collider experiment', *Nucl. Instr. Meth. Phys. Res. A*, 581, pp. 265–270.

Kobayashi, M. *et al.* (2019) 'Measurement of the electron transmission rate of the gating foil for the TPC of the ILC experiment', *Nucl. Instr. Meth. Phys. Res. A*, 918, pp. 41–53.

Kubo, H. *et al.* (2003) 'Development of a time projection chamber with micro-pixel electrodes', *Nucl. Instr. Meth. Phys. Res. A*, 513, pp. 94–98.

Kudryavtsev, V. N., Maltsev, T. V. and Shekhtman, L. I. (2017) 'Study of spatial resolution of coordinate detectors based on Gas Electron Multipliers', *Nucl. Instr. Meth. Phys. Res. A*, 845, pp. 289–292.

Kuger, F. *et al.* (2016) 'Mesh geometry impact on MICROMEGAS performance with an Exchangeable Mesh prototype', *Nucl. Instr. Meth. Phys. Res. A*, 824, pp. 541–542.

Kuger, F. (2017) 'Performance studies of resistive MICROMEGAS detectors for the upgrade of the ATLAS Muon spectrometer', *Nucl. Instr. Meth. Phys. Res. A*, 845, pp. 248–252.

Latronico, L. (2000) 'Status of the CMS MSGC Tracker', *Nucl. Instr. Meth. Phys. Res. A*, 446, pp. 346–354.

Lautner, L. *et al.* (2019) 'High voltage scheme optimization for secondary discharge mitigation in GEM-based detectors', *JINST*, 14, p. P08024.

Ledermann, B. *et al.* (2007) 'Development studies for the ILC: Measurements and simulations for a time projection chamber with GEM technology', *Nucl. Instr. Meth. Phys. Res. A*, 581, pp. 232–235.

Li, H. *et al.* (2015) 'Assembly and test of the gas pixel detector for X-ray polarimetry', *Nucl. Instr. Meth. Phys. Res. A*, 804, pp. 155–162.

Ligtenberg, C. *et al.* (2020) 'The gaseous QUAD pixel detector', *Nucl. Instr. Meth. Phys. Res. A*, 958, p. 162731.

Lippmann, C. (2016) 'A continuous read-out TPC for the ALICE upgrade', *Nucl. Instr. Meth. Phys. Res. A*, 824, pp. 543–547.

Llopart, X. *et al.* (2002) 'Medipix2: A 64-k pixel readout chip with 55-μm square elements working in single photon counting mode', *IEEE Trans. Nucl. Sci.*, NS-49, p. 2279.

Llopart, X. *et al.* (2007) 'Timepix, a 65k programmable pixel readout chip for arrival time, energy and/or photon counting measurements', *Nucl. Instr. Meth. Phys. Res. A*, 581, pp. 485–494.

Lombardi, M. and Lombardi, F. S. (1997) 'The leak microstructure, preliminary results', *Nucl. Instr. Meth. Phys. Res. A*, 392, pp. 23–27.

Lorikyan, M. P., Asryan, G. A. and Gary, C. K. (2007) 'Investigation of porous dielectric detectors at high intensity particles', *Nucl. Instr. Meth. Phys. Res. A*, 570, pp. 475–478.

Lu, C. and McDonald, K. T. (1994) 'Properties of reflective and semitransparent CsI photocathodes', *Nucl. Instr. Meth. A*, 343, pp. 135–151.

Lund, N. *et al.* (1999) 'JEM-X-The X-ray monitor on INTEGRAL', *Astro. Lett. Comm.*, 39, pp. 339–345.

Lund, N. *et al.* (2003) 'JEM-X: The X-ray monitor aboard INTEGRAL', *Astronomy and Astrophysics*, 411, pp. L231–L238.

Lyashenko, A. *et al.* (2009) 'High-gain continuous-mode operated gaseous photomultipliers for the visible spectral range', *Nucl. Instr. Meth. Phys. Res. A*, 610, pp. 161–163.

Maghrbi, Y., Verwilligen, P. and Maggi, M. (2020) 'Fast timing micropattern gaseous detector (FTM) simulations for future colliders and medical applications', *Nucl. Instr. Meth. Phys. Res. A*, 954, p. 161666.

Maia, J. M. *et al.* (2004) 'Avalanche-ion back-flow reduction in gaseous electron multipliers based on GEM/MHSP', *Nucl. Instr. Meth. Phys. Res. A*, 523, pp. 334–344.

Majewski, S. and Sauli, F. (1975) 'Support lines and beam killers for large size multiwire proportional chambers', *CERN-NP Int. Rep. 75-14*.

Majumdar, N. and Mukhopadhyay, S. (2007) 'Computation of electrostatic and gravitational sag in multiwire chambers', arXiv:physics/0703004.

Maltsev, T., Sauli, F. and Shekhtman, L. (2019) 'Study of discharge properties in cascaded gaseous detectors', *Proc. MPGD 2019 Conf. La Rochelle*.

Manjarrs, J. *et al.* (2012) 'Performances of anode-resistive MICROMEGAS for HL-LHC', *JINST*, 7, p. C03040.

Margato, L. *et al.* (2004) 'Performance of an optical readout GEM-based TPC', *Nucl. Instr. Meth. Phys. Res. A*, 535, pp. 231–235.

Margato, L. M. S. *et al.* (2011) 'Effect of the gas contamination on CF 4 primary and secondary scintillation', *Nucl. Instr. Meth. Phys. Res. A*, 695, pp. 425–428.

Martinengo, P. *et al.* (2011) 'R&D results on a CsI-coated triple thick GEM-based photodetector', *Nucl. Instr. Meth. Phys. Res. A*, 639, pp. 126–129.

Martinez-Davalos, A. *et al.* (1994) 'Evaluation of a low-dose digital X-ray system with improved spatial resolution', *Nucl. Instr. Meth. Phys. Res. A*, 348, pp. 241–244.

Martoff, C. J. *et al.* (2000) 'Suppressing drift chamber diffusion without magnetic field', *Nucl. Instr. Meth. Phys. Res. A*, 440, pp. 355–359.

Martoff, C. J. *et al.* (2005) 'Negative ion drift and diffusion in a TPC near 1 bar', *Nucl. Instr. Meth. Phys. Res. A*, 555, pp. 55–58.

Martoff, C. J. *et al.* (2009) 'A benign, low Z electron capture agent for negative ion TPCs', *Nucl. Instr. Meth. Phys. Res. A*, 598, pp. 501–504.

Masaoka, S. *et al.* (2003) 'Optimization of a micro-strip gas chamber as a two-dimensional neutron detector using gadolinium converter', *Nucl. Instr. Meth. Phys. Res. A*, 513, pp. 538–549.

Mathieson, E. and Smith, G. C. (1988) 'Charge Distribution in Parallel Plate Avalanche Chambers', *Nucl. Instr. Meth. Phys. Res. A*, 273, pp. 518–521.

Di Mauro, A. (2014) 'Status and perspectives of gaseous photon detectors', *Nucl. Instr. Meth. Phys. Res. A*, 766, pp. 126–132.

Mazzitelli, G. *et al.* (2018) 'A high resolution TPC based on GEM optical readout', *IEEE Nucl. Sci. Symp. Conf. Rec.*, pp. 3–6.

McCarty, R. *et al.* (1986) 'Identification of large transverse momentum hadrons using a ring-imaging Cherenkov counter', *Nucl. Instr. Meth. Phys. Res. A*, 248, p. 69.

Meinschad, T., Ropelewski, L. and Sauli, F. (2004) 'GEM-based photon detector for RICH applications', *Nucl. Instr. Meth. Phys. Res. A*, 535, pp. 324–329.

Miernik, K. *et al.* (2007) 'Optical time projection chamber for imaging nuclear decays', *Nucl. Instr. Meth. Phys. Res. A*, 581, pp. 194–197.

Minakov, G. D. *et al.* (1993) 'Performance of gas microstrip chambers on glass with ionic and electronic conductivity', *Nucl. Instr. Meth. Phys. Res.*, 326, pp. 566–569.

Miyamoto, J. *et al.* (2004) 'GEM operation in negative ion drift gas mixtures', *Nucl. Instr. Meth. Phys. Res. A*, 526, pp. 409–412.

Moleri, L. *et al.* (2017) 'The Resistive-Plate WELL with Argon mixtures — A robust gaseous radiation detector', *Nucl. Inst. Meth. Phys. Res. A*, 845, pp. 262–265.

Morello, G. *et al.* (2019) 'The micro-Resistive WELL detector for the phase-2 upgrade of the LHCb muon detector', *Nucl. Instr. Meth. Phys. Res. A*, 936, pp. 497–498.

Moreno, B. *et al.* (2011) 'Discharge rate measurements for MICROMEGAS detectors in the presence of a longitudinal magnetic field', *Nucl. Instr. Meth. Phys. Res. A*, 654, pp. 135–139.

Morishima, K. *et al.* (2017) 'Discovery of a big void in Khufu's Pyramid by observation of cosmic-ray muons', *Nature*, 552, p. 386.

Mrmann, D. *et al.* (2003) 'GEM-based gaseous photomultipliers for UV and visible photon imaging', *Nucl. Instr. Meth. Phys. Res. A*, 504, pp. 93–98.

Morozov, A. *et al.* (2010) 'Photon yield for ultraviolet and visible emission from CF4 excited with alpha-particles', *Nucl. Instr. Meth. Phys. Res. B*, 268, pp. 1456–1459.

Morozov, A. *et al.* (2011) 'Effect of the electric field on the primary scintillation from CF 4', *Nucl. Instr. Meth. Phys. Res. A*, 628, pp. 360–363.

Muller, T. (1998) 'The CMS tracker and its performance', *Nucl. Instr. Meth. Phys. Res. A*, 408, pp. 119–127.

Multiphysics (2019) *COMSOL*, https://www.comsol.com.

Muraro, A. *et al.* (2019) 'Directionality properties of the nGEM detector of the CNESM diagnostic system for SPIDER', *Nucl. Instr. Meth. Phys. Res. A*, 916, pp. 47–50.

Murtas, F. *et al.* (2010) 'Applications in beam diagnostics with triple GEM detectors', *Nucl. Instr. Meth. Phys. Res. A*, 617, pp. 237–241.

Nagayoshi, T. *et al.* (2003) 'Performance of large area micro pixel chamber', *Nucl. Instr. Meth. Phys. Res. A*, 513, pp. 277–281.

Nappi, E. (2020) 'Development of photocathodes for gaseous counters: From UV to visible', *Nucl. Instr. Meth. Phys. Res. A*, 970, p. 163424.

Nappi, E. and Sguinot, J. (2005) 'Ring imaging Cherenkov detectors: The state of the art and perspectives', *Riv. Nuovo Cimento*, 28(8–9), p. 2005.

Natal Da Luz, H. *et al.* (2018) 'Ion backflow studies with a triple-GEM stack with increasing hole pitch', *JINST*, 13, p. P07025.

Nath Patra, R. *et al.* (2018) 'Characteristic study of a quadruple GEM detector and its comparison with a triple GEM detector', *Nucl. Instr. Meth. Phys. Res. A*, 906, pp. 37–42.

Neyret, D. *et al.* (2012) 'New pixelized MICROMEGAS detector with low discharge rate for the COMPASS experiment', *JINST*, 7, p. C03006.

Nitti, M. A. *et al.* (2009) 'Influence of the substrate surface texture on the photon-sensitivity stability of CsI thin film photocathodes', *Nucl. Instr. Meth. Phys. Res. A*, 610, pp. 234–237.

Nygren, D. and Marx, J. (1978) 'The Time Projection Chamber', *Phys. Today*, 31, p. 46.

Ochi, A. *et al.* (2001) 'A new design of the gaseous imaging detector: Micro Pixel Chamber', *Nucl. Instr. Meth. Phys. Res. A*, 471, pp. 264–267.

Ochi, A. *et al.* (2002) 'Development of micro pixel chamber', *Nucl. Instr. Meth. Phys. Res. A*, 478, pp. 196–199.

Ochi, A. *et al.* (2009) 'A new MPGD design: Micro-mesh micro-pixel chamber', *Nucl. Instr. Meth. Phys. Res. A*, 604, pp. 11–14.

Oed, A. (1988) 'Position-sensitive detector with microstrip anode for electron multiplication with gases', *Nucl. Instr. Meth. Phys. Res. A*, 263, pp. 351–359.

Oed, A. *et al.* (1989) 'A new position-sensitive proportional counter with microstrip anode for neutron detection', *Nucl. Instr. Meth. Phys. Res. A*, 284, pp. 223–226.

Oed, A., Geltenbort, P. and Budtz-Jrgensen, C. (1991) 'Substratum and layout parameters for microstrip anodes in gas detectors', *Nucl. Instr. Meth. Phys. Res. A*, 301, pp. 95–97.

Ohnuki, T., Snowden-Ifft, D. P. and Martoff, C. J. (2001) 'Measurements of carbon disulfide anion diffusion in a TPC', *Nucl. Instr. Meth. Phys. Res. A*, 463, pp. 142–148.

de Oliveira, R. *et al.* (2010) 'First Tests of MICROMEGAS and GEM-like detectors made of a resistive mesh', *IEEE Trans. Nucl. Sci.*, NS-57, pp. 3744–3752.

De Oliveira, R., Maggi, M. and Sharma, A. (2015) 'A novel fast timing micropattern gaseous detector: FTM', arXiv:1503.05330v1.

Ortuiio-Prados, F. and Budtz-Jrgensen, C. (1995) 'The electron-conducting glass SCHOTT 523900 as substrata for microstrip gas chamber', *Nucl. Instr. Meth. Phys. Res. A*, 364, pp. 2–7.

Ostling, J. *et al.* (2003) 'Study of hole-type gas multiplication structures for portal imaging and other high counting rate applications', *IEEE Trans. Nucl. Sci.*, NS-50, pp. 619–623.

Ostling, J. *et al.* (2004) 'A radiation-tolerant electronic readout system for portal imaging', *Nucl. Instr. Meth. Phys. Res. A*, 525, pp. 308–312.

Ostling, J. (2006) 'New efficient detector for radiation therapy imaging using gas electron multipliers', *Doctoral Thesis Stockholm Univ.*

Pacella, D. *et al.* (2001) 'Ultrafast soft X ray 2D plasma imaging system based on Gas Electron Multiplier detector with pixel read-out', *Rev. Scient. Instrum.*, 72, p. 1372.

Pacella, D. *et al.* (2013) 'GEM gas detectors for soft X-ray imaging in fusion devices with neutron-gamma background', *Nucl. Instr. Meth. Phys. Res. A*, 720, pp. 53–57.

Pacella, D. *et al.* (2016) 'An hybrid detector GEM-ASIC for 2-D soft X-ray imaging for laser produced plasma and pulsed sources', *JINST*, 11, p. C03022.

Pancin, J. *et al.* (2004) 'Measurement of the n TOF beam profile with a MICROMEGAS detector', *Nucl. Instr. Meth. Phys. Res. A*, 524, pp. 102–114.

Pancin, J. *et al.* (2008) 'Piccolo MICROMEGAS: First in-core measurements in a nuclear reactor', *Nucl. Instr. Meth. Phys. Res. A*, 592, pp. 104–113.

Pansky, A. *et al.* (1995) 'The scintillation of CF, and its relevance to detection science', *Nucl. Instr. Meth. Phys. Res. A*, 354, pp. 262–269.

Papaevangelou, T. *et al.* (2018) 'Fast timing for high-rate environments with MICROMEGAS', *EPJ Web of Conferences*, 174, p. 02002.

Patra, R. N. *et al.* (2017) 'Measurement of basic characteristics and gain uniformity of a triple GEM detector', *Nucl. Instr. Meth. Phys. Res. A*, 862, pp. 25–30.

Pemler, P. *et al.* (1999) 'A detector system for proton radiography on the gantry of the Paul-Scherrer-Institute', *Nucl. Instr. Meth. Phys. Res. A*, 432, pp. 483–495.

Periale, L. *et al.* (2002) 'Detection of the primary scintillation light from dense Ar, Kr and Xe with novel photosensitive gaseous detectors', *Nucl. Instr. Meth. Phys. Res. A*, 478, pp. 377–383.

Peskov, V. *et al.* (2012) 'Development of novel designs of spark-protected micropattern gaseous detectors with resistive electrodes', *JINST*, 7, p. C01005.

Peskov, V. and Zichichi, A. (2007) 'A new supersensitive flame detector and it's use for early forest fire detection', *CERN-LAA-Preprint*, 863.

Pestov, Y. N. and Shekhtman, L. I. (1994) 'Influence of the bulk resistivity of glass with electronic conductivity on the performance of microstrip gas chamber', *Nucl. Instr. Meth. Phys. Res. A*, 338, pp. 368–374.

Phan, N. S. *et al.* (2016) 'GEM-based TPC with CCD imaging for directional dark matter detection', *Astroparticle Physics*, 84, pp. 82–96.

Phan, N. S. *et al.* (2017) 'The novel properties of SF6 for directional dark matter experiments', *JINST*, 12, p. P02012.

Pinci, D. *et al.* (2019) 'High resolution TPC based on optically readout GEM', *Nucl. Instr. Meth. Phys. Res. A*, 936, pp. 453–455.

Pitt, M. *et al.* (2018) 'Measurements of charging-up processes in THGEM-based particle detectors', *JINST*, 13, p. P03009.

Poli Lener, M. *et al.* (2017) 'The μ-RWELL: A compact, spark protected, single amplification-stage MPGD', *Nucl. Instr. Meth. Phys. Res. A*, 834, pp. 565–568.

Policarpo, A. (1977) 'The gas scintillation proportional counter', *Space Sci. Instr.*, 3, p. 77.

Posik, M. and Surrow, B. (2015) 'Optical and electrical performance of commercially manufactured large GEM foils', *Nucl. Instr. Meth. Phys. Res. A*, 802, pp. 10–15.

Procureur, S. *et al.* (2010) 'A Geant4-based study on the origin of the sparks in a MICROMEGAS detector and estimate of the spark probability with hadron beams', *Nucl. Instr. Meth. Phys. Res. A*, 621, pp. 177–183.

Procureur, S. *et al.* (2011) 'Discharge studies in MICROMEGAS detectors in a 150 GeV/c pion beam', *Nucl. Instr. Meth. Phys. Res. A*, 659, pp. 91–97.

Procureur, S. *et al.* (2012) 'Origin and simulation of sparks in MPGD', *JINST*, 7, p. C06009.

Procureur, S. (2018) 'Muon imaging: Principles, technologies and applications', *Nucl. Instr. Meth. Phys. Res. A*, 878, pp. 169–179.

Procureur, S. and Atti, D. (2019) 'Development of high-definition muon telescopes and muography of the Great Pyramid', *Comp. Rend. Phys.*, 20, pp. 521–528.

Procureur, S., Dupr, R. and Aune, S. (2013) 'Genetic multiplexing and first results with a 50×50 cm^2 MICROMEGAS', *Nucl. Instr. Meth. Phys. Res. A*, 729, pp. 888–894.

Radicioni, E. (2007) 'Design, construction and performance of a large GEM-TPC prototype', *Nucl. Instr. Meth. Phys. Res. A*, 572, pp. 195–197.

Radics, B. *et al.* (2015) 'The ASACUSA MICROMEGAS Tracker: A cylindrical, bulk MICROMEGAS detector for antimatter research', *Rev. Sci. Instr.*, 86, p. 83304.

Raether, H. (1964) *Electron Avalanches and Breakdown in Gases*. London: Butterworth.

Ratza, V. *et al.* (2018) 'Characterization of a hybrid GEM-MICROMEGAS detector with respect to its application in a continuously read out TPC', *EPJ Web of Conferences*, 174(01016).

Rehak, P. *et al.* (2000) 'First results from the micro pin array detector (MIPA)', *IEEE Trans. Nucl. Sci.*, 47, pp. 1426–1429.

Rice-Evans, P. (1974) *Spark, Streamer, Proportional and Drift Chambers*. London: Richelieu.

Roth, S. (2004) 'Study of GEM structures for a TPC readout', *Nucl. Instr. Meth. Phys. Res. A*, 518, pp. 103–105.

Rubin, R. *et al.* (2013) 'Optical readout: A tool for studying gas-avalanche processes', *JINST*, 8, p. P08001.

Rutherford, E. and Geiger, H. (1908) 'An electrical method of counting the number of α-particles from radio-active substances', *Proc. Royal Soc. A*, 81, p. 141.

Rzadkiewicz, J. *et al.* (2013) 'Design of T-GEM detectors for X-ray diagnostics on JET', *Nucl. Instr. Meth. Phys. Res. A*, 720, pp. 36–38.

Sachdeva, R. *et al.* (1994) 'Fast electronic readout of microstrip gas chambers', *Nucl. Instr. Meth. Phys. Res. A*, 348, pp. 378–382.

Salete Leite, M. (1980) 'Radioluminescence of Rare Gases', *Portugal Phys.*, 11, p. 53.

Sauli, F. (1997) 'GEM: A new concept for electron amplification in gas detectors', *Nucl. Instr. Meth. Phys. Res. A*, 386, pp. 531–534.

Sauli, F (1999) 'Micro-Pattern Gas Detectors', *Position Sensitive Detectors Conference London*, CERN-EP/99.

Sauli, F. (1999) 'Radiation detector of very high performance and planispherical parallax-free X-ray imager comprising such radiation detector', *Patent WO 99/21211*.

Sauli, F. (2003) 'Fundamental understanding of aging processes: review of the workshop results', *Nucl. Instr. Meth. Phys. Res. A*, 515, pp. 358–363.

Sauli, F. *et al.* (2004) 'Photon detection and localization with GEM', *IEEE Nucl. Sci. Symp. Conf. Rec.*, 41.

Sauli, F. (2005) 'Novel Cherenkov photon detectors', *Nucl. Instr. Meth. Phys. Res. A*, 553, pp. 18–24.

Sauli, F. (2010) *GEM framing and assemblage*, http://gdd.web.cern.ch/GDD/.

Sauli, F. (2014) *Gaseous Radiation Detectors: Fundamentals and Applications*. Cambridge University Press.

Sauli, F. (2016) 'The gas electron multiplier (GEM): Operating principles and applications', *Nucl. Instr. Meth. Phys. Res. A*, 805, pp. 2–24.

Sauli, F. (2018) 'Radiation imaging with gaseous detectors', *Nucl. Instr. Meth. Phys. Res. A*, 878, pp. 1–9.

Sauli, F. (2019) 'Six Concepts in search of an author', *Instruments*, 3, p. 51.

Sauli, F., Ropelewski, L. and Everaerts, P. (2006) 'Ion feedback suppression in time projection chambers', *Nucl. Instr. Meth. Phys. Res. A*, 560, pp. 269–277.

Savard, P. *et al.* (1993) 'An a-Si:H gas microstrip detector', *Nucl. Instr. Meth. Phys. Res. A*, 337, pp. 387–392.

Schade, P. and Kaminski, J. (2011) 'A large TPC prototype for a linear collider detector', *Nucl. Instr. Meth. Phys. Res. A*, 628, pp. 128–132.

Schlumbohm, H. (1958) *'Zur Statistik der Elektronenlawinen im ebenen Field'*, *Zeit. Phys.*, 151, p. 563.

Schmidt, S., Wertenbach, U. and Zech, G. (1994) 'Study of thin substrates for microstrip gas chambers', *Nucl. Instr. Meth. Phys. Res. A*, 337, pp. 382–386.

Segui, L. (2013) 'NEXT prototypes based on MICROMEGAS', *Nucl. Instr. Meth. Phys. Res. A*, 718, pp. 434–436.

Sguinot, J. *et al.* (1990) 'Reflective UV photocathodes with gas-phase electron extraction: solid, liquid, and adsorbed thin films', *Nucl. Instr. Meth. Phys. Res. A*, 297, pp. 133–147.

Sguinot, J., Tocqueville, J. and Ypsilantis, T. (1980) 'Imaging Cherenkov detectors: Photo-ionization of tri-ethyl-amine', *Nucl. Instr. Meth.*, 173, pp. 283–298.

Sguinot, J. and Ypsilantis, T. (1977) 'Photo-ionization and Cherenkov ring imaging', *Nucl. Instr. Meth.*, 142, pp. 377–391.

Sguinot, J. and Ypsilantis, T. (1994) 'A historical survey of ring imaging Cherenkov counters', *Nucl. Instr. Meth. Phys. Res. A*, 343, pp. 1–29.

Seravalli, E. *et al.* (2008) 'A scintillating gas detector for 2D dose measurements in clinical carbon beams', *Phys. Med. Biol.*, 53, pp. 4651–4665.

Seravalli, E. *et al.* (2009) '2D dosimetry in a proton beam with a scintillating GEM detector', *Phys. Med. Biol.*, 54, p. 3755.

Shah, A. *et al.* (2019) 'Impact of single-mask hole asymmetry on the properties of GEM detectors', *Nucl. Instr. Meth. Phys. Res. A*, 936, pp. 459–461.

Shalem, C. *et al.* (2006) 'Advances in thick GEM-like gaseous electron multipliers-Part I: atmospheric pressure operation', *Nucl. Instr. Meth. Phys. Res. A*, 558, pp. 475–489.

Sidiropoulou, O. *et al.* (2017) 'Performance studies under high irradiation and ageing properties of resistive bulk MICROMEGAS chambers at the new CERN Gamma Irradiation Facility', *Nucl. Instr. Meth. Phys. Res. A*, 845, pp. 293–297.

Silva, A. L. M. *et al.* (2013) 'Performance of a gaseous detector based energy dispersive X-ray fluorescence imaging system: Analysis of human teeth treated with dental amalgam', *Spectrochimica Acta — Part B*, 86, pp. 115–122.

Simon, A. *et al.* (2005) 'A scintillating triple GEM beam monitor for radiation therapy', *IEEE 2003 Nucl. Sci. Symp. Conf. Rec. IEEE*, 5, pp. 2770–2774.

Snyder, L. *et al.* (2018) 'Performance of a MICROMEGAS-based TPC in a high-energy neutron beam', *Nucl. Instr. Meth. Phys. Res. A*, 881, pp. 1–8.

Song, G. *et al.* (2020) 'Production and properties of a charging-up "Free" THGEM with DLC coating', *Nucl. Instr. Meth. Phys. Res. A*, 966, p. 163868.

Song, I. *et al.* (2016) 'Tomographic 2-D X-ray imaging of toroidal fusion plasma using a tangential pinhole camera with gas electron multiplier detector', *Cur. Appl. Phys.*, 16, pp. 1284–1292.

Stahl, H., Werthenbach, U. and Zech, C. (1990) 'First steps towards a foil microstrip chamber', *Nucl. Instr. Meth. Phys. Res. A*, 297, pp. 95–102.

Sugaya, Y. *et al.* (1996) 'Ionization Chamber as a p, d and alpha beam intensity Monitor', *Nucl. Instr. Meth. Phys. Res. A*, 368, pp. 635–639.

Suzuki, M. *et al.* (1987) 'The emission spectra of Ar, Kr and Xe + TEA', *Nucl. Instr. Meth. Phys. Res. A*, 254, pp. 556–560.

Takahashi, H. *et al.* (2013) 'Development of a glass GEM', *Nucl. Instr. Meth. Phys. Res. A*, 724, pp. 1–4.

Tamagawa, T. *et al.* (2006) 'Development of gas electron multiplier foils with a laser etching technique', *Nucl. Instr. Meth. Phys. Res. A*, 560, pp. 418–424.

Tamagawa, T. *et al.* (2009) 'Development of thick-foil and fine-pitch GEMs with a laser etching technique', *Nucl. Instr. Meth. Phys. Res. A*, 608, p. 390.

Terasaki, K. *et al.* (2013) 'R&D of thick COBRA GEM for the application of the GEM-based TPC', *IEEE Nucl. Sci. Symp. Conf. Rec. IEEE*, pp. 1–4.

Tessarotto, F. (2017) 'Status and perspectives of gaseous photon detectors', *Nucl. Instr. Meth. Phys. Res. A*, 876, pp. 225–232.

Thers, D. *et al.* (2001) 'MICROMEGAS as a large microstrip detector for the COMPASS experiment', *Nucl. Instr. Meth. Phys. Res. A*, 469, pp. 133–146.

Thuiner, P., Hall-Wilton, R., *et al.* (2015) 'Charge Transfer Properties Through Graphene Layers in Gas Detectors', arXiv:1503.06596v1.

Thuiner, P., Resnati, F., *et al.* (2015) 'Multi-GEM Detectors in High Particle Fluxes', *EPJ Web of Conferences*, 174, p. 05001.

Tikhonov, V. and Veenhof, R. (2002) 'GEM simulation methods development', *Nucl. Instr. Meth. Phys. Res. A*, 478, pp. 452–459.

Titov, M. and Ropelewski, L. (2013) 'Micro-pattern gaseous detector technologies and RD51 collaboration', *Modern Phys. Lett. A*, 28, pp. 1–25.

Titt, U. *et al.* (1998) 'A Time Projection Chamber with optical readout for charged particle track structure imaging', *Nucl. Instr. Meth. Phys. Res. A*, 416, pp. 85–99.

Tokanai, F. *et al.* (2009) 'Development of gaseous PMT with micropattern gas detector', *Nucl. Instr. Meth. Phys. Res. A*, 610, pp. 164–168.

Tserruya, I. (2006) 'Report on the hadron blind detector for the PHENIX experiment', *Nucl. Instr. Meth. Phys. Res. A*, 563, pp. 333–337.

Tserruya, I., Aoki, K., Woody, C. (2020) 'Hadron blind Cherenkov counters', *Nucl. Instr. Meth. Phys. Res. A*, 970, p. 163765.

Tsyganov, E. *et al.* (2008) 'Gas electron multiplying detectors for medical applications', *Nucl. Instr. Meth. Phys. Res. A*, 597, pp. 257–265.

Va'vra, J. (2003) 'Physics and chemistry of aging-early developments', *Nucl. Instr. Meth. Phys. Res. A*, 515, pp. 1–14.

Va'vra, J. *et al.* (1997) 'Study of CsI photocathodes: volume resistivity and ageing', *Nucl. Instr. Meth. Phys. Res. A*, 387, pp. 154–162.

Veenhof, R. (1998) 'GARFIELD, recent developments', *Nucl. Instr. Meth. Phys. Res. A*, 419, pp. 726–730.

Vellettaz, N., Oed, A. and Assaf, J. E. (1995) 'Two-dimensional neutron detector', in Contardo, D. and Sauli, F. (eds) *Proc. Int. Workshop on Micro-Strip Gas Chambers Lyon*, p. 73.

Veloso, J. F. C. A., Caldas, C. C., *et al.* (2007) 'High-rate operation of the micro-hole and strip plate gas detector', *Nucl. Instr. Meth. Phys. Res. A*, 580, pp. 362–365.

Veloso, J. F. C. A., Amaro, F. D., *et al.* (2007) 'PACEM: a new concept for high avalanche-ion blocking', *Nucl. Instr. Meth. Phys. Res. A*, 581, pp. 261–264.

Veloso, J. F. C. A. *et al.* (2011) 'THCOBRA: Ion back flow reduction in patterned THGEM cascades', *Nucl. Instr. Meth. Phys. Res. A*, 639, pp. 134–136.

Veloso, J. F. C. A., Santos, J. M. F. D. and Conde, C. A. N. (2000) 'A proposed new microstructure for gas radiation detectors: The microhole and strip plate', *Rev. Sci. Instrum.*, 71, p. 2371.

Veloso, J. F. C. A. and Silva, A. L. M. (2018) 'Gaseous detectors for energy dispersive X-ray fluorescence analysis', *Nucl. Instr. Meth. Phys. Res. A*, 878, pp. 24–39.

Villa, M. *et al.* (2011) 'Progress on large area GEMs', *Nucl. Instr. Meth. Phys. Res. A*, 628, pp. 182–186.

Volpe, G. (2020) 'PID performance of the High momentum particle identification (HMPID) detector during LHC-Run 2', *Nucl. Instr. Meth. Phys. Res. A*, 952, p. 161803.

Walenta, A. H., Heintz, J. and Schrlein, B. (1971) 'The multiwire drift chamber: a new type of Proportional Wire Chamber', *Nucl. Instr. Meth.*, 92, pp. 373–380.

Weisskopf, M. C. *et al.* (2016) 'The imaging X-ray polarimetry explorer (IXPE)', *Res. Phys.*, 6, pp. 1179–1180.

White, S. (2018) 'PICOSEC: Charged particle timing to 24 picosecond precision with MicroPattern gas detectors', *Nucl. Instr. Meth. Phys. Res. A*, 912, pp. 298–299.

Wotschack, J. (2012) 'Development of MICROMEGAS muon chambers for the ATLAS upgrade', *JINST*, 7, p. C02021.

Xie, Y. *et al.* (2013) 'Development of Au-coated THGEM for single photon, charged particle, and neutron detection', *Nucl. Instr. Meth. Phys. Res. A*, 729, pp. 809–815.

Yamane, F. *et al.* (2020) 'Development of the Micro pixel chamber with DLC cathodes', *Nucl. Instr. Meth. Phys. Res. A*, 951, p. 162938.

Ypsilantis, T. (1981) 'Cerenkov ring imaging', *Phys. Scripta*, 23, pp. 371–376.

Ypsilantis, T. and Sguinot, J. (1994) 'Theory of ring imaging Cherenkov counters', *Nucl. Instr. Meth. Phys. Res. A*, 343, pp. 30–51.

Yu, B. *et al.* (2005) 'A GEM based TPC for the LEGS experiment', in *IEEE Nucl. Sci. Symp. Conf. Records*, p. 924.

Zerguerras, T. *et al.* (2009) 'Single-electron response and energy resolution of a MICROMEGAS detector', *Nucl. Instr. Meth. Phys. Res. A*, 608, pp. 397–402.

Zerguerras, T. *et al.* (2015) 'Understanding avalanches in a MICROMEGAS from single-electron response measurement', *Nucl. Instr. Meth. Phys. Res. A*, 772, pp. 76–82.

Zeuner, T. (1997) 'Development of a 150000 channel MSGC tracking system for the experiment HERA-B', *Nucl. Instr. Meth. Phys. Res. A*, 392, pp. 105–108.

Zeuner, T. (2000) 'The MSGC-GEM inner tracker for HERA-B', *Nucl. Instr. Meth. Phys. Res. A*, 446, pp. 324–330.

Zhang, Z. *et al.* (2018) 'A high-gain, low ion-backflow double micro-mesh gaseous structure for single electron detection', *Nucl. Instr. Meth. Phys. Res. A*, 889, pp. 78–82.

Zhou, J. *et al.* (2020) 'Highly efficient GEM-based neutron detector for China Spallation Neutron Source', *Nucl. Instr. Meth. Phys. Res. A*, 953, p. 163051.

Ziegler, M. G. (1998) *Untersuchungen von Detektorprototypen fr das innere Spurkammersystem des HERA-B Experimentes*. Physikalishen Institut Univ. Heidelberg.

Zieliska, A. *et al.* (2013) 'X-ray fluorescence imaging system for fast mapping of pigment distributions in cultural heritage paintings', *JINST*, 8, p. P10011.

Index

www.ingramcontent.com/pod-product-compliance
Lightning Source LLC
Chambersburg PA
CBHW050537190326
41458CB00007B/1820